Nanoscale Science and Engineering Forum

Core Programming Topic at the 2012 AIChE Annual Meeting

Pittsburgh, Pennsylvania, USA
28 October - 2 November 2012

ISBN: 978-1-62276-734-2

Printed from e-media with permission by:

Curran Associates, Inc.
57 Morehouse Lane
Red Hook, NY 12571

Some format issues inherent in the e-media version may also appear in this print version.

Printed by Curran Associates, Inc. (2013)

AIChE
3 Park Avenue
New York, NY 10016-5991

Phone: (203) 702-7660
Fax: (203) 775-5177

www.aiche.org

Additional copies of this publication are available from:

Curran Associates, Inc.
57 Morehouse Lane
Red Hook, NY 12571 USA
Phone: 845-758-0400
Fax: 845-758-2634
Email: curran@proceedings.com
Web: www.proceedings.com

TABLE OF CONTENTS

(25a) Membrane Preconcentration Technique for the Colorimetric Detection of Trihalomethanes in Water

Evan K. Wujcik, Brad Vielhaber, Max Duckworth, G. G. Chase and Chelsea N. Monty, Department of Chemical and Biomolecular Engineering, The University of Akron, Akron, OH

We present a preconcentration and colorimetric sensing platform for detection of trihalomethanes (THMs) in water. The Environmental Protection Agency (EPA) regulates the four chief THM constituents (bromoform, chloroform, bromodichloromethane, and dibromochloromethane) to a limit of 80 parts per billion (ppb) [nL/L] in treated water. Currently, expensive and non-portable analytical techniques are used for detection of THMs in treated water samples. Here, we utilize the Fujiwara reaction which is an inexpensive colorimetric detection reaction that, in the presence of a THM and basic conditions, forms a chromophore that emits a red/pink color under visible light absorption. This reaction, however, is limited to a visible detection range of (μL/L). To increase sensitivity, we have developed a Fujiwara preconcentration system utilizing a superhydrophobic membrane to decrease the detectable limit of this reaction making the colorimetric technique applicable to real-world applications. This Fujiwara preconcentration system has the ability to allow THM vapor through the membrane allowing the Fujiwara reaction to occur with a minute amount of a THM. Two types of superhydrophobic membranes [polypropylene (PP) and poly(4-methyl-1-pentene) (P4M1P)] will be used. In this presentation, the optimization of the Fujiwara stoichiometry, membrane conditions, and geometry of the system will be discussed.

(25b) Hydrophobically Modified Biopolymer As Enhanced Carrier for in Situ Groundwater Remediation

Rubo Zheng[1], Jingjing Zhan[1], Srinivasa R. Raghavan[2], Bhanukiran Sunkara[3] and **Vijay T. John**[3], (1)Department of Chemical and Biomolecular Engineering, Tulane University, New Orleans, LA, (2)Department of Chemical and Biomolecular Engineering, University of Maryland, College Park, MD, (3)Chemical and Biomolecular Engineering, Tulane University, New Orleans, LA

Carbon submicrospheres as a carrier for nanoscale zerovalent iron particles are of much potential in the remediation of chlorinated compounds. Effective in situ groundwater remediation requires the successful delivery of these reactive systems through soil without aggregation. Here, we show that hydrophobically modified chitosan(HMC), with C12 alkyl groups stabilize carbon submicrospheres through hydropobic interactions. Detailed NMR characterizations indicate attachment of the alkyl groups onto the microspheres. The coating of the environmentally benign biopolymer therefore leads to suspension stability of these submicrospheres through electrostatic and steric repulsion. Compared with pristine chitosan and commonly used sodium carboxymethyl cellulose (CMC), HMC is more effective in stabilizing carbon submicrospheres, and this is even enhanced under normal groundwater iron strength condition. Our results indicate that this system has optimal transport characteristics in groundwater saturated sediments with carbon submicrospheres attachment efficiency of 0.04 calculated from breakthrough data of capillary transport experiments. The fundamental concepts will be detailed in the talk.

(25c) Photocatalytic Degradation of Acetaminophen in Dilute Aqueous Solutions

Dorothy W. Skaf[1], Amanda M. Grannas[2], Nicholas Natrin[1] and Christopher Bongo[1], (1)Chemical Engineering, Villanova University, Villanova, PA, (2)Chemistry, Villanova University, Villanova, PA

Photocatalytic degradation of acetaminophen in dilute aqueous solution was evaluated using both a catalyst slurry and catalyst immobilized on glass surfaces. Acetaminophen oxidation was studied under various operating conditions to better understand how the reaction conditions impact initial degradation rates and byproduct compositions. Reaction conditions, such as catalyst concentration, reagent concentration, metal loading on the catalyst, oxygen purging, solution pH, solution temperature, addition of oxidant, batch versus continuous operation, light intensity and semiconductor stabilization were investigated. Literature results for reaction intermediates and reaction kinetics were used to interpret experimental results. A main focus was optimization of reaction conditions that can make photocatalytic oxidation of dilute aqueous contaminants more sustainable and economical. Catalyst films may eliminate the need to separate fine catalyst particles from the reaction medium, yet few comparisons of their effectiveness to that of slurry reactors are available. Catalyst lifetime, stability on the support, consumption of additional reagents, value and/or safety of the degradation products and energy consumption were considered when comparing various reaction options.

(25d) Hexavalent Chromium Remediation by Electrokinetic Transport and Zero-Valent Iron Nanoparticle Injection: Effects of Organic and Inorganic Groundwater Constituents

Ryan Thacher and Massoud Pirbazari, Sonny Astani Department of Civil and Environmental Engineering, University of Southern California, Los Angeles, CA

Hexavalent chromium (Cr^{VI}) contamination is prevalent in aquifers around the world as a result of numerous anthropogenic activities. Chromium is used in many industrial processes for corrosion inhibition and also for adding pigment in dyes, paints, inks, and plastics. The highly toxic nature of Cr^{VI} makes it a target contaminant to remove from water sources, however its high solubility makes complete removal difficult. Treatment of Cr^{VI} in aquifers is generally limited to *ex situ* methods such as pump-and-treat or soil excavation, which are often costly and invasive. The growing field of nanotechnology however offers potential for *in situ* groundwater remediation. Nanoparticles can be injected directly into aquifers for remediation purposes, and can travel long distances with water flow through soil pore space treating contaminants as they move. Zero-valent iron nanoparticles (nFe^0) have the capacity to react with many common pollutants such as heavy metals and radionuclides, by reducing them to insoluble precipitates through an adsorption-reduction process mechanism. Mass transfer of Cr^{VI} to nFe^0 surface reactive sites is followed by the reduction to Cr^{III}, corresponding to the oxidation of Fe^0 to Fe^{2+}. Another source of Fe^{2+} is from the oxidation of Fe^0 by H^+, which can reduce Cr^{VI} as well during the oxidation of Fe^{2+} to Fe^{3+}. The production of Cr^{III} is advantageous due to its negligible toxicity and low solubility leading to precipitation as $Cr(OH)_3$, or Cr-Fe (oxy)hydroxides on the nano iron surface.

Once injected into groundwater, nFe^0 will react with most chemical and biological species present. Groundwater contains natural organic matter and minerals, which have been found to interact with nFe^0, and affect their reductive capacity. Nonetheless, it is important that nFe^0 may be toxic to microbial communities present in the aquifer materials. In many cases these microbial communities are naturally assisting in the remediation process.

In this study a groundwater treatment strategy of integrating electrokinetic contaminant transport with nFe^0 injection is evaluated. Through the application of an electric field across a contamination plume in an aquifer, Cr^{VI} will migrate towards the anode electrode due to the negative charge carried by the chromate ion. This technique is especially advantageous in aquifers that are difficult to access, such as below developed land, or in soils of low hydraulic conductivity. Upon concentration of Cr^{VI} at the anode, nFe^0 can be injected directly into the aquifer near the anode to reduce Cr^{VI} to Cr^{III}.

This study investigates 1) the capacity of surface stabilized nFe^0 for Cr^{VI} reduction under oxic and anoxic conditions, the effects of humic acid and calcium on Cr^{VI} reduction, as well as toxic effects of nFe^0 to microbial communities, and 2) optimizing the integration of electrokinetic and nFe^0 technologies for Cr^{VI} remediation in a soil column. The interactions between nFe^0 and organic and inorganic constituents under oxic and anoxic conditions will be discussed in terms of their effect on Cr^{VI} reduction. Adsorption isotherm studies will compare the effects of these constituents on Cr^{VI} removal.

Furthermore, laboratory-scale experiments were conducted with a continuous flow of simulated groundwater spiked with Cr^{VI} pumped through a 40-mesh silica sand soil column under an applied electric potential. The columns were oriented such that water flowed from anode to cathode, while Cr^{VI} migrated in the opposite direction towards the anode. The electric potential application was varied from 0.5-2.0 V/cm such that transport by electrokinetic phenomena counteracted the advective transport of Cr^{VI} towards the cathode by water flow to promote Cr^{VI} accumulation near the anode. Surface stabilized nFe^0 were injected into the sand column at specific time intervals directly adjacent to the anode reservoir where Cr^{VI} concentrations were highest. Samples were taken from multiple locations within the soil column, as well as from the cathode reservoir effluent to determine the degree of Cr^{VI} removal by electrokinetic and nFe^0 integration. The results from the aforementioned studies will be discussed during presentation.

(25e) Gas Liquid Mass Transfer of Volatile Organic Contaminants Using Magnetite Nanoparticles

Vasanta L. Pallem[1], Alexander P. Mathews[2], Dambar B. Hamal[3] and Kenneth J Klabunde[3], (1)Department of Civil Engineering, Kansas State University, Manhattan , KS, (2)Department of Civil Engineering, Kansas State University, Manhattan, KS, (3)Department of Chemistry, Kansas State University, Manhattan, KS

Gas-liquid mass transfer plays a critical role in the removal of volatile organic contaminants from water supplies and wastewaters using various air stripping technologies. However, high air to water ratios needed for removal of contaminants present high expenses for the water processing industry. Nanoparticle sizes are in the range of hydrodynamic boundary layers. Therefore, nanoparticles show potential for higher removal of organic contaminants and enhancement of gas liquid mass transfer due to increased interfacial surface area and turbulence in the boundary layer. The current research examined the effects of magnetite nanoparticles on gas-liquid mass transfer rates in the desorption of trichloroethylene and methyl tert-butyl ether from aqueous solutions. This paper will discuss the effects of different variables such as particle loading, air flow rate, ultrasonication amplitude, and surfactant addition on gas-liquid mass transfer rates of trichloroethylene and methyl tert-butyl ether, in control solutions and in the presence of magnetite nanoparticles.

(25f) Removal of Radioactive Uranium From Groundwater Using Nanoparticle Technology and Bioremediation Strategies

Hannah Gray[1], Ryan Thacher[1], **Varadarajan Ravindran**[2] and Massoud Pirbazari[1], (1)Sonny Astani Department of Civil and Environmental Engineering, University of Southern California, Los Angeles, CA, (2)University of Southern California, Los Angeles, CA

In recent times, water contamination from radiological materials has become a significant hazard to human health. Wastewaters containing uranium and other actinides generated in nuclear reactors in processes related to generating energy and manufacturing nuclear weapons have been discharged for over 50 years. Many sites are contaminated with toxic uranium owing to solubilization of natural uranium, production of nuclear energy, and mining of uranium. Consequently, uranium is one of the most common radionuclides in soils, sediments, and groundwater at several U.S. Department of Energy (DOE) sites and is therefore of particular environmental concern. Acid in situ mining technology has been adopted as the main approach for exploiting uranium deposits as nuclear fuel. After the leaching process, the groundwater is generally characterized by very low pH and high levels of radioactive uranium. The radionuclide occurs as hexavalent uranyl (UO_2^{2+}) that is extremely mobile, contaminating the groundwater resources. Generally, the long-lived radionuclides of the $^{238}U_{92}$ series are extremely hazardous owing to their chemical toxicity and high radioactivity. Ingestion of uranium from contaminated water can cause adverse health effects including several forms of cancer, severe kidney or liver damage, birth defects, and even death. The World Health Organization has recognized U(VI) as a human carcinogen, and has recommended that its concentration in drinking water be below 15 micrograms per liter. Some of the DOE sites have uranium contamination in both groundwater and soil. Traditional ex situ remediation approaches based on pump-and-treat methods such as lime neutralization, anion exchange, activated aluminum and biosorption are extremely expensive and limited due to poor extraction efficiency, inhibitory competing ions, and massive waste production. Furthermore, bringing the radioactive contaminants to the surface level shall increase the health and safety risk factors for the cleanup workers and general public. Therefore, efficient and economical strategies are needed for treatment of uranium contamination in groundwater and prevent its further spread into deep subsurface.

In nature, uranium is present in two valence states, namely, hexavalent uranium, U(VI), and tetravalent uranium, U(IV). In groundwater at low pH values, the hexavalent form prevails, forming a highly soluble and mobile ionic compound, uranyl (UO_2^{2+}); however, at higher pH levels U(VI) can form carbonate and hydroxide compounds. Meanwhile, tetravalent uranium is very insoluble in water, as it forms uraninite, UO_2. Reduction of uranium from U(VI) to the U(IV) form can prevent its migration through groundwater and effectively protect our drinking water sources. The present work investigates the removal of uranium from groundwater by two methods: use of zero-valent iron nanoparticles (nZVIs or nFe^0) and bioremediation by microbial reduction using sulfate reducing bacteria (SRB).

Nano-scale zero-valent iron (nZVI) is of increasing interest for use in a variety of environmental remediation, water and waste water treatment applications. The nZVI particles are highly mobile with high surface-to-volume ratio, and various types of nZVI have been widely studied as chemical reductants and adsorbents for environmental applications. This can be attributed to their low production costs and high efficiencies for the removal of a wide range of contaminants including heavy metals and chlorinated compounds. Their use for next-generation remediation technologies is directed at improving the efficacy, versatility and economics of treatment. The application of nZVI particles shall not only reduce U (VI) to insoluble U(IV), but will also result in the adsorption and precipitation of uranyl ions onto these particles. The fundamental concept is that nZVIs are a source of aqueous Fe(II), Fe(III), hydrogen, free radicals, and other species including precipitates of iron-hydroxy complexes that can significantly contribute to uranium removal. It must be noted that the bulk of the U(VI) is in contact with the reactive sites of the nZVI (nFe^0) particles, so that reduction of U(VI) to U(IV) occurs as the UO_2^{2+}species is transformed to UO_2(s), and the nZVI (nFe^0) is oxidized to the ferrous iron (Fe^{2+}). Another reaction occurs between the nZVI and the protons (H^+) in solution to yield the ferrous ions (Fe^{2+}), which in turn undergoes oxidation to the ferric form (Fe^{3+}). Simultaneously, the U(VI) is reduced to U(IV) in the redox coupling reactions. The products UO_2(s) and $Fe(OH)_3$ (s) are synchronously coprecipitated and uranyl-ferric metal complexes are formed resulting in uranium immobilization.

Biological remediation technologies using microbial populations have proven efficient and cost-effective in the treatment waters contaminated with uranium. Among various subsurface microorganisms, the sulfate reducing bacteria (SRB) are a major group responsible for heavy metal immobilization and groundwater detoxification. It is well documented that various types of SRB including several trains of *Desulfovibrio* have been successful in the reduction of uranium from the U(VI) to U(IV) form by transforming uranyl (UO_2^{2+}) to uranite (UO_2).. Additionally, when SRB are artificially stimulated in the subsurface using nutrient injection, multiple benefits might be derived besides uranium reduction. Other heavy metals and radionuclides such as chromium Cr(VI), U(VI), and technetium Tc(VII) could also undergo reduction to less soluble forms. Bioremediation of uranium using SRB essentially involves microbial reduction of U(VI) to U(IV) through the release of enzymes, as well as through chemical reduction by interaction with microbial by-products. Therefore, the reduction of U(VI) to U(IV) using SRB involves at least these three important processes: (1) U(VI) binding to the cell surface and to extracellular biopolymers (biosorption); (2) chemical reduction of U(VI) by microbially generated hydrogen sulfide (H_2S); and (3) bioreduction of U(VI), which is enzymatic dissimilatory metal reduction with U(VI) acting as a terminal electron acceptor. Additionally, sorption onto bio-surfaces, including polymers and dead cells can contribute to the removal and immobilization of uranium. Chemical reduction by the H_2S produced by SRB is an important aspect of this research, as these bacteria can substitute uranium (VI) as an electron acceptor to produce energy. However, certain process variables including pH, nutrients, temperature, and uranium concentration must be investigated and carefully controlled for achieving maximum removals of uranium. One of the major challenge sin the implementation of the technology is to make the SRB reduction of uranium a more efficient and economical process by optimally using carbon sources and electron donors such as methanol, ethanol, etc.

The experimental methodology for this study uses two parallel columns packed with aquifer sand media – one containing nZVI particles, and the other a microbial biofilm of SRB. The uranium concentrations in the aqueous phase were determined by liquid scintillation counting (LSC). This analytical technique achieved a low uranium detection limit of 2 micrograms per liter. The continuous flow column experiments determined uranium removal rates as a function of time, and evaluated the overall process dynamics and efficiencies for the two scenarios (using nZVI particles and employing a a SRB biofilm). The presentation shall involve detailed investigation of the mechanisms associated with uranium removal in the two different processes – sand column containing nZVIs, and sand column with a biofilm constituted by SRB. It will also include the optimization of the sand column with SRB biofilm using appropriate carbon source and electron acceptor for optimizing the process efficiency and economics. Furthermore, the discussion will include the presentation of a mathematical model for forecasting the removal of uranium in abiotic (nZVI) and biotic (SRB) soil columns.

(25g) Facile Fabrication of Functional TiO2 Nanoarrays On Different Substrates

Xinghua Meng[1], K. Y. Simon Ng[2] and Da Deng[1], (1)Chemical Engineering and Materials Science, Wayne State University, Detroit, MI, (2)Department of Chemical Engineering and Material Science, Wayne State University, Detroit, MI

Facile Fabrication of Functional TiO_2 Nanoarrays on Different Substrates

Xinghua Meng,1 K. Y. Simon Ng,1 Da Deng1,

1. Department of Chemical Engineering and Materials Science, Wayne State University, Detroit, MI 48202

Email: da.deng@wayne.edu

There is increasing interest in fabrication of functional nanoarrays on substrates, which can find important functional applications in catalysis, energy storage, solar cells, absorption, photonics, and chemical sensors. In particular, semiconductor TiO_2 nanoarrays, fabricated and employed for energy and environmental applications, attract much attention. However, it is still challenging to fabricate TiO_2 nanoarrays with desired structures in a controlled manner on arbitrary substrates. In this presentation, our recent efforts on the design and preparation of TiO_2 nanoarrays on different substrates with structure control will be reported. The structure and composition are thoroughly characterized by field emission scanning electron microscopy (FESEM), high-resolution transmission electron microscopy (HRTEM), x-ray diffraction (XRD). A number of proof-of-concept applications will be demonstrated, and the improved performance could be attributed to structure and chemical functionalized surface control.

(26a) Enantioselective Separation On Chiral Au Nanoparticles

Nisha Shukla, Institute for Complex Engineered Systems, Carnegie Mellon Univeristy, Pittsburgh, PA and Andrew J. Gellman, Chemical Engineering, Carnegie Mellon University, Pittsburgh, PA

Adsorption of chiral compounds on chiral surfaces is the initial step in enantioselective processes such as separations and catalysis. There has been a significant effort over the past decade aimed at the preparation of chiral nanoparticles based on metallic cores modified by chiral ligands. In principle, these can serve as the basis for enantioselective chemical processing. In this work we demonstrate a simple measurement of enantioselective adsorption on chiral metal nanoparticles using a method that can yield quantitative measures of the enantiospecific adsorption equilibrium constants [1].

The surfaces of chemically synthesized Au nanoparticles have been modified with D- or L-cysteine to render them chiral and enantioselective for adsorption of chiral molecules. Their enantioselective interaction with chiral compounds has been probed by optical rotation measurements when exposed to racemic propylene oxide. The ability of optical rotation to detect enantiospecific adsorption arises from the fact that the specific rotation of polarized light by R- and S-propylene oxide is enhanced by interaction Au nanoparticles. This effect is related to previous observations of enhanced circular dichroism by Au nanoparticles modified by chiral adsorbates. More importantly, chiral Au nanoparticles modified with either D- or L-cysteine selectively adsorb one enantiomer of propylene oxide from a solution of racemic propylene oxide, thus leaving an enantiomeric excess in the solution phase. Au nanoparticles modified with L-cysteine (D-cysteine) selectively adsorb the R-propylene oxide (S-propylene oxide). A simple model has been developed that allows extraction of the enantiospecific equilibrium constants for R- and S-PO adsorption on the chiral Au nanoparticles.

[1] N. Shukla, M.A. Bartel, A.J. Gellman "Enantioselective separation on chiral Au nanoparticles" *Journal of the American Chemical Society*, *132*(25), (**2010**), 8575–8580

(26b) Specific Targeting Stealth Nanoparticles Coated with Ultra-Low Fouling Peptide

Ann K. Nowinski, Lei Zhang, Wei Yang and Shaoyi Jiang, Chemical Engineering, University of Washington, Seattle, WA

Nanoparticle-based biotechnology is relevant to many biomedical applications including drug delivery, diagnosis, and biosensing. A large obstacle to these applications is nonspecific protein adsorption, which can result in cellular uptake, nanoparticle aggregation, or an immune system response. Here we present an ultra-low fouling all -natural peptide coating based on the design principle of alternating positively and negatively charged residues. These mixed charge peptides exhibit high resistance to nonspecific protein adsorption. Furthermore, they can be extended with a functional peptide sequence to achieve specific targeting. Integrating the ultra-low fouling peptide sequence with a functional peptide sequence allows replacement of nonfouling synthetic materials and avoids the complex chemistries used in bioconjugation. These peptide-coated gold nanoparticles display stability in salt solution, protein solutions, and even undiluted human blood serum. Particles lacking specific targeting moieties possess stealth properties, whereas particles containing specific targeting moieties (RGD) are specifically uptaken by cells. This talk will discuss the assembly conditions, characterization, stability, and versatility of the peptide-based gold nanoparticle system.

(26c) Development of Magnetic Field-Responsive Surfactants

Sun Hae Ra Shin[1], Rajiv Misra[2], Peter Schiffer[2] and Kyle J. M. Bishop[3], (1)Chemical Engineering, The Pennsylvania State University, University Park, PA, (2)Physics, The Pennsylvania State University, University Park, PA, (3)Chemical Engineering, Penn State University, University Park, PA

Development of magnetic field-responsive surfactants

Sun Hae Ra Shin[1], Rajiv Misra[2], Peter Schiffer[2], Kyle Bishop[1]*

[1]Department of Chemical Engineering, The Pennsylvania State University, University Park, PA 16802

[2] Department of Physics, The Pennsylvania State University, University Park, PA 16802

*Address correspondence to kjmbishop@engr.psu.edu

This talk presents the development of magnetic field-responsive surfactants that can be used for dynamic and continuous tuning of interfacial properties. The magnetic surfactants consist of inorganic nanoparticles with amphiphilic surfaces. Particles of magnetic materials such as cobalt ferrite exhibit large magnetic moments and can be oriented and positioned by externally applied fields. The preparation of amphiphilic surface is readily achieved by spontaneous self-organization of a mixed monolayer of hydrophobic and hydrophilic ligands on the particle surface. Strong coupling between the particles' internal magnetic moments and their asymmetric surface chemistry allows external magnetic fields to directly influence the properties of the interface through the competition of magnetic and surface forces. The field-induced response of liquid interfaces containing these responsive surfactants is investigated with pendant drop tensiometry.

(26d) Deposition of Coated Titanium Dioxide Nanoparticles with Controllable Wetting Properties

Mikko Aromaa[1], Janne Haapanen[1], Hannu Teisala[2], Mikko Tuominen[2], Jurkka Kuusipalo[2], Milena Stepien[3], Jarkko J. Saarinen[3], Martti Toivakka[3] and Jyrki M. Mäkelä[1], (1)Aerosol Physics Laboratory, Tampere University of Technology, Tampere, Finland, (2)Paper Converting and Packaging Technology, Tampere University of Technology, Tampere, Finland, (3)Laboratory of Paper Coating and Converting, Abo Akademi University, Turku, Finland

Liquid Flame Spray (LFS) method can be used in synthesis and deposition of nanoparticles. The deposited nanoparticle coating forms a functional surface on top of the substrate. We have previously demonstrated that the coating can be applied on paperboard in a roll-to-roll process in a pilot scale with the substrate speed up to 150 m/min [1]. Titanium dioxide nanoparticles are hydrophilic by nature but with a careful selection of the coating parameters the substrate, especially paper board, can participate in the process. Even without affecting the visual or mechanical properties of the substrate some combustion fumes might be released from the substrate. The combustion fumes condense on top of the LFS deposited titania nanoparticles forming a coating on top of the particles [2]. The coating layer is only a few nanometres thick and thus transparent. The photoactive properties of titanium dioxide are preserved as the coating on top is transparent and therefore UV-light can be used in switching the wetting properties of the surface from superhydrophobic to superhydrophilic [3].

[1] Mäkelä, J.M., Aromaa, M., Teisala, H., Tuominen, M., Stepien, M., Saarinen, J.J., Toivakka M. and Kuusipalo, J., Nanoparticle deposition from liquid flame spray onto moving roll-to-roll paperboard material, Aerosol Science and Technology, 45, 827-837, 2011.

[2] Aromaa, M., Arffman, A., Suhonen, H., Haapanen, J., Keskinen, J., Honkanen, M., Nikkanen, J.-P., Levänen, E., Messing, M.E., Deppert, K., Teisala, H., Tuominen, M., Kuusipalo, J., Stepien, M., Saarinen, J.J., Toivakka, M. and Mäkelä, J.M., Atmospheric Synthesis of Superhydrophobic TiO_2 Nanoparticle Deposits in a Single Step using Liquid Flame Spray, Journal of Aerosol Science, 17 pages, accepted 2012.

[3] Stepien, M., Saarinen, J.J., Teisala, H., Tuominen, M., Aromaa, M., Kuusipalo, J., Mäkelä, J.M. and Toivakka, M., Surface chemical analysis of photocatalytic wettability conversion of TiO_2 nanoparticle coating, Surface Coatings Technology, 17 pages, submitted 2011.

(26e) Nanoparticles Surface Design Process for Aggregation and Dispersion Behavior Control in Liquid to Apply Functional Material

Hidehiro Kamiya and Motoyuki Iijima, Institute of Engineering, Tokyo University of Agriculture and Technology, Koganei, Japan

Nanoparticle and nanomaterials dispersion and aggregation behavior control is one of the most important fundamental processes to apply the functional properties development of energy related devices, such as polymer composite and electrode of Li battery. Surface modification and treatment by using various surfactants and coupling agents with different molecular structure on prepared nanoparticle surfaces in organic solvents were applied to control surface interaction between particles. After surface treatment on nanoparticles in suspension, if surface structure on nanoparticles was almost completely covered and modified by surfactant or coupling agent, hard aggregate or agglomerate of nanoparticles were not formed after evaporating and drying process. Such completely surface modified nanoparticles were able to be re-dispersed into different organic solvent. Uniform nanoparticles dispersed polymer composite or electrode for battery were prepared from the above prepared nanoparticle's suspension without aggregation. Furthermore, molecular and nanometer scaled surface structure and interaction after various modification and treatment was characterized by using a colloid probe atomic force microscope, AFM, method. Based on the measurement results, the action mechanism of surface modification was discussed.

Reference

Iijima, M., Kobayakawa, M., Yamazaki, M., Ohta, Y., Kamiya, H., (2009). *Anionic Surfactant with Hydrophobic and Hydrophilic Chains for Nanoparticle Dispersion and Shape Memory Polymer Nanocomposites,* J. Am. Chem. Soc. 131(45) 16342-16343.

Iijima, M., Kamiya, H., (2010). *Layer-by-Layer Surface Modification of Functional Nanoparticles for Dispersion in Organic Solvents*, Langmuir, 26, 17943–48.

(87a) Activation of NALP3 Inflammasome by Gold Nanoshell/Silica Core Nanoplasmonics

Hai Nguyen, Chemical Engineering, University of Washington, Seattle, WA and Hong Shen, Department of Chemical Engineering, University of Washington, Seattle, WA

Gold nanoshells are particles consisting of a silica core coated in a thin gold shell used for the photothermal ablation of tumors and are currently in clinical trials. The benefit of using a nanoparticle system consisting of a silica core and gold shell layer is the relative biocompatibility of the two materials and lack of toxicity in clinical use. Nanoshell accumulation in tumors relies on the enhance permeability and retention(EPR) effect in which circulating nanoparticles extravasate into tumors due to the leaky nature of the vasculature within the rapidly expanding tumor mass. Circulating nanoparticles also accumulate in other areas as well, such as the liver, kidneys, spleen, and lungs.

Particulate has been shown to activate NALP3 inflammasome complex. Inflammasomes form high molecular weight complexes that lead to the activation of caspase-1 to cleave precursors of proinflammatory cytokines, such as pro-IL-1β and pro-IL-18. The generation of IL-1β, a potent proinflammatory cytokine, is believed to be the key mediator in the generation of a cascade of immune responses. It can recruit neutrophils and other inflammatory cells to the site of injury, promote the maturation of dendritic cells (DCs), and prime $CD8^+$ T-cells, induce the differentiation of type 17 T-helper cells, and stimulate the production of various downstream molecules such as nitric oxide(NO) and proinflammatory cytokines such as IL-6 and IL-12. Generation of proinflammatory cytokines at sites of nanoparticle accumulation leads to recruitment of other immune cells to clear the particulate accumulation, but can also lead to persistent inflammation and possible tissue damage.

In this study we show that gold nanoshells thought to be biocompatible induce inflammatory responses through NALP3 inflammasome. Surface functionalization with polyethylene glycol(PEG) attenuated their activity of stimulating inflammasomes. Some mechanisms involved in the inflammasome complex formation in response to gold nanoshell/silica core nanoplasmonics are elucidated. Implications of our results for nanoshell enabled photothermal therapy will be discussed.

(87b) Development of in-Vivo Screening Benchmark for Complex Engineered Silica Nanoparticles

Michelle Najera, Department of Chemical Engineering, University of Pittsburgh, Mascaro Center for Sustainable Innovation, University of Pittsburgh, Pittsburgh, PA, Qing Bai, Department of Neurology, University of Pittsburgh, Edward Burton, Department of Neurology, University of Pittsburgh, Pittsburgh and Götz Veser, Chemical and Petroleum Engineering, University of Pittsburgh, Pittsburgh, PA

Silica nanomaterials have emerged as facile, easily modifiable templates for a wide range of applications, including biomedical imaging, drug delivery, catalysis, separations, and sensors. More broadly, new applications of many types of functional nanomaterials has led to the development of engineered nanomaterials at an ever accelerating pace, and to a rapid increase in their use in consumer products, including cosmetics, sporting goods, clothing, and consumer electronics. Concurrently, increasing evidence is indicating elevated toxicity of some nanomaterials, suggesting an urgent need to establish appropriate toxicity tests and thresholds for this emerging technology.

Studies to-date have mostly concentrated on air contamination as the exposure route. However, increasing use of nanomaterials-containing products is likely to result in an accumulation of these materials in the environment and hence increased risk from other exposure routes. This motivates the need for the development of high-throughput screenings for nanomaterials toxicity along other exposure routes. In our work, we are using a zebrafish (*Dario rerio*) model for toxicity screenings of complex nanomaterials in aqueous environments. Several factors make zebrafish excellent candidates for testing nanoparticle toxicity compared to more traditional *in vivo* models, including prolific breeding, embryo tissue transparency, rapid development time, and the simplicity of media-dosed toxin exposures.

In the present contribution, we are specifically investigating the possibility to utilize encapsulation of nanoparticles in nanostructured silica shells to simultaneously enhance material functionality and mitigate toxic effects. We present results from a study in which the toxicity of carefully synthesized Ni-silica nanomaterials with different, well-controlled nanostructures is investigated. Multi-component materials of this type, containing nanoparticles of one or more metal in combination with an inert matrix (here: silica), are expected to be especially important structures in emerging nanomaterial applications which widely rely on the use of embedded (rather than free) nanoparticles.

The toxicity of surface-deposited Ni nanoparticles is compared to that of embedded and encapsulated metal nanoparticles. Since porous silica nano- and microparticles have been shown to be non-toxic, it was hypothesized that embedding or encapsulation with this inert support could reduce or entirely mitigate nanoparticle toxicity while preserving the accessibility of the metal nanoparticle surface. Our results demonstrate that porous silica coatings can indeed significantly lower the toxicity of metal nanoparticles compared to the equivalent dosing of analogous metal salts. The toxicity trends can be explained by the high stability of the nanoparticles as indicated by a limited dissolution of metal ions from the structured materials into the media. However, measurement of metal uptake into zebrafish tissues also indicates that particle transport and hence exposure can be *increased* as a result of silica structuring, suggesting the presence of a "Trojan horse"-like mechanism. Thus, while nanostructuring can be useful for mitigating the toxicity of nanoparticles, it may also enable an entirely new mechanism of toxicity.

(87c) Disruption of Tethered Lipid Bilayers by Silica-Core Nanoparticles: Effect of Surface Functional Group

Ying Liu[1], Quanxuan Zhang[2], Gregory Baker[2], Zhen Zhang[3] and R. Mark Worden[4], (1)Chemical Engineering and Materials Science, Michigan State University, East Lansing, MI, (2)Chemistry, Michigan State University, East Lansing, MI, (3)Department of Statistics and Probability, Michigan State University, East Lansing, MI, (4)Chemical Engineering and Materials Science, Michigan State Univeristy, East Lansing, MI

Disruption of tethered lipid bilayers by silica-core nanoparticles: effect of surface functional group

Ying Liu, Quanxuan Zhang, Gregory L. Baker, Zhen Zhang, and R. Mark Worden

In this paper, we use present a method that uses a tethered bilayer lipid membrane (tBLM) to study the interaction of engineered nanomaterials (ENM) with biomembranes. Highly insulating tBLM were formed on gold electrodes using 1, 2-dipalmitoyl-*sn*-glycero-phosphothioethanol (DPPTE) to form the lower leaflet and 1, 2-dioleoyl-*sn*-glycero-phosphocholine (DOPC) to form the upper leaflet. The tBLM were then exposed to silica nanoparticles having different surface properties (unmodified, amine terminated, and carboxylic-acid terminated). Electrochemical impedance spectroscopy was used to measure time-dependent changes in the tBLM's impedance following ENM exposure. The data were fit to an exponential model and then analyzed using a hierarchical clustering algorithm. The resulting dendrograms confirmed that ENM having different surface functional groups induced statistically different changes in tBLM impedance. The amine terminated ENM reduced tBLM impedance considerably faster than did the carboxylic-acid terminated and unmodified ENM. Based on dynamic light scattering data, exposure to the tBLM triggered an increase in average particle size for the unmodified ENM but not the amine-terminated and unmodified ENM. Results of this study could provide insight into fundamental mechanisms by which ENM interact with biomembranes and may lead to high-throughput methods to assess the health risk of ENM based on their interactions with biomembranes.

Keywords: tBLM; silica nanoparticles; electrochemical impedance spectroscopy; hierarchical clustering; lipid bilayer

(87d) Effect of Nanoparticle Shape and Charge On Cytotoxicity: Molecular Dynamics Simulations

Shikha Nangia, Department of Biomedical and Chemical Engineering, Syracuse University, Syracuse, NY, Stephen DeSalvo, Biomedical and Chemical Engineering, Syracuse University, Syracuse and **Radhakrishna Sureshkumar**, Department of Biomedical and Chemical Engineering and Department of Physics, Syracuse University, Syracuse, NY

Nanoparticles (NPs) that exhibit plasmon resonance in the IR regime are promising candidates for biomedical applications in targeted vivo imaging, diagnostics, and cancer therapy since the biological tissue is transparent to near IR radiations. However, these engineered nanoparticles can be highly toxic to living tissue and lead to cell-death. To better understand the molecular mechanisms underlying cytotoxicity caused by NPs, we perform MD simulations of various gold nanoparticles such as nanocube, nanocage, nanocylinder, nanocone, and nanorice among others and, study their interaction with a model lipid bilayer. It is known that charged nanoparticles penetrate the lipid bilayer by forming holes leading to cyotoxicity. We will provide estimates of the cell-membrane damage and kinetic data to quantify the penetration of the various nanoshapes. Combined effect of nanoshape and charge on the translocation of these particles and membrane damage will also be discussed.

Acknowledgements: Finacial support from NSF grants CDI-1049489 and EFRI- 1137186 are gratefully acknowledged.

(87f) Size-Dependent Cytotoxicity of Nanosilver: Effect of Released Ag+ Ions

Anna Pratsinis[1], Pablo Hervella[2], Jean-Christophe Leroux[2], Sotiris E. Pratsinis[3] and **Georgios A. Sotiriou**[4], (1)Particle Technology Laboratory, ETH Zurich, Zurich, Switzerland, (2)Department of Chemical and Applied Biosciences, Drug Formulation and Delivery Laboratory, Institute of Pharmaceutical Sciences, Zurich, Switzerland, (3)Particle Technology Laboratory, Institute of Process Engineering, Department of Mechanical and Process Engineering, ETH Zurich, Zurich, Switzerland, (4)Department of Mechanical and Process Engineering, Particle Technology Laboratory, ETH Zurich, Zurich, Switzerland

Silver nanoparticles (nanosilver) are broadly used today in textiles, food packaging, household devices and in bioapplications prompting for a better understanding of their toxicity and biological interactions [1]. However, the cytotoxicity mechanism of nanosilver is still not clear as limitations regarding its controlled synthesis hinder such investigations. Here, nanosilver with closely controlled size (from 5.7 to 16.8 nm) immobilized on nanostructured silica without any coating was produced by flame synthesis [2,3]. Its dispersibility in aqueous solutions was facilitated by the presence of silica, and therefore, the Ag^+ ion release of this nanosilver was not inhibited by surface coatings [4,5]. Detailed physicochemical characterization of these nanoparticles with X-ray diffraction, N_2 adsorption and electron microscopy revealed the size tunability. Smaller nanosilver released larger fractions of its mass as Ag^+ ions upon its dispersion water. This strongly influenced the cytotoxicity of such nanosilver upon incubation with murine macrophages, which was assessed by monitoring their mitochondrial activity. The nanosilver size dictated its mode of cytotoxic activity. The toxicity of small nanosilver that released high fractions of Ag^+ was mostly mediated by the released ions. In the case of large nanosilver, which released only minimal fractions of Ag^+, the toxicity was attributed to both the released ions but also to the direct nanoparticle-cell contact. The nanosilver behavior when incubated with murine macrophages was also monitored with dark-field and hyperspectral imaging microscopy. It was found that nanosilver tends to agglomerate upon interacting with cells.

[1] Sotiriou, G. A. & Pratsinis, S. E. Engineering nanosilver as an antibacterial, biosensor and bioimaging material. *Curr. Opin. Chem. Eng.* **1**, 3-10 (2011).

[2] Sotiriou, G. A. & Pratsinis, S. E. Antibacterial activity of nanosilver ions and particles. *Environ. Sci. Technol.* **44**, 5649-5654 (2010).

[3] Sotiriou, G. A., Teleki, A., Camenzind, A., Krumeich, F., Meyer, A., Panke, S. & Pratsinis, S. E. Nanosilver on nanostructured silica: Antibacterial activity and Ag surface area. *Chem. Eng. J.* **170**, 547-554 (2011).

[4] Sotiriou, G. A., Hirt, A. M., Lozach, P. Y., Teleki, A., Krumeich, F. & Pratsinis, S. E. Hybrid, silica-coated, Janus-like plasmonic-magnetic nanoparticles. *Chem. Mater.* **23**, 1985-1992 (2011).

[5] Sotiriou, G. A., Sannomiya, T., Teleki, A., Krumeich, F., Vörös, J. & Pratsinis, S. E. Non-toxic dry-coated nanosilver for plasmonic biosensors. *Adv. Funct. Mater.* **20**, 4250-4257 (2010).

(87g) Studies of Protein Expression Upon Cellular Exposure to Nanomaterials

Gabrielle Rogers-Nieman, Department of Chemical Engineering, West Virginia University, Morgantown, WV

Studies of protein expression upon cellular exposure to nanomaterials

Gabrielle Rogers-Nieman[1], David Lowry [2],

Michael L. Kashon[2], Linda M. Sargent[2], and Cerasela Zoica Dinu[1]*

[1]Department of Chemical Engineering, West Virginia University, Morgantown WV, 26506, USA

[2]National Institute for Occupational Safety and Health, Morgantown WV, 26505, USA

* Corresponding author

Carbon nanotubes (CNTs) have many potential applications in industry because of their extraordinary mechanical, electrical and structural properties. However, recent studies demonstrate that CNTs can enter the working environment as suspended particulate matter of respirable sizes, which could pose an occupational inhalation exposure hazard and thus create challenges for exposure control. Research has also shown that CNTs associate with cellular elements, namely cytoskeletal filament microtubule and the DNA and interfere with cell progression to normal division, most notably by disruption of the centrosome. This leads to aneuploidy, an early event in the progression of lung mesothelioma. Such results further suggest that CNT cellular uptake and the mechanisms of nanotube cellular transport may play a role in both tumorigenesis and tumor progression. In our research we are unraveling the mechanisms associated with the nanotube uptake and their cellular transport in order to understand if the nanotube cellular interactions affect key protein expression involved in mitotic spindle formation. Using real time PCR complemented by optical microscopy and fluorescence activated cell sorting, our preliminary results show that CNTs affect the cell cycle and this is correlated with fluctuations in the expression level of integral motor proteins responsible for nanotube transport inside the cell. These observations are time and dose dependent. The quantification of expressional differences in messenger RNA of integral molecular motors during a time course of CNT exposure can provide a more detailed explanation of genotoxicity associated with cellular exposure to CNTs.

References:

Sargent LM, Hubbs AF, Young SH, Kashon ML, Dinu CZ, Salisbury JL, Benkovic SA, Lowry DT, Murray AR, Kisin ER, Siegrist KJ, Battelli L, Mastovich J, Sturgeon JL, Bunker KL, Shvedova AA, Reynolds SH. Single-walled carbon nanotube-induced mitotic disruption. Mutat Res. 2011. Dec 8. (Epub ahead of print)

Hubbs AF, Mercer RR, Benkovic SA, Harkema J, Sriram K, Schwegler-Berry D, Coravanahally MP, Nurkiwicz TR, Castranova V, Sargent LM. Nanotoxicology-a pathologist's perspective. Toxicol Pathol. 2011 Feb; 39(2): 301-24/

Quintyne NJ, Schroer TA. Distinct cell cycle dependant roles for dynactin and dynein at centrosomes. J Cell Biol. 2002 Oct 28: 59(2): 245-54.

Kapitein LC, Peterman EJG, Kwok BH, Kim JH, Kapoor TM, Schmidt CF. The bipolar mitotic kinesin Eg5 moves on both microtubules that it crosslinks. Nature. 5 May 2005: 453, 114-118.

Gable A, Qiu M, Titus J, Balchand S, Fernez NP, Ma N, Collins ES, Fagerstrom C, Ross JL, Yang G, Wadsworth P.

Dynamic reorganization of Eg5 in mammalian spindle throughout mitosis requires dynein and TPX2. Mol Biol Cell. 2012 Apr;23(7) 1254-66.

WiltshireC, SinghBL, Stockley J, Fleming J, Doyle B, Barnetson R, Robson CN, Kozielski F, Leung HY. Docetaxel-resistent Prostate cancer cells remain sensitive to S-trityl-L-cysteine-mediated Eg5 inhibition. Mol Cancer Ther. 2010 Jun; 9(6): 1730-3.

(102a) Electrochemical Fabrication of Organic Nanorods On Nanoparticle-Decorated Electrode

Li Li and Guangzhao Mao, Chemical Engineering and Materials Science, Wayne State University, Detroit, MI

Our previous studies have shown that nanoparticles are capable of nucleating and confining organic crystallization into nanorods based on seed-mediated nucleation. This unique mechanism promises a versatile, room-temperature, solution-based process to the manufacturing of organic nanorods. This presentation describes electrochemical deposition of gold nanoparticles on graphite electrode and their use as a confinement medium for electrochemical crystallization of tetrathiafulvalene bromide charge transfer salt (TTF) nanorods and partially oxidized potassium tetracyanoplatinate sesquihydrate (TCP) nanorods. Cyclic voltammetry, AFM, and SEM are used to characterize the nanostructure. The size, shape, and particle density of gold nanoparticles are varied by gold salt concentration, applied overpotential, and deposition time. The nanoparticle-decorated graphite electrode is used for the synthesis of TTF and TCP nanorods. Both compounds show a preferred nucleation on the gold nanoparticle and a confined nanorod growth as compared to the crystals formed on bulk graphite electrode. The results contribute to the understanding of seed-mediated nucleation and manufacturing of nanomaterials.

(102b) Synthesis of Gold Nanokites by Temperature Controlling Method

Chunrong Wang[1], Jiang Lai[1] and Yun Fang[2], (1)School of Food Science and Engineering, Jiangxi Agricultural University, Nanchang, China, (2)School of Chemical and material engineering, Jiangnan University, Wuxi, China

This article describes a simple approach to gold nanokites prepared by the reduction of $HAuCl_4$ in the solution of *β*-cyclodextrin (*β*-CD) through controlling reaction temperature. In this approch, *β*-CD served as both a mild reducing agent and protective agent. However, the reacted temperature was the more important factor that controlled the anisotropic gold nanostructures with various shapes. With a fixed concentration of 0.2 mM $HAuCl_4$ and 0.4% *β*-CD, as the reaction temperature was 90°C, there were some gold nanotadpoles (about 50 nm long) formed; as 80°C, some branched gold nanostructures with equivalent diameter around 150 nm; as 70 °C, gold nanowires as long as 1μm; as 60 °C, gold nanokites (more than 1μm). Electron diffraction pattern (ED) suggested that these nanokites were single crystalline. UV-Vis absorption spectrum of gold nanoparticles indicated that there were anisotropic nanostures formed in the reacted solutions. The preferred noble metal nanokites may be obtained easily by this simple method.

(102c) Organic Nanoparticle Deposition in Organosilane Nano-Arrays

Sunxi Wang, Pedram Jahanian, Juxhin Xhahysa and Guangzhao Mao, Chemical Engineering and Materials Science, Wayne State University, Detroit, MI

The ability to make nano-arrays enables high-throughput screening and nanofluidics. This paper describes a method to generate octadecyltrichlorosilane (OTS) nano-arrays on oxidized silicon wafers for nanoconfining small organic molecules. Particle lithography was carried out on 300 and 900 nm polystyrene particle array templates by chemical vapor deposition of OTS, which reacted with the substrate in the water menisci to form a ring pattern. We discovered reaction conditions that favor bulk polymerization resulting in ~14 nm thick rings for the 900 nm particles and ~9 nm thick rings for the 300 nm particles, both of which exceed the OTS monolayer thickness of 2.6 nm. It is estimated that a 1×1 cm^2 wafer (900 nm particles) contains more than 100 million ordered "nano-flasks" with volume $\sim3\times10^{-10}$ nL. The nano-array pattern is stable against common organic solvents thus allowing repeated use in solution crystallization screening experiments. Several hydrophobic compounds including docosane and aspirin were deposited on the nano-array pattern for the study of nanoconfinement effect on molecular crystallization. The conditions for preferential deposition of the compounds insider the "nano-flasks" will be discussed. The resultant organic nanoparticle arrays on the nanopattern were characterized by AFM and field-emission SEM.

(102d) Transformers: Colloidal Nanocrystals Converting Into Globules On Graphene Oxide

Doh C. Lee and Chaewon Pak, Department of Chemical and Biomolecular Engineering, Korea Advanced Institute of Science and Technology (KAIST), Daejeon, South Korea

Because of its high carrier mobility and mechanical strength, graphene, a one-atom thick carbon sheet, is a promising two-dimensional (2-D) microcontact for nanometer length scale objects, such as nanocrystals. However, tethering nanocrystals on graphene remains challenging, as graphene has few functional sites to adhere nanocrystals to. Graphene oxide (GO), an oxidized form of graphene, has oxygen-rich moieties that can anchor the nanocrystals on the surface of the 2-D microcontact. We exchanged the ligand on the surface of nanocrystals so that the nanoparticles can bind electrostatically to GO. Interestingly, when blended with GO in aqueous solution for 12 h, the nanocrystals undergo substantial size increase and crystalline transformation. For example, 3.5 nm CdSe nanocrystals turn into ~500 nm amorphous Se globules. The crystalline transformation was investigated with nanocrystals of PbSe, Co, and $CoPt_3$ each blended with GO; notably, the crystalline transformation by acidic sites of GO turn out to be dictated by the bonding strength in the nanocrystals. We also identified the reaction conditions that allow uniform dispersion of intact nanocrystals on GO.

(102e) Automated Capillary Force Layer-by-Layer (CF-LBL) Assembly of Polyelectrolytes

Wei Li, Steven Castleberry and Paula T. Hammond, Chemical Engineering, Massachusetts Institute of Technology, Cambridge, MA

In the layer-by-layer (LBL) process, alternating charged or other complementary interacting species are deposited onto a substrate in sequences. LBL assembly enables the tunable design and fine control of functional materials into nanoscale thin films, which has found diverse applications in reactive membranes, drug delivery systems, and electrochemical and sensing devices. However, as the field continues to expend to pursue new discoveries in cell biology and commercial translation in pharmaceutical industry, several engineering challenges need to be overcome, such as low viability of film architectures and morphologies, over-consumption of building materials, and non-sterile environment during film preparation.

Herein, we present a simple and accessible procedure of capillary force LBL (CF-LBL) assembly to build microstrips of thin films at sufficiently high throughput and low cost. Two examples are discussed as follows: 1) pH-dependent thickness behavior of sequentially adsorbed layers of weak polyelectrolytes, and 2) development and investigation of LBL film libraries for the delivery of DNA and RNAi from surfaces. Furthermore, we are in the process of developing this CF-LBL system into a fully programmable and automated platform by using computer-assisted liquid handling robot.

(102f) Tunable Porosity Gradients in Photocatalytic Titania Nanostructures for Air and Water Purification

Michael Riley, Chemical and Biological Engineering, Rensselaer Polytechnic Institute, Troy, NY and **Joel L. Plawsky**, The Isermann Department of Chemical and Biological Engineering, Rensselaer Polytechnic Institute, Troy, NY

Titanium dioxide nanoparticles have been used extensively a photocatalyst to breakdown organic compounds when exposed to ultraviolet light. However, it is difficult to control the morphology and structure of a collection of nanoparticles. Oblique angle deposition allows for the growth of catalytic nanostructures on solid surfaces and the ability to tailor the shape, structure, and density of the material as a function of deposition conditions. The efficiency of porous TiO2 films developed for this purpose is limited by light intensity losses due to absorption as well as the ability of the reactant to diffuse into the pores. The interplay of light absorption and reactant diffusion in films grown by oblique angle deposition was investigated to find the optimal reaction conditions. *High surface area nanorods of titanium dioxide were grown on a transparent substrate to investigate their effectiveness as photocatalytic agents for the destruction of organic contaminants in air and water. Optical transmission measurements were made that allowed for an estimation of the porosity of the film (75%-78%). The photocatalytic degradation of an indigo carmine dye solution and a gas mixture containing isobutylene on the porous films was shown to depend on film thickness and annealing conditions. The effectiveness of the film was assessed by observing the change in absorbance of the dye at 610nm over time and quantifying the film performance using a pseudo-first-order reaction rate model. Reaction rates increased as the film thickness increased from 600nm to 1000nm, but leveled out or decreased at thicknesses beyond 1500nm. A transport/reaction model was used to show that there exists an optimal geometry that maximizes the overall reaction rate and that such a geometry can be simply produced using glancing angle deposition. The nanorod films were benchmarked against nanoparticle films and were shown to perform as well as 0.73g/L of 25nm diameter anatase nanoparticles having a surface area of 50m^2/g.*

(102g) Unique Regimes in Nanofiber Formation From Pendant Precursor Drops Using Gas Jet

Rafael E. Benavides, Polymer Engineering, The University of Akron, Akron, OH, Sadhan C. Jana, Department of Polymer Engineering, University of Akron, Akron, OH and Darrell H. Reneker, Department of Polymer Science, The University of Akron, Akron, OH

A large body of literature covers formation of nanofibers from a pendant drop of polymer solution in electrospinning process. The nanofibers result from electrically charged jet of polymer solution emanating from the pendant drop by the action of electric fields of about 1 kV/cm. In this work, we present a novel approach for obtaining nanofibers of similar diameter as in electrospinning from a pendant drop by the action of compressed air jet. This novel and cost effective method to produce nanofibers is demonstrated experimentally by producing fibers from different polymer solutions. Fibers ranging from a few tens of nanometers to a few micrometers are found. Two different regimes of fiber formation are reported – (1) a *surface drainage regime* where fibers are formed from a single cone formed in the drop similar to the electrically formed Taylor cone in electrospinning, and (2) a *volumetric drainage regime* where the drop is continuously deformed to form the liquid jet that is stretched by the gas jet. These fibers find an array of applications including scaffolds, encapsulated materials, and controlled drug delivery systems.

(102h) Micro/Nanofiber Assemblies with Controlled Fiber Morphology

Ji Wang, Engineering Science and Mechanics, Virginia Tech, Blacksburg, VA and Amrinder S. Nain, Mechanical Engineering, Virginia Tech, Blacksburg, VA

Micro/nanofiber Assemblies with Controlled Fiber Morphology

Ji Wang[1], Amrinder S. Nain[2,3]

[1] Engineering Science and Mechanics, [2] Mechanical Engineering, [3] Biomedical Engineering, Virginia Tech, Blacksburg, VA

Introduction

Polymeric micro/nanofibers are increasingly gaining importance due to their versatile application in diverse fields of tissue engineering, smart textiles, sensors and actuators. However, aligned deposition and continuous production of long fibers with sub-micron and nanoscale diameters with desired morphologies has been challenging. Recently, techniques of electrospinning, template synthesis and sequential micro dry spinning have been used to deposit a wide variety of micro/nano scale polymeric fibers in aligned configurations [1-6]. Among all these methods, electrospinning is the most popular process, which allows for the production of polymer/ceramic fibers ranging from tens of nanometers to a few microns in diameter [7-11]. Recent research has been devoted to control electrospun fiber morphology and surface porous /wrinkled fibers have been demonstrated [12-13]. In addition, co-axial electrospinning has been well established as an easy and cost effective method to fabricate tubular structures [14]. Using specialized techniques, electrospinning methods are able to achieve control on fiber alignment in single layer. However, fiber spacing and three-dimensional stacking of fibers in precise locations has not been demonstrated using electrospinning methods. Our recently reported non-electrospinning fiber manufacturing strategy based on Spinneret based Tunable Engineering Parameter (STEP) technique is able to deposit fibers and fiber arrays in aligned configurations with control on spacing in single and multiple layers. This allows fabrication for the very first time of fiber hierarchical structures with each layer containing fibers of customized diameters (sub 100nm-micron), spacing (sub 100 nm-microns), orientation (0 deg. -90 deg.) and variety of polymeric materials. In this study, we extend this capability to include control on surface morphology of fibers through manipulation of solvent and ambient atmosphere humidity.

Experimental

All chemicals were used as received without further purification.

Polystyrene(PS) Two molecular weights PS (650K and 2000K $g.mol^{-1}$, Polymer scientific, USA) were dissolved in xylene and DMF mixture with varying wt% concentrations (from 1% to 18%) for one week prior to experiments.

Poly(methyl methacrylate). PMMA(540K $g.mol^{-1}$, Polymer scientific, USA) dissolved in chlorobenzene at 14%wt was used in this study.

Polyurethane (PU5719, Thermal Plastic, Lubrizol Advanced Material, USA)was dissolved in DMF at 30%wt and heated to 80 °C stirring for 24 hours to prepare homogenous solution.

Polyacrylonitrile(PAN,150K $g.mol^{-1}$,SIGMA ALDRIC, USA) was dissolved in DMF at 11%wt overnight to prepare homogeneous solutions

STEP spinning. All experiments were performed at 20°C. As shown in Fig 1(a), the viscous polymer solutions were pumped through the micropipette spinneret (ID=100mm) and formed a solution droplet at the nozzle. The substrate (Thermanox, USA) was mounted to a DC motor, which in turn was mounted onto a motorized three degree of freedom micro-positioning stage (VP-25XA, Newport Inc., USA). After the substrates contacted the solution droplet, fibers were collected on the substrate in aligned configurations. By adjusting the linear movement speed of the substrate and the angle between micropipette and the rotating axis, fiber spacing and orientation was tightly controlled.

Fiber Fusion: PMMA fiber network were treated with chlorobenzene vapor at 20°C in an enclosed chamber for 2 minutes.

Carbonization: PAN fibers were calcinated at 900°C in nitrogen atmosphere for 3 hours to produce carbon fibers.

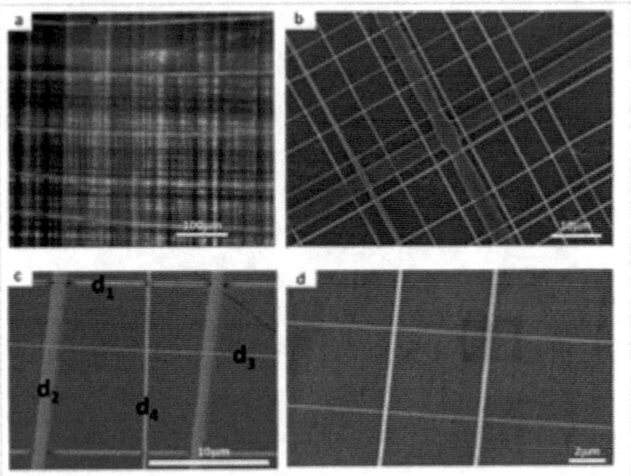

Figure2. (a) 4 μm diameter PS fiber(red) and 400nm diameter(green) PS suspended network (b)PU network deposited over crisscross PS

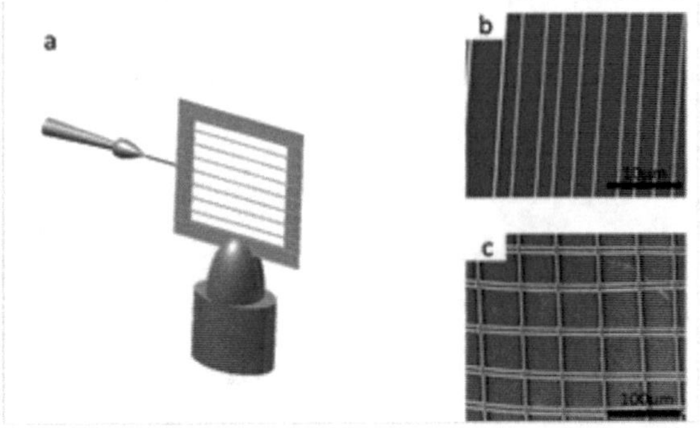

Figure1. (a) Schematic illustration of STEP technique (b) PS fiber arrays with equal fiber spacing. (c) A fused PMMA fiber network.

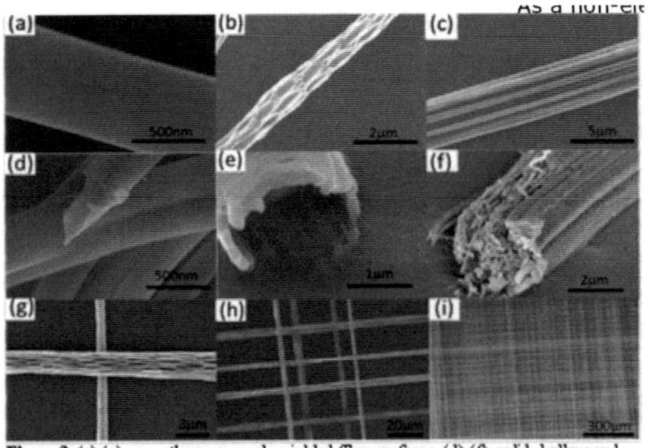

Figure 3. (a)-(c): smooth, porous and wrinkled fiber surfaces. (d)-(f): solid, hollow and co-continuous fiber inner structures.(g) A 2D crisscross of porous fibers with different pore size (h) A 3D assembly(4 layers of 0/90° alternating layup) of surface wrinkled fibers. (i) A scaffold of 16 layers 0/90° alternating layup of smooth fibers.

Results and Discussion

As a non-electrospinning technique, STEP relies on the rotating substrate to transform the polymer solution filament into nano/micro scale fibers. Since no voltage was used in this technique, the classic jet bending and spraying in electrospinning were avoided in STEP, and as a result, the STEP fiber diameter is narrowly distributed (within 20%) and STEP fibers maintain aligned configurations(angles between paralleled fibers are less than 1°) . In Fig 1(b), arrays of uniform diameter PS fibers were deposited with equal inter-fiber space; in Fig 1(c), two PMMA fiber arrays were deposited orthogonally, forming a crisscross fiber network, which was then treated by solvent vapor using a custom in-house solvent evaporator, creating an inter-connected structure composing of "unit cells".

Using the STEP technique, we have successfully fabricated aligned nanofibers from a variety of polymers other than

PS and PMMA; typical examples include polyurethane(PU), polyacrylonitrile(PAN), poly(ethylene oxide)(PEO), poly(lactic-co-glycolic acid)(PLGA) and fibrinogen. Using these uniform diameter fibers as building blocks, hierarchical fiber assemblies were achieved. Fig 2(a) shows a suspended PS fiber assembly composing of 4 mm and 400 nm diameter fibers. Fig 2(b) is a hierarchical assembly of PU fibers (300nm in diameter) and PS fiber (4mm in diameter). Fig 2(c) demonstrates a $10\times10mm^2$ PS fiber assembly, composing of 4 different diameters (d_1=400nm, d_2=800nm,d_3=80nm, d_4=160nm). In addition, we demonstrated that inorganic nanofibers could be fabricated by calcination of polymer precursors. In Fig 2 (d), crisscross carbon nanofibers were obtained through calcination of PAN fibers at 900° C in nitrogen atmosphere for 3 hours.

By changing the solvent and adjusting the humidity of atmosphere, we demonstrate that both the surface and inner morphology of STEP fibers can be altered. As shown in Fig 3(a)-(c), at 60% relative humidity (RH), by gradually adding DMF in xylene (from 0% wt to 50%wt and 75%wt), STEP fiber surfaces evolve from smooth to porous and finally wrinkled. The mechanism for surface attributes formation is solvent induced phase separation—the solvent rich phase turns into pores while the solvent poor phase turns into the matrix. As DMF concentration becomes dominating in the mixed solvent, the solvent becomes more hydrophilic and the polymer-solvent interaction is weakened, leading to larger dispersed phases. As a result, surface features evolve from discontinuous pores (Fig 3(b))to parallel channels(Fig 3(c)). Fig 3(d)-(f) demonstrate the corresponding inner fiber structures, which depend on the relative rates of solvent evaporation to phase separation rate. When solvent evaporation rate dominates over the phase separation rate (using solvent of 0% wt DMF), a uniformly solidified cross section is obtained (Fig3 d). When the solvent evaporation rate is faster(using solvent of 50%wt DMF), a thin skin consisting of almost pure polymer is formed first, which subsequently develops towards the core of the fiber, leading to the formation of a hollow fiber (Fig 3 e). To the best of our knowledge, this is first time tubular structures were obtained through control of evaporation process. In electrospinning, the tubular structures usually collapse into flat ribbons because of the lateral pressure caused by cumulated electrical charges[15]. In STEP, tubular structures were well preserved due to abscence of electrical field. The formation of a bi-continuous core (Fig 3 f) is attributed to the phase separation process when the evaporation rate is slowed (using solvent of 75% wt DMF). We go further to demonstrate that both the pore size and tube wall thickness are function of solution concentrations. By changing polymer concentration in the solvent, fibers with different diameters, surface morphologies and inner structures were obtained. Finally, fibers of different morphologies were incorporated into 2D/3D assemblies as shown in Fig 3(g) to (i).

Conclusion:

STEP turns out to be a very versatile method, which allows controlled fiber deposition of various materials. The unique advantage of STEP in spacing and orientation control realizes the 3D nano/micro fiber assembly. Fiber morphology is subjected to both processing parameters (humidity) and material parameters(concentration and solvent). Compared with electrospinning, STEP offers an alternative to achieve nano/micro fibers with desired position and attributes.

Reference:

[1] D. H.Reneker, A. L.Yarin, H.Fong , S.Koombhongse, *J. Appl. Physics*. **2000**,87,4531.

[2] J. Doshi , D. H.Reneker, *J. of Electrostatic*. **1995**,35,151.

[3] A.Theron, E.Zussman, A.Yarin, *Nanotechnology* **2001**,12,384.

[4] E.Zussman, A .Theron, and A.Yarin, *Appl. Phys. Lett.* **2003**,82,973.

[5] C. Martin ,*Science*. 1994,266,**1961**.

[6] L.Feng, S. Li, H.Li, J.Zhai, Y.Song, L.Jiang, D. A. Zhu, *Chem. Int. Ed.* **2002**, 114,1269.

[7] P. X. Ma , R. J. Zhang, *Biomed. Mater. Res.* **1999**,46,60.

[8] F.Yang, R.Murugan, S. Ramakrishna, X.Wang, Y. X.Ma, and S. Wang, *Biomaterials.***2004**,25,1891.

[9] A. S.Nain, C.Amon, M. Sitti, *IEEE Trans. on Nanotechnology.* **2006**,5, 499.

[10] A. S.Nain, J. C.Wong, C.Amon and M. Sitti , *Appl. Phys. Lett.* **2006**,89,183105.

[11] Dan Li, YuLiang Wang, Younan Xia. *Adv.Mater.***2004**,16,361.

[12]Lifeng Wang, Chia Ling Pai, Mary C.Boyce, and Greory C.Rutledge. *Appl. Phys. Lett.* **2009**,94,151916

[13] Pratyush Dayal and Thein Kyu. *J.Appl.Phys* 2006, 100,043512.

[14] Y.Dror, W.Salalha, R.Avrahami, E.Zussman, A.L.Yarin, R.Dersch, A.Greeiner, and J.H.Wendorff. *Small.* **2007**, 3,1064.

[15] Sureeporn, Koombhongse, Wenxia Liu, Darrell H. Reneker. *Journal of Polymer Science: Part B: Polymer Physics.* **2001**.39,2598.

(113a) Reaxff Reactive Force Field Study of Oriented Attachment of TiO2 Nanocrystals in Vacuum and Humid Environments

Muralikrishna Raju[1], Sung-Yup Kim[2], Ya Zhou[3], Kristen Fichthorn[3] and Adri van Duin[4], (1)Dept. of Physics, The Pennsylvania State University, University Park, PA, (2)Dept. of Mechanical and Nuclear Engineering, Pennsylvania State University, University Park, PA, (3)Dept. of Chemical Engineering, Pennsylvania State University, University Park, PA, (4)Dept. of Mechanical and Nuclear Engineering, The Pennsylvania State University, University Park, PA

We use a ReaxFF reactive force field in molecular dynamics simulations to study the aggregation of various titanium dioxide (anatase) nanocrystals in vacuum and humid environments. The nanocrystals are in the 2-6 nm size range and they generally possess shapes dictated by the Wulff construction. Wulff-shaped nanocrystals are terminated by {101} and {001} facets. To mimic the effects of HCl in the hydrothermal synthesis of anatase[1], we also create {112} facets on the surfaces of some of the particles. In vacuum, the nanocrystals tend to merge along their direction of approach, resulting in a polycrystal[2]. By contrast, in the presence of water vapor, the nanocrystals tend to reorient themselves and aggregate via the oriented attachment mechanism to form a single or twinned crystal.

We find that adsorbed water molecules and hydroxyl groups play multiple roles in oriented attachment. As the nanocrystals approach one another closely, adsorbed water molecules and surface hydroxyls prevent their immediate aggregation. These adsorbed species create a hydrogen bonding network, which aligns the nanoparticles in registry. Upon the eventual elimination of these species, the nanoparticles fuse into a single-crystal or twinned aggregate.We observe this aggregation mechanism for anatase(101), anatase(112), and anatase(001) surfaces, as is also seen experimentally[1]. This indicates the important role that solvent plays in nanocrystal aggregation and how solvent can be a powerful tool for directing and controlling nanocrystal growth to fabricate nanostructures with desired shapes and sizes.

1. R.L. Penn, J.F. Banfield, Geochim. Cosmochim. Acta 63, 1549 (1999).
2. M. Alimohammadi, K.A. Fichthorn, Nano. Lett. 9, 4198 (2009).

(113b) Analysis of the Dynamics and Interactions of Magnetic Nanoparticles in Gradient and Time-Varying Magnetic Fields

Xiaozheng Xue and Edward P. Furlani, Dept. of Chemical and Biological Engineering, University at Buffalo (SUNY), Buffalo, NY

The interest in magnetic nanoparticles and ferrofluids has grown substantially in recent years as their applications continue to proliferate. Current applications are diverse and include the transport of biomolecules and therapeutic drugs, gene transfection, bioseparation, ferrofluidic seals and pumps, microfluidic mixers and magnetoresistive-based sensors, among others. However, despite the widespread and growing use of magnetic nanoparticles, there are many fundamental aspects of their collective behavior that remain unknown. Of particular interest is the self-assembly of the particles in a gradient field and the response of assembled particle-based microstructures in a time-varying field. In this presentation, a model is presented for predicting the field-directed assembly of colloidal magnetic particles under the influence of a high-gradient magnetic field and the manipulation of extended magnetic structures in a time varying field. The model is based on a modified discrete element method and takes into account several competitive effects. These include the applied magnetics force, induced magnetic dipole-dipole interactions, Brownian dynamics, Van der Waals interaction, a repulse contact force, viscous damping and hydrodynamic interactions among the particles. A dynamic time-stepping approach is used in the model to stabilize and accelerate the computation. The model is useful for predicting the structure of assembled particles in a given field and the stability of such structures in a time varying field. The model is demonstrated via application to various nanoparticle systems.

(113c) The Structural Ordering of Ionic Liquids Nanoconfined Between Charged Walls

Christopher R. Iacovella[1], Hugh Docherty[1], Matthew A. Gebbie[2], Markus Valtiner[3], Xavier Banquy[3], Jacob N. Israelachvili[3] and Peter T. Cummings[1,4], (1)Chemical and Biomolecular Engineering, Vanderbilt University, Nashville, TN, (2)Materials Department, University of California, Santa Barbara, Santa Barbara, CA, (3)Chemical Engineering, University of California, Santa Barbara, Santa Barbara, CA, (4)Center of Nanophase Materials Sciences, Oak Ridge National Laboratory, Oak Ridge, TN

In recent years, room temperature ionic liquids (RTILs) have been widely studied due to their unique properties such as negligible volatility and high electrochemical and thermal stability. The behavior of RTILs at surfaces and in confined geometries is also of great interest for use in various electrochemical applications, such as super-capacitors[1]. However, the structure and behavior of the RTILs in small, confined geometries is still not well understood, especially when the RTIL is interfaced with a charged surface[2]. The ordering of RTILs confined between charged walls is dictated by several competing factors, e.g., interactions between the charged species in the fluid, interactions between charged fluid species and the charged walls, aggregation behavior of alkyl side-chains, and geometric effects of the pore. In this work, we perform molecular dynamics (MD) simulations that examine the behavior and trends of model confined RTILs. We perform atomistically detailed simulations of several commonly used RTILs (e.g., [CnC1im][NTf2], where n is the length of the side-chain) confined between mica walls. We combine these results with MD simulations of a coarse-grained, minimal model, to efficiently establish trends over a wide range of statepoints. Specifically, we investigate the impact of wall charge, pore separation, and side-chain interactions on the self-assembly within the pore. Our atomistic and coarse-grained simulations provide similar results, where we observe several clear structural ordering motifs. Under certain conditions, we find close agreement with ordering schematics previously proposed in the literature[2]. We conclude by comparing our simulated results to surface force apparatus measurements of the intersurface forces across the RTIL film for matching [CnC1im][NTf2] systems.

[1] G. Feng and P.T. Cummings, J. Phys. Chem. Lett., 2011, 2, 2859-2864

[2] S. Perkin, et al. Chem. Commun., 2011, 47, 6572–6574

(113d) Molecular Dynamics Simulations of Structure and Dynamics of Cylindrical Micelles and Micelle-Nanoparticle Complexes

Ashish Sangwai[1], **Abhinanden Sambasivam**[2], Yutian Yang[2] and Radhakrishna Sureshkumar[3], (1)Intel Corporation, Hillsboro, OR, (2)Department of Biomedical and Chemical Engineering, Syracuse University, Syracuse, NY, (3)Department of Biomedical and Chemical Engineering and Department of Physics, Syracuse University, Syracuse, NY

Coarse-grained Non-equilibrium Molecular Dynamics (MD) simulations are performed to study the structure alignment effects of a cationic cylindrical micelle in shear flow. Micelles undergo cycles of flow alignment, elongation and tumbling when subjected to shear flow. The onset of flow alignment occurs at Weissenberg number (*Wi*) » 1 where *Wi* is the product of shear rate and angular relaxation time. The frequency of tumbling has a power law relationship with *Wi* with an exponent of 0.7. Simulations of relaxation of micelle from its elongated state provide a linear relationship between the micelle stretching energy calculated from pair potential energy and the length. This provides a constant force F needed to elongate the micelle to be about 653 kJ mol^{-1} nm^{-1}. Micelle scission was observed at a very high shear rate corresponding to Wi = 76.2 where the micelle elongated by ~3.5nm and crossed an energy barrier of ~3700 kJ/mol.

Self-assembly of metal nanoparticles with surfactant micelles is known to produce stable nanogels with rich rheological and optical properties. Coarse Grained MD simulations are performed to explore the molecular mechanism underlying this self-assembly process. It is observed that charged nanoparticles prefer a vesicular structure with a double layer of surfactants whereas uncharged, hydrophobic particles prefer a monolayer with the tail groups adsorbing onto the surface of the particle. Interaction of cylindrical micelles with these supramolecular structures results in the formation of stable junctions primarily though the opening up of the end cap of the micelle. The energetics of the bridging process and the effect of flow shear on their stability will be discussed.

References

1) A. V. Sangwai & R. Sureshkumar, Langmuir, 27(11), 2011.

2) A. V. Sangwai & R. Sureshkumar, Langmuir, 28(2), 2012.

3) T. Cong, S. N. Wani, P.A Paynter and R. Sureshkumar, Applied Physics letters, 99 043112 (2011)

(113e) How Particle Geometry Controls the Properties and Assembly of Nanoobjects

David A. Walker[1], Emily Leitsch[1] and Bartosz A. Grzybowski[2], (1)Chemical and Biological Engineering, Northwestern University, Evanston, IL, (2)Chemical Engineering, Northwestern University, Evanston, IL

While there has been significant progress in controlling the size, shape and properties of nanomaterials, their controlled assembly into nanostrucutres remains a significant challenge. To date, many nanoparticles (NPs) of unusual shapes have been reported - but the influence of geometry on the surface chemistry of these particles remains a mystery. We recently found that the pK_a of carboxylic acids bound to the surface of NPs deviates drastically compared to the values for free ligands in solution. This deviation can be described by the coupling the acid/base chemical equilibrium taking place at the particle surface with the surrounding electrostatic environment. Interestingly, by controlling the NP geometry, one can control how charges are distributed on the surface of NPs as well as other chemical properties (such as shifts in redox potentials). The gradients in chemical properties across particle surfaces can be exploited to create higher order structures which would not be possible without the influence of externals forces (such as the use of electronic or magnetic fields). A mixture of experimental and theoretical results will be presented.

(113f) Computational Investigation of Organic Molecule Assemblies On Cu(111)

Rees B. Rankin[1], Esmeralda Yitamben[2], Erin Iski[2], Joe Smerdon[2], Nathan Guisinger[2], Jeffrey Guest[2] and Jeff Greeley[1], (1)Center for Nanoscale Materials, Argonne National Laboratory, Argonne, IL, (2)Center for Nanoscale Materials, Argonne National Laboratory

The ability to understand and control assembly of molecular-surface interactions, and hence overall assembled molecular adlayers on surfaces, is critical for future design of nanomaterials and nanodevices in areas including electronics and biochemistry. Copper has shown promise as a relatively low-cost metal which promising surface electronic structure to interact with various molecular species. Based on intriguing results seen via STM investigations, we will present work from recent density functional theory (DFT) calculations aimed at characterizing the local surface adlayer structure of directed assembled adlayers of several organic molecules on the Cu(111) surface. Specifically, we will present results related to three distinct assembled adlayers on the Cu(111) surface. The first system studied is high-coverage adlayers of alanine on Cu(111). Current STM studies show that alanine/Cu(111) can form highly ordered hexagonally tiling local quantum corrals. The calculations performed help lead insight into the possible local atomic structure and hydrogen bonding networks present in the observed quantum corrals of alanine on Cu(111); understanding how to engineer nano-scale confined quantum materials can have applications in practical nanodevice design. The second system studied is isoleucine on Cu(111). Isoleucine represents a candidate amino acid with an appreciably large and predicted non-surface-binding side chain for investigation of coverage effects in adlayer assembly on Cu(111). In particular, this work seeks to determine whether energetic for differing isoleucine adlayer chiral handedness exist, as well as the possibility for higher adlayer density itself between chiral hands of isoleucine. The final system examined involves the directed assembly of buckyball-pentacene chiral pinwheels on Cu(111). In addition to examining the possibility of using this system for applications in electronic heterojunctions, we perform calculations to determine the local geometry, charge transfer in the chiral pinwheel network as well as the overall thermodynamic driving force for its formation.

(113g) In Situ Structural, Mechanical and Electrical Property Development During the Self-Assembly of Conjugated Polymer Organogels

Greg Newbloom, Kathleen Weigandt and Danilo Pozzo, Chemical Engineering, University of Washington, Seattle, WA

Organogels made from p-type conjugated polymers have shown promise as a platform for developing morphologies that benefit charge transport in organic photovoltaic devices.[1] However, previous studies of conjugated polymer organogels have largely been limited to the equilibrium state. Engineering specific gel structures requires an understanding of the driving forces behind self-assembly as well as the progression of structural intermediates. The latter can be difficult to probe because studying the progression of self-assembly requires non-invasive techniques that can probe evolving structures over all relevant length-scales. Small angle scattering is an ideal technique for this and is employed in this work to study self-assembling P3HT organogels.

Complementary small angle neutron scattering (SANS), AC dielectric spectroscopy and rheology are utilized simultaneously to study the self-assembly of poly(3-hexylthiophene) (P3HT). This talk will shows that nanoscopic structural features of fibrillar P3HT organogels, which are crucial for the optimization of organic photovoltaic devices, evolve throughout the gelation process. By performing cooling (i.e. gelation) and heating (i.e. dissolution) cycles we find substantial structure-property hysteresis that suggests the formation and dissolution mechanisms are drastically different, both of which are proposed in this talk. We also find that P3HT organogels formed in different solvents show differences of more than two orders of magnitude in conductivity. The similarity in nanoscale structural features between these gels suggests that the mesoscale network structure plays a critical role in the determination of charge transport efficiency. This talk demonstrates the importance of controlling self-assembly towards improving charge transport properties and demonstrates the potential to rationally design specific network structures using organogels as a generally applicable platform.

Reference: [1] – Newbloom, G. M.; Weigandt, K. M.; Pozzo, D. C. *Macromolecules* **2012**, *45*, 3452-3462.

(141a) Multimedia Environmental Distribution of Nanomaterials

H. Haven Liu and Yoram Cohen, Chemical and Biomolecular Engineering, University of California, Los Angeles, Los Angeles, CA

Nanotechnology has brought about many benefits and improvements to commercial and industrial applications, but its potential risks to the environment must also be systematically assessed. In this regard, exposure levels (i.e. concentrations, exposure time, etc.) of engineered nanomaterials (eNMs) in the environment are crucial to assessing and managing risks due to potential release of eNMs to the environment. Although models are available for predicting the multimedia environmental distribution of organics chemical contaminants, such models cannot be directly applied to eNMs due to the differences in the fate and transport (F&T) behavior of eNMs. Although "substance flow analysis" estimations have been employed in recent years as part of life cycle analysis (LCA), such analyses are not a priori predictive and do not consider mechanistic descriptions of fundamental transport processes. Accordingly, in the present study, a generalized model of the multimedia environmental distribution of engineered nanomaterial (Mend-Nano) is developed via multimedia fate and transport model to estimate the potential exposure levels to ENMs. The multimedia environment is considered as a collection of environmental compartments and subcompartments, with each compartment treated as either well-mixed (e.g. atmosphere, water), or non-uniform compartments (e.g. soil, sediment) that uses convection-diffusion models to increase spatial resolution. Fundamental intermedia transport processes (e.g. dry deposition, rain scavenging, resuspension, runoff) are then used to link various compartments while considering the particle size distribution (PSD) of eNMs in both the atmospheric and aquatic compartments. In order to enable rapid "what-if?" scenario analyses, a web-based graphical user interface (GUI) was developed for the model as part of a cloud-based modeling system. In the present model, the user defines the modeling scenario via the web-based GUI by specifying relevant properties of the eNM under consideration, geographical and meteorological parameters (i.e., size of regions, temperature, wind speed, rain rates, etc.), transport process parameters and source emission estimates (i.e., release rates, etc.). The model then calculates the dynamic change of ENM concentration and mass in each environmental compartments, and displays the result within the GUI. It is envisioned that the present multimedia analysis tool will assist regulators, industrial, and academic users in rapidly assessing the potential environmental implications of ENMs releases to the environment.

(141b) Molecular Weight Fraction of Natural Organic Matter Determines Nanoparticle Stability Against Aggregation and Attachment to Hydrated Metal-Oxide Surfaces

Stacey M. Louie and Gregory V. Lowry, Department of Civil & Environmental Engineering, Carnegie Mellon University, Pittsburgh, PA

Nanoparticles have become of interest as a potential emerging contaminant due to their novel properties and increasing use in commerce. Quantitative prediction of nanoparticle transport in the environment is necessary to assess exposure risks resulting from nanoparticle releases. Adsorption of macromolecules, especially natural organic matter (NOM), has been repeatedly demonstrated to greatly affect a nanoparticle's surface properties and its transport in the environment by imparting steric or electrosteric forces. However, predicting the effect of adsorbed NOM on a nanoparticle's properties is greatly complicated by the heterogeneous nature of NOM. NOM is polydisperse in terms of molecular weight, hydrophobicity, and functional groups present.

Adsorbed NOM has been shown to either stabilize nanoparticles against aggregation, or to destabilize them. In most studies assessing the effect of adsorbed NOM on nanoparticle transport, polydisperse NOM samples are used in whole. These samples are rarely fully characterized, so it is not possible to determine if particular components of the NOM may have adsorbed preferentially to the nanoparticle surface to produce the observed effects. Hence, only a qualitative explanation of the effects is currently possible. A more complete characterization of the adsorbed NOM coating is required to determine the specific NOM properties that most influence nanoparticle behavior.

In this study, we provide molecular weight characterization of NOM by size exclusion chromatography coupled with multi-angle light scattering (SEC-MALS). For Suwanee River NOM, two distinct molecular weight populations are observed: a smaller fraction (M_w<16 kDa) comprising ~98 wt-% of the unfractionated NOM, and a persistent large fraction (M_w~600 kDa) comprising ~2 wt-%. Preparative fractionation of these two NOM components is achieved by ultrafiltration. The effect of these fractions on the stability of 15 nm citrate-stabilized gold nanoparticles is studied. The large NOM fraction is observed to stabilize the nanoparticles against aggregation in 100 mM NaCl, whereas the small NOM fraction provides no or poorer stabilization, even at similar and significantly higher mass concentrations. A suite of characterizations is used to assess differences in the coating properties that may explain differences in stability, including adsorbed mass, chemical functionalities by ^{13}C NMR and comparison of UV adsorption ratios, and charge and extended layer thickness by an electrokinetic method. These results provide a deeper understanding of the effects of adsorbed NOM on nanoparticle transport behavior, and will be useful for interpretation of prior and future studies of NOM-coated nanoparticles.

(141c) CFD-Based Modeling of Nanopaint Application and Sustainability Assessment

Rohan Uttarwar and Yinlun Huang, Department of Chemical Engineering and Materials Science, Wayne State University, Detroit, MI

Nanocoating is a material in which inorganic nano-size filler particles are incorporated in the coating matrix. Over the past decade, nanocoatings have found significant application areas in the automotive, OEM, chemical, architectural and many other industries, as they may be able to provide advanced performance properties, such as significantly improved resistance to heat, chemical, scratch, corrosion, or to incorporate newer functionalities, such as self-cleaning, self-healing etc. With the advancement in the coating systems, there is a growing concern over the issues related to material-product-manufacturing sustainability. Among those, environmental and health impact of nanocoating manufacturing has caught a special attention. In the automotive coating manufacturing process, thermoset nanocoating film is developed through paint spray and curing. The emission of nanoparticles from the coating films in these operations can cause severe environmental and health problems, if the paint application processes are not designed and operated by following sustainability principles.

In this work, a computational fluid dynamics (CFD)-based system modeling and simulation approach is introduced to investigate the emission and flow patterns of nanoparticles and paint material during the paint spray operation. In addition, a set of carefully selected sustainability metrics are proposed to quantify the product and process sustainability. The CFD spray model can characterize the dynamics of paint film formation on substrate, and the sustainability performance is evaluated based on nanopaint formulation data and rheological parameters, operational data about paint spray ejection speed, downdraft air velocity, positioning of spray guns, and the paint booth geometry and ventilation system design data. These parameters are quantified and analyzed using the CFD models for air flow, species transport and energy consumption. Model-based simulation facilitates the identification of optimal operation parameters for superior coating quality, minimum nanoparticles emission, and minimum energy consumption. The design is further studied for improving its energy efficiency by modifying the geometry of the spray booth and positioning of spray guns. A comprehensive case study will demonstrate its application potential for achieving a significant improvement of sustainability performance in nanopaint applications.

(141d) Dependence of Surface Chemistry On Environmental Fate and Transport of Metallic Nanoparticles

Ashley E. Hart[1], Christopher L. Kitchens[1], Brian A. Powell[2], Hilary Emerson[2], O. Thompson Mefford[3] and Dan D'Unger[3], (1)Department of Chemical and Biomolecular Engineering, Clemson University, Clemson, SC, (2)Environmental Engineering and Science, Clemson University, Clemson, SC, (3)Materials Science and Engineering, Clemson University, Clemson, SC

As research in nanotechnology becomes more prominent, nanoparticles have developed into a commercial market and are produced in large quantities. Silver nanoparticles are currently available in over 1000 consumer products and gold nanoparticles have potential in many biomedical applications such as drug delivery and bioimaging. As the amount of nanoparticles used commercially increases, the concern of their environmental consequences becomes more prudent. To date there have been many studies conducted on nanoparticle toxicity under lab conditions, but very few studies have been conducted on nanoparticle s in environmental settings. Many factors contribute to the behavior of engineered and natural nanoparticles in soil, including surface chemistry, elemental composition, size and concentration. The properties of the soil also contribute to nanoparticle transport properties in soil. Soil inherently contains natural molecules such has humic acid, fulvic acid, citric acid, and desferrioxamine-B; each of which are components of natural organic matter, (NOM). Ligand exchange is a common reaction that occurs on the surface of nanoparticles when a molecule with a higher affinity to the nanoparticle core material replaces the original surface chemistry. Ligand exchange on nanoparticle surface with natural molecules found in soil can either promote or inhibit nanoparticle transport in soil, and this is being investigated in our study. An accelerated study was also conducted to obtain information on adsorption properties of nanoparticles to soil under saturated conditions. In short, citrate stabilized silver nanoparticles were mixed with 0.25g of sandy loam soil in 10 mL of water in concentrations of 20, 200, and 200 ppb. The pH of each solution was adjusted to 5.5. The silver content of the water phase and soil phase was measured by ICP-MS after one week to determine the fate of the nanoparticles. The transport of nanoparticles through soil was monitored using miscible displacement column experiments. A sandy loam soil was packed in 8.3 cm x 1.5 cm (OD) columns and nanoparticle suspensions were added as a finite step of one column volume then followed by nanoparticle free groundwater simulant. The concentration of nanoparticles in the effluent was monitored and the columns were segmented after the experiments to determine nanoparticle distribution within the column. Studies will also be conducted on silver nanoparticles with a variety of surface chemistries to determine the role of surface chemistry in nanoparticle fate and transport in soil systems.

(141e) Surfactant-Nanoparticle Interactions: Order of Addition Matters

Amir M. Farnoud, Chemical and Biochemical Engineering, University of Iowa, Iowa City, IA and Jennifer Fiegel, Pharmaceutics and Translational Theraputics; Chemical and Biochemical Engineering, The University of Iowa, Iowa City, IA

Interactions of nanoparticles and Langmuir monolayers of surfactants are important for several fields including food, cosmetics and pharmaceutical sciences. Several reports have shown that nanoparticles can disrupt the function of surfactant monolayers. However, results have been inconsistent and the underlying mechanisms are not fully understood. Considering the variability of results, it is of great importance to understand how the differences in methods are affecting the mechanisms of interaction. The present study focuses on interaction protocols and their effects on the mechanisms of interaction between carboxyl modified polystyrene nanoparticles (200 nm) and Langmuir monolayers of dipalmitoylphosphatidylcholine (DPPC).

Two interaction protocols were used to study nanoparticle effects on DPPC function: addition of nanoparticles after spreading the DPPC monolayer (subphase injection) and spreading DPPC on a subphase that already contains nanoparticles (DPPC addition). Surfactant function was probed by measuring the dynamic surface tension of the surfactant films during compression and expansion cycles using a computer-controlled Langmuir trough equipped with a Wilhelmy plate balance. A surface potentiometer was used to measure surface potential and provide information on surfactant packing density. Also, surfactant microstructure and topology were studied using fluorescence and atomic force microscopy.

The two interaction protocols resulted in significantly different surface pressure isotherms. Subphase injection resulted in higher surface pressures in the liquid expanded-liquid condensed coexistence phase but did not affect the maximum surface pressure. Subphase injection of nanoparticles reduced the size of the condensed surfactant phases without nanoparticle penetration to the air-water interface suggesting that electrostatic screening of surfactant dipoles by negatively charged nanoparticles is the mechanism of effect. On the contrary, DPPC addition resulted in premature monolayer collapse and lower maximum surface pressure in the following cycles. DPPC addition protocol did not show a significant electrostatic screening effect but resulted in an early rise in surfactant dipoles suggesting that space competition between nanoparticles and DPPC molecules has resulted in a more compact packing of DPPC molecules. These results show that interaction protocols greatly affect the mechanisms of interaction between nanoparticles and surfactant and are important in understanding mechanisms proposed in the literature.

(152a) Synthesis and Applications of Low Melting Point Tin/Indium (Sn/In) Nanosolders

Yang Shu, Karunaharan Rajathurai, Fan Gao, Qingzhou Cui and Zhiyong Gu, Chemical Engineering, University of Massachusetts Lowell, Lowell, MA

Many electronics manufacturing processes require low temperature processing, such as flexible electronics, electronics or devices involving thermal sensitive components, etc. To meet the requirement of low temperature processing as well as to satisfy the need for energy saving, low melting point solder particles are necessary for electronics assembly and joining of certain products. Herein we report the synthesis of tin/indium (Sn/In) alloy nanoparticles as a new nanosolder material for low temperature applications. The nanosolders are synthesized by a surfactant-assisted chemical reduction method. The parameters that affect the synthesis process such as solution pH, stirring speed and surfactant concentration are studied to control the size and shape of the Sn/In nanosolders. It is found that oxidation plays a significant role in nanosolder formation. SEM, TEM, XRD, and EDS are used to determine the structure and composition of the nanosolders. DSC test has been conducted to determine the melting points of the nanosolders with various compositions. The nanosolders have been used to join different components. Our results show that the nanosolder particle size and shape can be well controlled under optimum conditions. These low melting point nanosolders have potential for such applications as micro- or nanojoining techniques, bonding and interconnection on flexible electronics.

(152b) Electron Optics of Self-Assembled Nanocomposite Metamaterials

D. Keith Roper, Chemical Engineering, University of Arkansas, Fayetteville, AR; Microelectronics/Photonics Graduate Program, University of Arkansas, Fayetteville, AR, Drew Dejarnette, Microelectronics/Photonics Graduate Program, University of Arkansas, Gyoung-Gug Jang, Ralph E. Martin Department of Chemical Engineering, University of Arkansas, Fayetteville, AR, Aaron G. Russell, Department of Chemical Engineering, University of Akansas, Fayetteville, AR, Phillip Blake, Department of Chemical Engineering, University of Arkansas, Fayetteville, AR and Keith Berry, Chemical Engineering, University of Arkansas

Actively enhanced opto-electronic coupling would fundamentally impact nanoscale systems for sustainable energy, synthetic biology, and biomedical theranostics. Quantized photon-exciton coupling is the source of electron-hole pairs in semiconductor photovoltaics,induced charge transfer in ordered DNA chains, and distance-dependent Forster resonance energy transfer in spectroscopic molecular rulers. Recent advances in our lab identify relative contributions of photon diffraction and quantized plasmon polarizabilities to extraordinary opto-electronic coupling in ordered metalloceramic nanocomposite metamaterials. Metamaterials exhibit tunable electromagnetic functionality -- from simple iridescence in butterfly wings to radiofrequency cloaking -- due to coherent interference from multidimensional structuring of suitable condensed-matter composites. Surface energy driven self-assembly of these metamaterials has been demonstrated in our lab by electroless nanosphere synthesis on nanolithographed anti-gecko lattices. We showed that by modulating nanoscale photodynamic, hydraulic, and redox potentials, ionic precursors can be directed to nucleate into various nanostructured isoforms including spheres, island films, and clusters in ceramic and polymeric substrates. Energy dissipation from metamaterials via interaction with lattice phonons can induce predictable optoplasmonic profiles and phase changes in nanocomposites to drive microscale fractionation of 2nd generation biofuels. Microspectroscopies (x-ray photoelectron; transmission UV; and Raman) and electron/optical microscopies are critical to evaluate such effects. These advances in modeling, fabrication, characterization, and systems integration are important milestones toward integrating electromagnetically active nanocomposite metamaterials for next-generation energy harvesting, molecular engineering, disease diagnosis, and personalized therapies.

(152c) Electrostatic Interaction Mechanism for near-Surface Defect Redistribution

Prashun Gorai, Department of Chemical & Biomolecular Engineering, University of Illinois at Urbana-Champaign, Urbana, IL and Edmund G. Seebauer, Department of Chemical & Biomolecular Engineering, University of Illinois at Urbana-Champaign

Concentration and diffusion of point defects in metal oxides strongly influence the manufacture and performance of these materials in nanoelectronic, gas sensing, photonic, and photocatalytic applications. Near-surface effects are particularly important in nanoscale devices because most of the bulk is located in the vicinity of the surface. The electric fields that naturally exist near many semiconductor surfaces influence the motion of charged defects, resulting in their redistribution. In metal oxides, the redistribution of charged oxygen defects affects the performance of nanoelectronic devices such as memory resistors and Schottky barriers. Near-surface electric fields can arise from electrostatic charge buildup due to dangling bonds, surface polarity and adsorption. These effects are controllable in principle. Another important contributor that is less controllable is the chemical potential difference between defects residing deep in the bulk compared to near the surface. With a view toward developing improved methods for defect engineering, we have employed a combination of diffusion measurements, surface potential measurements and simulations to estimate the relative contributions of electrostatic and thermodynamic forces to the surface potential V_S. Oxygen defect behavior was monitored indirectly via isotopic oxygen self-diffusion measurements made by exposing single-crystal rutile titania and zinc oxide to isotopically-labeled oxygen gas. The resulting profiles were measured by secondary ion mass spectrometry and subsequently simulated with continuum equations describing the reaction, Fickian diffusion and electric field-driven diffusion of the key point defects.

(152d) Opening and Tuning of Band Gap by Formation of Nanodiamond Superlattices in Twisted Bilayer Graphene

Andre R. Muniz, Department of Chemical Engineering, Federal University of Rio Grande do Sul, Porto Alegre, Brazil and Dimitrios Maroudas, Department of Chemical Engineering, University of Massachusetts Amherst, Amhest, MA

We report results of first-principles density functional theory (DFT) calculations, which introduce a novel class of carbon nanostructures formed due to the creation of interlayer covalent C-C bonds in twisted bilayer graphene (TBG). In TBG, the two graphene planes of the bilayer are rotated with respect to each other by a specified twist angle over the range from 0 to 30 degrees. Interlayer C-C bonding in TBG is the result of hydrogenation of the graphene layers according to certain hydrogenation patterns that generate stable interlayer-bonded configurations. The fully relaxed hydrogenated interlayer-bonded structures consist of two-dimensional (2D) superlattices of diamond-like nanocrystals embedded within the graphene layers; the superlattices have the same periodicity as that of the Moiré pattern corresponding to the rotational layer stacking in TBG and the 2D diamond nanodomains may resemble the cubic or the hexagonal diamond phase. The detailed structure of these superlattice configurations is determined by the twist angle, the effects of which we have examined over the range from 0 to about 15 degrees; the twist angle is an important parameter in this study. Other important parameters in the analysis include the hydrogen coverage and hydrogenation pattern, which determine the number of interlayer C-C bonds formed and, consequently, the size of the embedded diamond nanodomains.

We have calculated the electronic band structure of all of the relaxed atomic configurations we generated and we demonstrate that the formation of interlayer-bonded finite domains causes the opening of a band gap, which depends on the density and spatial distribution of the interlayer C-C bonds. We have found that the band gap increases monotonically with increasing size of the embedded diamond nanodomain in the unit cell of the superlattice and predicted band gaps as wide as 1.2 eV. These results suggest that this approach is very promising for opening a tunable band gap in bilayer graphene, which can be controlled precisely in order to tailor graphene's electronic properties for applications in nanoelectronics.

(152f) Titanosillicates, Are They Natural Host for Quantum Wires?

Al Sacco Jr.[1], Nina Bordeaux[2], Önnaz Özkanat[3], Peter Ryan[4] and Nicol E. McGruer[4], (1)Box 43103, Engineering Center Rm 100, Texas Tech University, Lubbock, TX, (2)Chemical Engineering, Northeastern University, Boston, MA, (3)Delft University of Technology, Delft, Netherlands, (4)Electrical and Computer Engineering, Northeastern University, Boston, MA

To develop micro/nano optoelectronics, sustain miniaturization trends as well as develop photon/electron detectors at the cellular level, methods to produce quantum wire arrays are being evaluated. Quantum wire arrays are predicted to be superior when compared to quantum wells and quantum dots in that they have more prominent energy quantization effects, extremely high electron and photon mobility and are less temperature and are less polarization dependent than quantum wells. In addition, arrays 'may' allow some defects to be present and preserve device performance through electron tunneling. An investigation was performed utilizing the natural occurring titania chains in titanosilicate materials (diameter ~ 7 Å, intrinsic quantum wires) to produce "quantum arrays". A method to grow large (≥ 200 μm, ETS-4; ≥80 μm, ETS-10), oriented, reduced-defect crystals was developed. These crystals were incorporated into "microcircuits", and current versus voltage (IV) curves were measure at 293 K, 194 K, 64 K, and 15 K. Results indicated a non-Ohmic behavior, which was not initially understood. Low temperature results showed higher resistance relative to 293 K, and the resistance saturated around 65 K. The IV curve at 15 K was found to be almost identical with that at 65 K. Additionally, in these initial tests, there appeared to be a hysteresis effect, which was determined to be due to structure breakdown due to resistance heating. Two probe tests showed current peaks consistent with the theory of resonant tunneling for quantum superstructures, showing both rapidly increasing current peaks and negative differential resistances. Similar behavior has been observed on a metal-insulator-nanoparticle-insulator-metal (MINIM) capacitor. The observed IV curve was reported to be due to single electron traveling through each particle, suggesting by analogy possible quantum effects in ETS-4 and ETS-10. However, additional two probe tests using Zeolite A (NaA) and Zeolite Y (NaY) but without titania chains present showed similar IV behaviors. These findings suggest that quantum behavior does not dominate electron transport in ETS-10 and ETS-4, but rather the observed IV behavior likely occurs due to cation movement within the framework.

(152g) Selective Nanomanufacturing of Particle-Specific Oligomeric Clusters

Benjamin Robinson[1], Rajasekhar Anumolu[1] and Leonard F. Pease III[2], (1)Chemical Engineering, University of Utah, Salt Lake City, UT, (2)Chemical Engineering and Pharmaceutics & Pharmaceutical Chemistry, University of Utah, Salt Lake City, UT

Selective Nanomanufacturing of Particle-Specific Oligomeric Clusters

Benjamin Robinson, [1] **Rajesekhar Anumolu,**[1] **Leonard F. Pease III**[1,2*]

1. Department of Chemical Engineering, University of Utah, Salt Lake City, UT 84112

2. Departments of Internal Medicine and Pharmaceutics & Pharmaceutical Chemistry, University of Utah, Salt Lake City, UT 84112

ABSTRACT

Here we demonstrate the ability to selectively fabricate nanoparticle clusters composed of two or more diverse types of particles. Quantitatively predicting and controlling nanoparticle aggregation within drying droplets remains challenging despite recent advances in the fabrication of nanoclusters from nanoelectronic and other materials. Here, we report our modeling effort to statistically predict the cluster distribution for two or more distinct particle types and compare these distributions to those obtained experimentally using electrospray differential mobility analysis (ES-DMA). In this technique, droplets encapsulate two or more random particles and then dry to form the nanoclusters. To predict the formation of oligomeric nanoparticle clusters formed from monomers in solution, we use previously developed theory that predicts the probability of cluster formation for one type of particle and then multiply it by the mixing probability for two or more distinct particles. The predicted probabilities are in good agreement with measured cluster distribution from ES-DMA measurements. These results now predict the conditions necessary to generate nanoparticle clusters from any combination of two or more particles using droplet drying techniques. The resulting, easily accessible correlations may be used to minimize undesirable clustering along with maximizing yield of particle specific nanophotonic heteroclusters and others. Complex nanoclusters such as coupled quantum dots are important to nanoenergy, nanophotonics, and nanomedicine applications as well as fundamental studies of heterocluster properties.

(155a) Invited Talk: A Review of the Use of Nanotechnology in Tissue Engineering

Thomas J. Webster, School of Engineering, Brown University, Providence, RI

Inspired from biological systems, nanotechnology is beginning to revolutionize medicine including improving the prevention, diagnosis, and treatment of numerous diseases. This talk will summarize efforts over the past decade that have synthesized novel nanoparticles, nanotubes, and other nanomaterials to improve medicine. Efforts focused on the use of nanomaterials to minimize immune cell interactions, inhibit infection, and increase tissue growth will be especially emphasized. Tissue systems covered will include the nervous system, orthopedics, bladder, cardiovascular, vascular, and the bladder. Due to complications translating in vitro to in vivo results, only in vivo studies will be emphasized here. Materials to be covered will include ceramics, metals, polymers, and composites thereof. Self-assembled nano-chemistries will also be emphasized. As the FDA has now approved several nanomaterials for medical applications, recent results from FDA trials will also be discussed. In summary, this talk will provide the latest information concerning the design and use of numerous nanomaterials in regenerative medicine while highlighting what is necessary for this field to continue to grow.

(155b) Invited Talk: Plant-Derived and Inspired Nanostructured Scaffolds for Tissue Engineering

Mingjun Zhang, The University of Tennessee, Knoxville, Knoxville, TN

Much research has focused on the development of scalable scaffolds that are highly compatible with tissue engineering applications. The goal of this research is to demonstrate and propose a naturally occurring nano-scaffold from the adhesive of the Sundew, and bio-inspired scaffold coating using naturally occurring nanoparticles isolated from ivy adhesive, for applications in tissue engineering.

First, we will show how we evaluate and propose the naturally occurring nano-structured scaffold, from Sundew (*Drosera*) adhesive, for tissue engineering applications. It is well-known that the Sundew uses a highly elastic adhesive for insect capturing. We recently discovered that this adhesive forms a nanoparticle and nanofiber-based highly elastic and porous scaffolds when deposited on a surface. Through experimental studies using atomic force microscopy, we were able to visualize the complex network of nanofibers, identify uniform nanoparticles, and determine cursory material properties from the scaffold. Further experimental studies indicated that the scaffold was capable of supporting cell growth, and showed that PC12 cells as well as ostebolasts strongly adhered to the scaffold. We have determined the complete physical properties, biocompatibility, and biostability of the Sundew adhesive derived scaffold and are attempting to modify the nano-scaffold for potential tissue engineering applications.

Our group recently discovered that uniform nanoparticles, in high abundance were secreted from aerial rootlets of ivy for surface affixing. These nanoparticles were found to contribute to the adhesive strength of the adhesive discs, through the formation of a concrete-like matrix. Similar studies by other groups have also revealed the presence of nanoparticles in barnacle and marine mussel adhesives. More recently, we have further characterized the nature of ivy nanoparticles – size, shape and chemical composition. We have developed a highly robust and cost-effective approach for nanoparticle isolation and purification. This talk will discuss how the ivy nanoparticles have been used for scaffold coating in tissue engineering applications.

Through the above case studies, we hope to demonstrate an alternative approach for developing nanostructured scaffolds for tissue engineering. The bio-inspired or nature-derived approach seems offer unique merit for biomedical applications.

(155c) Invited Talk: Engineering Nanostructured Interface for Next-Generation Biodegradable Implants

Huinan Liu, Bioengineering, University of California at Riverside, Riverside, CA; Materials Science and Engineering, University of California at Riverside, Riverside, CA

A novel class of biodegradable metal, i.e. magnesium (Mg) alloys, has attracted growing interest as a promising alternative for next-generation biodegradable medical implants due to their advantageous mechanical and physiological properties. The implants/devices made from magnesium alloys can have the desirable modulus and mechanical strength to prevent breakage during implantation and to provide good initial fixation. Controlling the interface of magnesium with the biological environment is the key challenge that currently limits this biodegradable metal for broad applications in medical devices and implants. This talk will particularly focus on how to create a nanostructured interface between the biodegradable metallic implant and surrounding tissue for the dual purposes of (1) enhancing tissue integration and regeneration and (2) simultaneously mediating the degradation of the metallic implant.

(155d) Invited Talk:Differentiation of Cardiac Progenitor Cells Using Fibrous Scaffolds

Jianjun Guan, Materials Science Engineering, Ohio State, Columbus, OH

Matrix mechanical properties-induced stem cell differentiation has attracted tremendous interest in the tissue engineering community, as cell differentiation can be simply controlled by matrix physical properties, without using complicated biochemical stimulation [1]. We investigated whether mechanical properties of the extracellular matrix-mimicking fibrous scaffolds can be modulated to direct stem cell differentiation into a cardiac lineage. The scaffolds were fabricated by electrospinning a polymer blend based on a highly soft poly(N-isopropylacrylamide) hydrogel and a relatively stiffer polyurethane. By varying ratio of the hydrogel and the polyurethane, scaffolds with three different stiffnesses were fabricated, i.e., 14 ± 3, 223 ± 33 and 463 ± 38 kPa. To investigate stem cell differentiation in these scaffolds, cardiosphere-derived cells (CDCs) - a type of emerging cardiac progenitor cell for cardiac therapy, were incorporated to the scaffolds, by simultaneously electrospinning polymer fibers and electrospraying cells. Cell growth and differentiation were characterized after a 7-day culture in a spinner flask. dsDNA (for live cells) results showed that CDCs remained alive in the scaffolds. CDC differentiation was characterized at the mRNA level by real time RT-PCR, and the protein level by immunohistochemistry. At the mRNA level, the expressions of cardiac markers cardiac troponin T (cTnT) and cardiac myosin heavy chain (MYH6) were significantly upregulated in all scaffolds. The CDCs also developed calcium channels (CACNA1c) that are necessary for imparting electrophysiological properties. The scaffold with modulus of 14.3 kPa most efficiently stimulated the differentiation. Immunohistochemical analysis demonstrated that CDCs in the scaffold with modulus of 14.3 kPa expressed both troponin I (cTnI) and connexin43 (CX43) markers at protein level. This study suggests that stiffness of 14.3 kPa may be ideal for CDCs to differentiate into a cardiac linage in a fibrous scaffold.

(155e) Synthesis of Magnetic Particles Using Co-Precipitation Method for Biomedical Applications

Jaclyn Lock[1] and Huinan Liu[1,2], (1)Bioengineering, University of California at Riverside, Riverside, CA, (2)Materials Science and Engineering, University of California at Riverside, Riverside, CA

Magnetic iron oxide particles are highly advantageous for biological applications (such as drug delivery) due to their magnetic properties coupled with their low toxicity and biocompatibility. We synthesized poly(vinyl alcohol)-ferrite magnetic particles using the co-precipitation method and characterized these particles using scanning electron microscopy (SEM) with energy dispersive X-ray spectroscopy (EDX) analysis. Furthermore, we showed that the amount of ferrite encapsulated in the resulting particles is dependent upon the concentrations of: (1) poly(vinyl alcohol) polymer, and (2) iron chloride. The SEM micrographs demonstrated that these magnetic particles have nano-structural features, which have been previously shown to improve cell adhesion and function.[1] We observed that these particles were non-toxic to rat bone marrow stromal cells, i.e., cells adhered and spread-out on the magnetic particles after a 24-hour incubation period. We believe that these synthesized magnetic particles serve as promising platform for applications in targeted drug delivery and tissue engineering.

1. Liu, H.; Webster, T., Enhanced biological and mechanical properties of well-dispersed nanophase ceramics in polymer composites: from 2D to 3D printed structures. *Materials Science and Engineering* **2010**.

(155f) Nanofiber Scaffolds Mimicking Structural Organization of Tendon-to-Bone Insertion Site

Jingwei Xie, Marshall Insititute for Interdisciplinary Research, Marshall University, Huntington, WV, Bing Ma, Marshall Insistitute for Interdisciplinary Research, Marshall University, Huntington, WV and Franklin Shuler, Orthopaedic Surgery, Marshall University, Huntington, WV

Nanofiber Scaffolds Mimicking Structural Organization of

Tendon-to-Bone Insertion Site

Jingwei Xie[1*], Bing Ma[1], Shuler D. Franklin[2]

[1]Marshall Institute for Interdisciplinary Research and Center for Diagnostic Nanosystems, Marshall University, Huntington, WV, 25755 USA

[2]Department of Orthopaedic Surgery, Joan C. Edwards School of Medicine, Marshall University, Huntington, WV 25701 USA

[*]Correspondence should be addressed to: xiej@marshall.edu

Introduction

It is known that tendons are attached to bones across a specialized transitional tissue with varying structures (i.e., collagen fibers are less oriented at the insertion compared to the tendon) and compositions (i.e., the relative mineral concentration vs. distance across the insertion site shows an approximately linear increase across the interface.[1] Efforts have been devoted to the development of scaffolds which could mimic the structure (i.e., collagen fiber orientation) and/or composition (i.e., mineral content) of tendon-to-bone insertions for the repair of rotator cuff injury. A number of studies demonstrated the fabrication of scaffolds with a gradient of mineral content for mimicking the composition and mechanical function of the interface between soft tissue to hard tissue (i.e., tendon-to-bone, ligament-to-bone, and cartilage-to-bone).[2-4] However, few studies have centered on the creation of scaffolds capable of exhibiting the unique structural organization of collagen fibers seen at the native tendon-to-bone insertion site except that a recent study attempted to fabricate "aligned-to-random" nanofiber scaffolds by making use of a specially-designed collector during electrospinning for mimicking the structure of the tendon-to-bone insertion site.[5] But some issues still remain in this study: the nanofiber scaffold has no gradual transitions from aligned portion to random portion; and there is apparent difference of the thicknesses between the two portions, which could easily cause breaking at the interface upon exerting a force. Therefore, so far no such a scaffold can fully recreate the gradients in structure found at the uninjured tendon-to-bone attachment.

In this study, we report a new and simple method for fabrication of nanofiber scaffolds with gradations in fiber organization by depositing random fibers on the uniaxially-aligned nanofiber mat in a gradient manner. These surfaces mimic the change in fiber orientation which presents at the tendon-to-bone insertion site. Tailoring the surface features in a user-specified manner can provide spatial control over cell response like cell morphology. Hence, we hypothesize that adipose-derived stem cells (ADCSs) could respond differently to different locations of such nanofiber surfaces. An adipose tissue is one of the sources of mesenchymal stem cells and ADSCs are very similar in nature to bone mesenchymal stem cells (BMSCs) in terms of multipotency.[6] Adipose tissue is abundant in most individuals and can be harvested using a simple liposuction procedure which is less invasive and cause less discomfort and donor-site damage. In addition, adipose tissue has a significantly higher stem cell density than does bone marrow (5% vs 0.01%). ADSCs have been demonstrated for use in bone, cartilage, and tendon tissue engineering and are used in this study.

Materials and Methods

The uniaxially-aligned nanofibers were produced by electrospinning using an air-gap collector based on previous studies. Poly(ε-caprolactone) (PCL) (Mw=80,000 g/mol; Sigma-Aldrich, St. Louis, MO) was dissolved in a solvent mixture consisting of dichloromethane (DCM) and N, N-dimethylformamide (DMF) (Fisher Chemical, Waltham, MA) with a ratio of 4:1 (v/v) at a concentration of 10% (w/v). Polymer solution was pumped at a flow rate of 0.5 mL/h using a syringe pump. A dc high voltage of 12 kV was applied between the nozzle and a grounded air-gap collector made of aluminum foil (with an open void 2 cm × 5 cm). Subsequently, the uniaxially-aligned nanofibers were easily transferred to a glass slide. Nanofiber scaffolds were fabricated using the setup shown in Figure 1. Briefly, a glass slide covered with uniaxially-aligned nanofibers served as a collector. A movable plastic mask was connected to a second syringe pump and placed 2 mm above the collector during deposition of random fibers.

Human ADSCs were obtained from Cellular Engineering Technologies (Coralville, IA) and were cultured in α-modified Eagle's medium (α-MEM) supplemented with 10% fetal bovine serum (FBS, Invitrogen) and 1% gentalmycine/streptomycin (Invitrogen) at 37 °C in an atmosphere of 95% air / 5% CO_2. Cell culture medium was changed every 2 days. Prior to cell seeding to nanofiber scaffolds, cells were trypsinized and counted. Around 1×10^4 cells were seeded on each nanofiber scaffold without encapsulation of coumarin 6.

Results and Discussion

Nanofiber scaffolds with gradations in fiber organization were generated by deposition of random fibers on an uniaxially-aligned nanofiber mat. It is observed that more and more random fibers were deposited from 0 mm to 6 mm along the nanofiber scaffold. The Fourier fast transfer (FFT) patterns suggest progression of random fiber generation from 0 to 6 mm. In particular the FFT pattern at 6 mm suggests randomly oriented nanofibers.

Once we demonstrated successful generation of random fiber deposition, we studied whether this gradient fiber orientation can direct morphological changes following incubation with ADSCs. Scaffolds generated using the techniques above were incubated with ADSCs for 3 and 7 days. Live cells were detected using a fluorescein diacetate (FDA) assay (Figure 3, A-H), The response of pre-seeded ADSCs to underlying nanofiber scaffolds was quantified by measuring the orientation of individual cells relative to the long axis of the fibers (Figure 3I). The distribution of cell orientations demonstrated that ADSCs seeded on random nanofibers lacked organization and directional specificity, as the cells projected in all possible directions. In contrast, ADSCs seeded on the aligned scaffolds demonstrated orientation and alignment along the fibers in contact with the cells. In these cases, approximately 70% of the cells were oriented within 20° of the axis of nanofiber alignment. Comparison of the distributions of ADSCs orientations on the various portions of scaffolds using the Kolmogorov-Smirnov *test further confirmed that the pattern of cellular orientation observed at different portions (distance > 4 mm) on the nanofiber scaffolds was significantly different (p < 0.05).*

To broaden potential biomedical applications, chemical gradients were established on this novel scaffold using nanoencapsulation. Coumarin 6-loaded random PCL nanofibers were generated in gradient fashion. We quantified the fluorescence intensity by measuring the grey scale using Image J software. It is observed that the fluorescent intensity increased along the distance. This method could be readily extended to fabricate two-component gradients on the surface by depositing random fibers twice in two different directions.

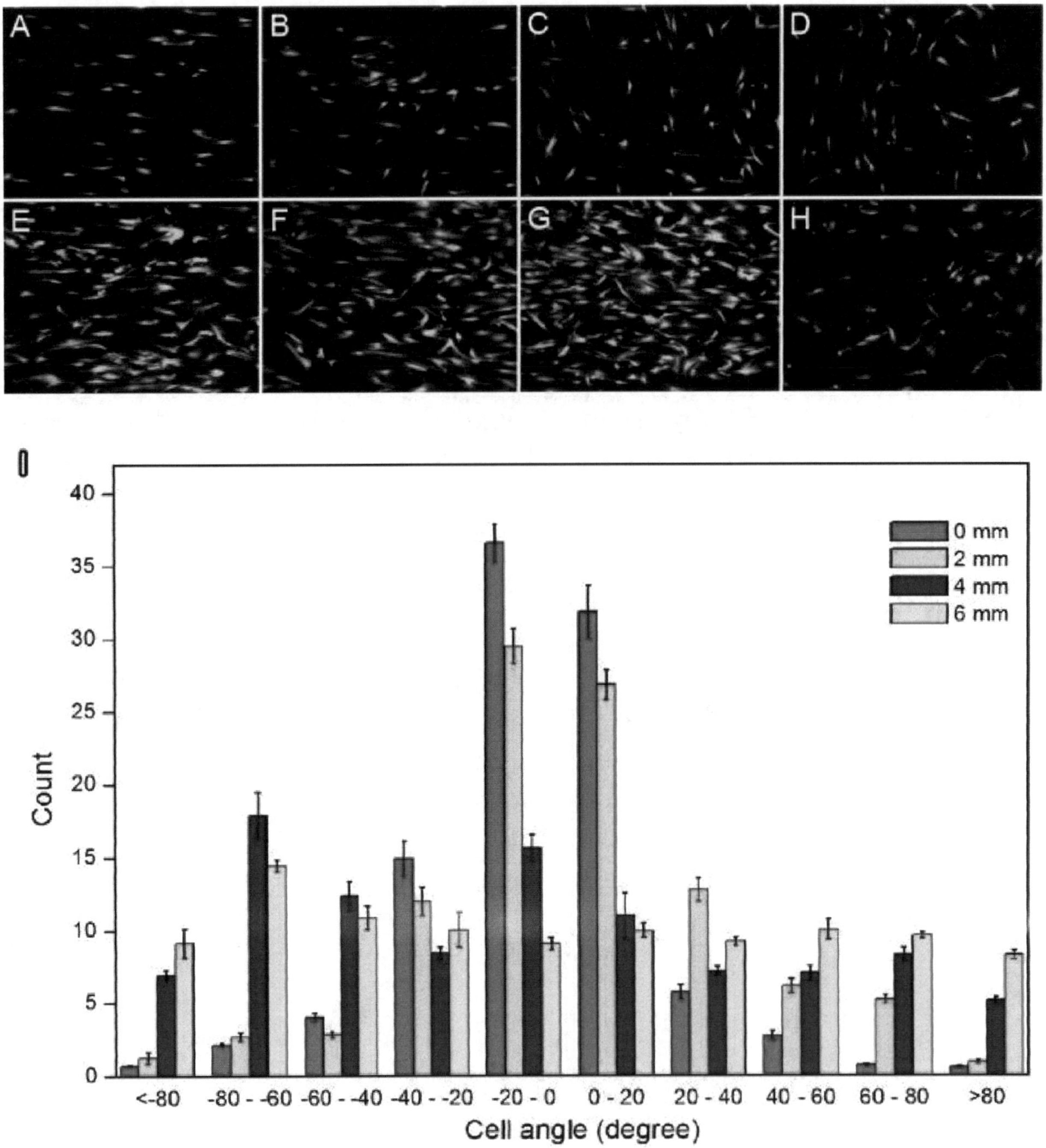

Figure 1. Fluorescein diacetate assay: Adipose-derived stem cells after incubation for 3 days (A-D) and 7 days (E-H) showing different morphologies on different locations of nanofiber scaffolds with gradations in fiber organizations (A, E): 0 mm; (B, F): 2 mm; (C, G): 4 mm; and (D, H): 6 mm. (I): Quantitative analysis of morphology of adipose-derived stem cells on different locations of nanofiber scaffolds after incubation for 3 days. Compared to the distributions of cell angle at different locations using the Kolmogorov-Smirnov *test,* p values of 0.007 (between 0 mm and 6 mm), 0.675 (between 0 mm and 2 mm), 0.031 (between 0 mm and 4 mm), 0.031 (between 2 mm and 6 mm), 0.11 (between 2 mm and 4 mm), and 0.111 (between 4 mm and 6 mm) were obtained. $p < 0.05$ indicates significant difference.

Conclusion

We have demonstrated a simple, robust and facile method to fabricate nanofiber scaffolds with gradations in fiber organization, which could potentially mimic the structure (i.e., collagen fiber organization) at tendon-to-bone insertion site. We also demonstrated that ADSCs seeded on such a scaffold exhibited different morphologies at different locations. In addition, we demonstrated the formation of nanofiber surfaces with chemical gradients through encapsulation of desired materials inside deposited nanofibers. By incorporating mineral gradient to this scaffold, it could be used to recapitulate both structure and composition of tendon-to-bone insertion site to a considerable degree, which may be promising for the repair of tendon-to-bone insertion.

References

1. G. M. Genin, A. Kent, V. Birman, B. Wopenka, J. D. Pasteris, J. P. Marquez, S. Thomopoulos, *Biophy. J.* **2009**, *97*, 976.

2. X. Li, J. Xie, J. Lipner, X. Yuan, S. Thomopoulos, Y. Xia, *Nano Lett.* **2009**, *9*, 2763.

3. C. Lu, Z. Han, J. T. Czernuszka, *Acta Biomater.* **2009**, 5, 661.

4. S. Samavedi, C. O. Horton, S. A. Guelcher, A. S. Goldstein, A. R. Whittington, *Acta Biomater.* **2011**, *7*, 4131.

5. J. Xie, X. Li, J. Lipner, C. N. Manning, A. G. Schwartz, S. Thomopoulos, Y. Xia, *Nanoscale* **2010**, *2*, 923.

6. J.M. Gimble, F. Guilak, *Cytotherapy* **2003**, *5*, 362.

(155g) Biomimetic and Biphasic Nanocomposite Scaffold with Growth Factor-Encapsulated Nanospheres for Repairing Osteochondral Defects

Nathan Castro, Mechanical and Aerospace Engineering, The George Washington University, Washington, DC, Christopher O'Brien, Electrical and Computer Engineering, The George Washington University, Washington, DC and Lijie Grace Zhang, Department of Mechanical and Aerospace Engineering and Institute for Biomedical Engineering, The George Washington University, Washington, DC

Biomimetic and Biphasic Nanocomposite Scaffold with Growth Factor-Encapsulated Nanospheres for Repairing Osteochondral Defects

Nathan J. Castro[1], Christopher O'Brien[2], Lijie Grace Zhang[1]

[1]Department of Mechanical and Aerospace Engineering

[2]Department of Electrical and Computer Engineering

The George Washington University, Washington, DC 20052

Email addresses:

Nathan J. Castro: njcastro@gwmail.gwu.edu

Christopher O'Brien: cobrien2@gwmail.gwu.edu

Lijie Grace Zhang: lgzhang@gwu.edu

Abstract: Over 25 million people in the United States alone are afflicted with osteoarthritis or some form of degenerative joint disease. People with this disease experience reduced joint mobility and severe pain due to the gradual loss or traumatic injury to the articular cartilage and subchondral bone which is collectively known as osteochondral tissue. Current treatment methods used to address these defects including autografts, allografts, and mosaicplasties contain their own inherent limitations. Donor site morbidity, infection, poor tissue integration, and neovascularization continue to prevent the clinical success of traditional methods. Extensive research employing tissue engineering approaches have focused on regenerating both tissues independently with good results. Due to the complex nature of the osteochondral site both, in composition and mechanical properties, single-material constructs lack the dynamic range necessary to address such distinctly different tissues. Therefore, this work aims at fabricating a biphasic and biomimetic nanostructured construct for osteochondral tissue regeneration. Highly porous poly(caprolactone) (PCL) with bone morphogenetic protein-2 (BMP-2)-encapsulated poly(dioxanone) (PDO) nanospheres as well as nanocrystalline hydroxyapatite is fabricated via porogen-leaching through the combination of water soluble poly(ethylene glycol) diacrylate (PEG-DA) and sodium chloride salt particles, serving as the bone layer of the biphasic scaffold. The use of a co-porogen system has been shown to provide greater control over pore size and architecture. The cartilage layer is composed of PEG-DA with transforming growth factor-β1 (TGF-β1)-encapsulated PLGA nanospheres and is cast on the porous PCL bone layer then cured under ultraviolet (UV) radiation physically joining the two layers by crosslinking of infiltrated PEG-DA. In this manner, a biomimetic nano osteochondral scaffold with sustained chondrogenic and osteogenic growth factor distributions is created. Human bone marrow derived mesenchymal stem cells will be seeded onto each layer and evaluated for directed osteogenic and chondrogenic differentiation *in vitro*. We expect that sustained growth factor release in combination with a more biomimetic biphasic nanostructured construct will yield a more clinically relevant tissue-engineered construct for improved osteochondral regeneration.

(155h) Changes in Cellular Behavior with Implementation of Aligned PLGA Fibers in Wound Healing Model

C. Ng, Mechanical Engineering, Virginia Polytechnic Institute and State University, Blacksburg, VA and Amrinder S. Nain, Mechanical Engineering, Virginia Tech, Blacksburg, VA

Changes in Cellular Behavior with Implementation of Aligned PLGA fibers in Wound Healing Model

C. Ng[1], and A. S. Nain[1]

1 Virginia Tech, Blacksburg, VA

Introduction

The body has a natural process to heal wounds or breaks to the skin surface. This process known as wound healing is a complex sequence of events that are broken down into inflammation, tissue repair and tissue remodeling. [1] Depending upon the severity, this process can take a few days or months to heal and even after healing the skin may never regain its original strength [2]. Much work has been done through wound healing assays to understand wound repair, specifically trying to enhance our current knowledge in cell polarization, matrix-remodeling and cell migration in an attempt to optimize the wound healing process. [3]

New biopolymers and fabrication techniques are being constantly explored as a means of enhancing this natural process with the ultimate aim of decreasing healing time while maintaining the native aesthetics. Current wound healing strategies based on fibrous networks try to maintain an ideal environment to facilitate cell regeneration without causing further damage to the wound or patient. [4] However, a limitation to these techniques is their lack of ability to obtain high levels of alignment and much work has been devoted to creating highly aligned polymeric fibers for wound healing, specifically with electro-spinning. [5][6]

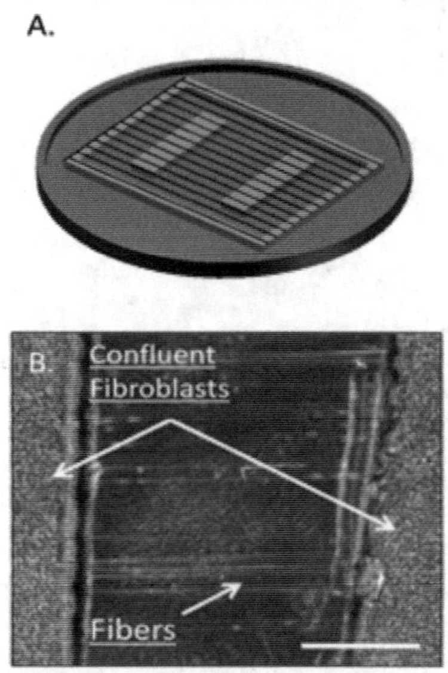

Figure 1 shows the conceptual layout of the experiment (A) and the actual setup (B). Two flat substrates cultured with fibroblasts until confluent were bridged with aligned PLGA fibers (scale bar=500μm)

With advancement in fiber deposition platforms, it is now possible to deposit fibers and fabricate scaffolds of a variety of polymeric systems of different length scales mimicking the native extra-cellular matrix (ECM) upon which cells can be seeded for wound healing. Our previously reported STEP (Spinneret based Tunable Engineered Parameters) platform is uniquely suited to fabricate biomedical scaffolds, as it provides superior control on fiber dimensions (diameter: sub 100nm –micron, length: mm-cm, spacing: sub-micron-microns) and deposition (parallelism $\leq$ 2.5 degrees).[7][8][9] Using this technique, we developed a new wound healing assay for analyzing the wound healing process. In this study, biophysical cues, specifically alignment and geometry are analyzed for their effect on wound closure. Specifically, poly-lactic glycolic acid (PLGA) aligned fibers are deposited between two confluent 3T3 monolayers mimicking a scratch test (Fig 1). We envision that understanding these cues will allow for the design of new sutures and the development and enhancement of wound healing devices.

Materials and Methods:

Polymer solution of 3%(w/w) PLGA dissolved in Chloroform was prepared and spun using STEP technique to achieve deposition of highly aligned uniform suspended fibers (app. 400nm diameter). Scaffolds were coated with fibronectin to improve cellular adhesion. Two plastic coverslips were seeded with NIH3T3 fibroblasts and cultured in DMEM for several days until a monolayer formed covering the entire plastic surface. The PLGA suspended scaffolds were then applied to the flat substrates to bridge the gap between the two separate monolayers shown in Figure 1. Time-lapse microscopy was started once cells from the monolayer began migrating into the gap now bridged by the nanofibers. These fiber scaffolds were imaged for approximately a week until a confluent layer of cells covered the entire fiber scaffold.

Results

With fiber implementation, Figure 2a, three major stages of wound closure (axial and lateral migration, and fiber gap closure) were observed depending on cell confluency.

Axial Migration: First, cells from the stationary monolayer (confluent) began to accelerate out at an average speed of ~50um/hr onto the (nonconfluent) fibers shown in Figure 2b. Each subsequent cell following the leading cell generally showed a decreased migration velocity and were found to be attached to trailing cells. The leading migratory cells as they reached the center were unstable, frequently changing its polarization and speed. However whenever individual cells came into contact with dense trailing groups of cells (along the advancing boundary), they migrated toward the less confluent area. Over the next 48 hours, the cells completely covered the fibers and migration along the fiber was inhibited by the high cell density, leading to lateral migration.

Lateral Migration: The cells continued to migrate along the fiber axis but with no room to attach to the actual fibers, the cells began to migrate outwards and along the inner attached cells. With a confluent fiber, the outer cells were no longer directly attached to the fiber but to its cell neighbor, stacking as many as three cells in either direction of the fiber. This stacking was found to be higher along junctions between the fiber scaffold and the confluent boundary leading to fiber gap closure.

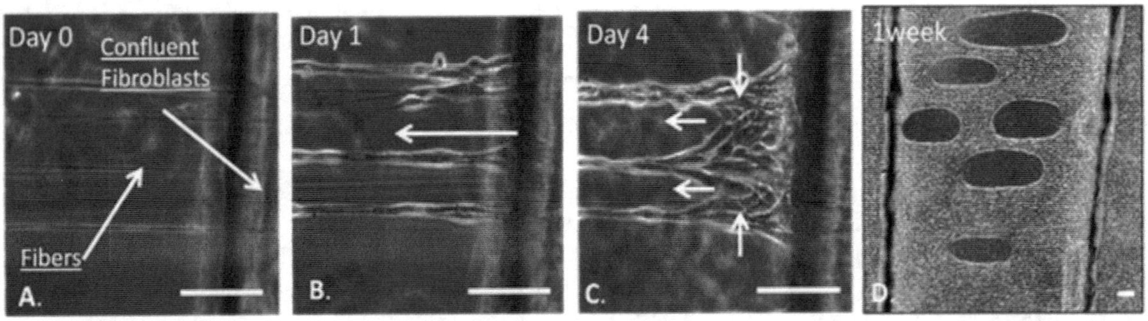

Figure 2(a) Fibers applied to fibroblast monolayers. (b)Cells are migrating along fibers showing axial migration (c) Fibroblasts with limited space to migrate begin closing fiber gap through lateral migration (perpendicular motion to the fiber). (d) After 1 week, fibroblasts almost completely cover the suspended fibers (all scale bars = 100μm)

Gap Closure: Migrating cells closed the gap between the suspended fibers by utilizing the perpendicular intersection between the fibers and the confluent boundary. Interestingly, cells were observed to continuously bridge the corners of the intersections leading to filling of gaps between fibers. This process was capable of closing fiber gaps as large as 100 microns shown. In Figure 2c, the white arrows show the migration of cells in between the fibers only supported by neighboring cells. The speed of this gap closure was found to be a function of fiber separation. In the case of large fiber separations (*100* microns or more) the cell boundary only at migrated about 2 microns per hour requiring many cells to be recruited before the growth could continue to advance. However, as fiber separation decreased, migratory speed increased exponentially where fibers separated by 20 microns or less could be bridged by single cells without support and closed the gap at 40microns/ hr. Within a week, almost all the fiber gaps were completely closed and are shown in Figure 2d.

Conclusion

STEP enabled wound closure model reveals fibroblast behavior in the wound healing process based solely on biophysical cues. . The confluent boundary represents the healthy tissue along the edges of a wound. With the implantation of aligned fibers onto a wound, the healthy tissue was able to cells into the wound as shown by the fibroblast migration. The wound healing process leading to wound closure was found to be dependent upon cellular axial and lateral migration. The aligned fiber migration wound healing scaffold provides the importance of biophysical cues and when coupled with biochemical cues opens new directions in design of wound healing strategies.

References

[1]R. A. F. Clark, The Molecular and Cellular Biology of Wound Repair, R. A. F. Clark, Ed., pp. 3–50, Plenum Press, New York, NY, USA, 2nd edition, 1996.

[2]J Yarrow, BMC Biotechnol. 2004 Sep 9;4:21.

[3] J. N. Percival, Surgery 2002, 20, 114.

[4] Payam , Zahedi, Polymers for Advance Technology 2009

[5]Huan Pan, Luming Li, Long Hu, Xiaojie Cui, Polymer, Volume 47, Issue 14, 28 June 2006, Pages 4901-4904,

[6]Seeram Ramakrishna, Kazutoshi Fujihara, Wee-Eong Teo, Thomas Yong, Zuwei Ma, Ramakrishna Ramaseshan, Electrospun nanofibers: solving global issues, Materials Today, Volume 9, Issue 3, March 2006,

[7] Sasha Bakhru, Amrinder S. Nain, et al., Integr. Biol., 2011, 3, 1207–1214

[8] Nain, A.S., Kowalewski, T., Jacobson, A., Sitti, M., and Amon, C., Macromolecular Rapid Communications, 2009. 30, pp. 1406-1412, 2009.

[9]Nain, A.S., Phillippi J. A., Sitti, M., MacKrell, J., Campbell, P. G., and Amon, C., Small (Weinheim an der Bergstrasse, Germany), 2008. 4(8), 1153-1159.

(194a) Invited: The Complex Natural Bionanotechnology of Autoimmune Disease, Atherosclerosis, and Plaque Diseases Including Alzheimer's and Parkinson's Disease

Annelise E. Barron, Bioengineering, Stanford University, Stanford, CA

My lab developed and studied host defense peptide (**HDP**) mimics based on oligo-*N*-substituted glycines (*peptoids*). Peptoid HDP mimics were designed to emulate the cationic, helical HDPs (or “antimicrobial peptides”) such as magainin (frog) and LL-37 (human). We have examined mechanisms of action and selectivity (or lack of it) in many peptoid sequence variants, comparing with HDPs of interest. Here, selectivity is defined the propensity to incapacitate pathogens, with less (or, acceptable) levels of toxicity to host cells. For peptoid mimics as for HDPs, activity and selectivity depend upon net positive charge, hydrophobicity, amphipathicity, and self-association into meta-stable coiled coils that dissemble on binding to electronegative cell membranes. We undertook mechanistic studies with peptoid variants having high (or low) activity, and/or high (or low) selectivity, comparing to selective and non-selective HDPs. Vesicle leakage, membrane depolarization, and TEM yield surprising and we believe paradigm-shifting new information about HDP and HDP mimic mechanisms of killing.

Unexpected insights into biology and physiology followed a realization that *the most active antibiotic peptoids always carry concomitant cytotoxicity—**as do HDPs, the human cathelicidin LL-37 in particular***. I was struck in particular by similarities between antibacterial and cytotoxic activities of our HDP mimics and LL-37, a phylogenetically ancient innate immune effector that is unique in the human proteome and that plays indispensible, pleiotropic roles in vivo. Misregulated LL-37 is known, already, to play a role in the autoimmune diseases rosacea, eczema, psoriasis, and lupus erythematosus, and most likely also rheumatoid arthritis.

In this lecture, I hypothesize a key role for LL-37 in the etiology of other human degenerative diseases, including Alzheimer’s and amyloid diseases including atherosclerosis, Type II diabetes, and other maladies, as consequences of LL-37’s intrinsic biophysical mechanisms of action, which can be both beneficial and damaging in a strongly concentration-dependent fashion. Dearth of LL-37 prevents normal autophagic cell maintenance, crippling cells over time. Yet disease can also result from overactivation of physiological triggers of LL-37's expression. Chronic nervous or traumatic hyperactivation of HDP expression, overstimulating the normally protective and beneficial HDP detoxification mechanisms, lead to hemolysis and trigger the deposition of atherosclerotic and amyloid plaque, which normally can be rapidly degraded by macrophages. However, frequent overexpression of LL-37 as a result of acute phase responses, occurring against a background of chronic underexpression and/or improper CAMP gene regulation (for instance as a result of inadequate Vitamin D3) can disrupt the balance between the two systems. Amyloid, initially formed around host cells to serve as "short-term" protective barrier that binds up endogenous cytotoxic HDPs preventing their disruption of essential cell processes, accumulates, engendering a "snowballing" of damaging, cytotoxic inflammatory processes.

(194b) Invited: "Multilayered" Approaches to the Design of Stable, Unstable, and Reactive Biointerfaces

David M. Lynn, Department of Chemical and Biological Engineering, University of Wisconsin-Madison, Madison, WI

Methods for the stepwise, layer-by-layer deposition of polymers on surfaces have had an enormous impact on the design of nanostructured surfaces and interfaces for biomedical and biotechnological applications. Most conventional approaches to layer-by-layer assembly exploit interactions between oppositely charged polymers to design thin polyelectrolyte-based films ('polyelectrolyte multilayers') held together by multivalent weak interactions. This approach allows the fabrication of polymer-based films on an incredibly varied range of surfaces (including the topographically complex surfaces of common biomedical devices), and it provides precise, nanometer-scale control over both film thickness and composition. These methods are also particularly well suited for the incorporation of bioactive species (such as proteins, peptides, and DNA) and, thus, also provide opportunities to design functional biointerfaces that provide new levels of control over the immobilization, presentation, and release of these agents from surfaces. The first part of this presentation will describe our work on the design and fabrication of nanostructured and ultrathin coatings useful for the delivery of DNA, with a focus on (i) fundamental aspects related to the design and characterization of films that erode and disintegrate in physiological media, and (ii) the ability of these ultrathin coatings to release or locally transfer therapeutic DNA constructs from the surfaces of interventional devices and promote beneficial physiological responses *in vivo*. We have also developed new methods for the "reactive" or "covalent" layer-by-layer assembly of polymer multilayers. These methods exploit interfacial reactions between mutually reactive polymers (as opposed to electrostatic interactions between oppositely charged polymers) and, thus, lead to polymer multilayers that are covalently crosslinked and physically stable. These methods preserve many of the key practical advantages of layer-by-layer assembly but, importantly, also result in films containing reactive groups that can be used to further decorate or pattern surfaces with chemical or biological functionality. The second part of this presentation will describe our work on the design of azlactone-functionalized multilayers as reactive platforms for the design, fabrication, and selective functionalization of new nano-biointerfaces that control, confine, and manipulate cell growth, provide new substrates for the automated and multi-step synthesis of complex biomolecules, and enable the design of reactive superhydrophobic surfaces that open new opportunities for the application of these extremely non-wetting materials in biotechnological contexts.

(194c) Bionanotechnology to Guide Vessel Sprouting

Tatiana Segura, Chemical and Biomolecular Engineering, University of California Los Angeles, Los Angeles, CA

Vascular endothelial growth factor (VEGF) is known to activate proliferation, migration, and survival pathways in endothelial cells through phosphorylation of VEGF receptor-2 (VEGFR-2). VEGF has been incorporated into biomaterials through encapsulation, electrostatic sequestration, and covalent attachment, but the effect of these presentation strategies on VEGF signaling and endothelial cell sprouting has not been thoroughly investigated. In our laboratory, we investigate if the manner in which VEGF is presented from a scaffold to endothelial cells influences the phosphorylation of VEGFR-2 and the architecture of the blood vessels formed. We used three different VEGF presentations to study their role of VEGFR-2 signaling and vessel sprouting: covalently bound (Vc), covalently bound and clustered (cVc), and sustained release from nanocapsules (nV). We covalently bound VEGF to either self-assembled monolayers on gold to make Vc or to polymeric nanoparticles to make cVc. To make nV we used in situ polymerization to polymerize a plasmin degradable nanogel around the protein. We found that covalently bound VEGF (Vc) is able to phosphorylate VEGFR-2 (VR-2) to the same extent as soluble VEGF (Vs) in endothelial cells (EC), but that the mode of VEGF presentation alters the tyrosine residues that are phosphorylated, the time course of phosphorylation, and the resulting downstream signaling. Using cVc and an EC branching assay or CAM assay we found that VEGF clusters led to an increase in EC sprout branching, thickness, and total vessel network length compared to soluble VEGF or less clustered VEGF. Last, sustained released VEGF from nV resulted in sustain receptor phosphorylation and enhanced EC branching without VEGF replenishing.

Short Bio

Professor Tatiana Segura received her Bachelors of Science degree from the University of California Berkeley and her doctorate from Northwestern University. She pursued post-doctoral training at the Swiss Federal Institute of Technology, Lausanne. Professor Segura's Laboratory studies hydrogel materials for stem cell culture and drug/gene delivery. On this topic she has published over 20 peered reviewed publications. She has been recognized with the Outstanding Young Investigator Award from the American Society of Gene and Cell Therapy, the American Heart Association National Scientist Development Grant, and the CAREER award from National Science Foundation.

(209a) Human Health and Nanomaterials: Occupational Exposure and Toxicity In Context

Randy L. Vander Wal, Energy and Mineral Engineering & The EMS Energy Institute, Pennsylvania State University, University Park, PA

At the forefront of particulate health concerns are nanomaterials. This talk will propose some considerations for evaluating nanomaterials and their health effects.

1. Is the material as prepared truly "nano"? For example, even though SWNTs satisfy the accepted definition of being nanoscale in 2-dimensions, in practice and certainly as synthesized, they are highly coalesced into aggregates of micron in size.

2. What distinguishing characteristics apply to the particular nanomaterial? High surface area is a universal hallmark of a nanomaterial. Yet fumed silica with a surface area of > 200 m^2/g is rated biologically safe and even used a food additive.

3. What are the synthesis methods? Even if synthesis is conducted within a closed system, subsequent processing steps of harvesting and processing can create far greater exposure risk.

3. Do the current markets and/or applications warrant concern? For example quantum dots contain heavy metals, yet they are a commercial product used in limited quantities for biological imaging.

4. What are the exposure routes and levels likely to be encountered in the manufacture, product development and consumer use of nanomaterial based products? Examples in which the nanomaterial, manufacturing process and product are generally considered safe include nanoclays used in various commercial polymers, TiO_2 in sunscreen and carbon black in car tires.

5. Is there any relevant information on related materials and/or nanoparticles and if so, is it applicable?

The talk will balance these considerations with those nanomaterials derived accidentally such from combustion-based stationary power generation or transportation. For the general population these sources overwhelm their potential exposures to nanomaterials, even after decades of study and emission controls. In fact recent studies have linked soot with asthma, heart arrhythmias and even death.

(209b) Structure-Activity Relationship Predicted Toxicity Metric for Metal Oxide Nanoparticles

Rong Liu[1], Haiyuan Zhang[1], Zhaoxia Ji[1], Robert Rallo[2], Andre E. Nel[3] and Yoram Cohen[4], (1)1. Center for the Environmental Implications of Nanotechnology, University of California, Los Angeles, Los Angeles, CA, (2)Departament d'Enginyeria Informatica i Matematiques, Universitat Rovira Virgili, Tarragona, Spain, (3)Division of NanoMedicine, University of California, Los Angeles, Los Angeles, (4)Chemical and Biomolecular Engineering, University of California, Los Angeles, Los Angeles, CA

The rapidly growing use of Engineered Nano-Materials (ENMs) in modern industrial products and processes has increased public concern regarding their potential of adverse environmental and health impacts. As a result, toxicity screening, which is critical for characterization of the potential hazard of ENMs for subsequent risk assessment and development of environmental and health regulatory policies, is confronted by a formidable task of (*in vitro* and *in vivo*) testing the large number and diversity of current and expected future ENMs. In order to tackle this challenge, there is a need for *in silico* methods (computational approaches) that enable virtual but rapid toxicity assessment. Structure-Activity Relationships (SARs) is such an approach that intends to generate predicted toxicity metric for studied materials from their physicochemical (structural-dependent) properties. In the present work, classification SAR were developed using a biological profile that depicts the cytotoxicity of twenty-four metal oxide nanoparticles. The cytotoxicity of these nanoparticles was assessed using human bronchial epithelial (BEAS-2B) as well as murine myeloid (RAW 264.7) cell lines with both single parameter assays (including MTS, ATP and LDH) and multi-parameter assays (comprising ROS production (MitoSox red fluorescence), intracellular calcium flux (Fluo4 fluorescence), mitochondrial membrane potential (JC1 fluorescence), and surface membrane permeability (PI uptake)) over a range of concentrations of 0.375-200 $mg \cdot L^{-1}$ and exposure times of up to 24 h. The cytotoxicity profile was statistically processed via dose-response analysis and rank analysis, which labeled seven nanoparticles as toxic and seventeen as nontoxic. The labeling result was then used as an endpoint for developing SARs with an initial pool of thirty nanoparticle descriptors. In order to identify the most suitable SAR models, five different models (including Naïve Bayesian Classifier (NBC), Linear Regression (LIR), Linear Discriminate Analysis (LDA), Logistic Regression (LGR), and Support Vector Machine (SVM)) were evaluated. For each model, the most suitable nanoparticle descriptors were identified via an exhaustive feature (descriptor) selection. In order to ensure the simplicity and generalization capability of SARs, simple models (including NBC, LIR, LDA, and LGR) were restricted to having up to three descriptors, while having no more than two descriptors were enforced for more complex models (i.e., second order LGR and SVM). The models were assessed and compared using a validation approach which provided a weighted average of a re-substitution classification accuracy and a 200-round bootstrapping/re-sampling classification accuracy to estimate model performance. The second order LGR and SVM were identified as the best performing models, both demonstrating a good classification accuracy above 90% as well as Y-randomization validated robustness. More importantly, both SAR models selected conduction band energy and hydration energy as the most suitable descriptors, which agreed well with recent studies suggesting that the electron transfer from biological system to nanoparticles and the metal ion shedding from nanoparticle surface are two important mechanisms of cellular oxidative stress generation. Relying on the posterior class probabilities (i.e., the likelihoods of a nanoparticle being toxic or nontoxic) inferred in the model development, the SARs can assist in environmental and health regulatory decision making for nanoparticles under different acceptance levels of false negative and false positive. In addition, the developed SARs can also be used to guide the design of safer nanomaterials as well as prioritizing nanomaterials for confirmatory studies via expanded toxicity testing.

(209c) Toxicity of Non-Cd-Based Fluorescent Nanoparticles for Biomedical Imaging

Kristi Olesik[1], Olivia Wetta[1], C. Jenny Dorcéna[1], Kalpesh D. Mahajan[2], Jianqian Xu[1], Qirui Fan[1], Gang Ruan[1], Andrew Herrington[3], Peter Kner[3] and Jessica O. Winter[4,5], (1)William G. Lowrie Department of Chemical and Biomolecular Engineering, The Ohio State University, Columbus, OH, (2)William G. Lowrie Dept. of Chemical and Biomolecular engineering, The Ohio State University, Columbus, OH, (3)Faculty of Engineering, University of Georgia, Athens, GA, (4)William G. Lowrie Department of Chemical and Biomolecular Engineering, the Ohio State University, Columbus, OH, (5)Department of Biomedical Engineering, The Ohio State University, Columbus, OH

Fluorescent nanoparticles have a unique potential for biomedical imaging because of their small size and unique fluorescent properties, including resistance to photobleaching, narrow emission bandwidth, and broad excitation spectra. However, the current gold standard nanoparticles for biomedical imaging, quantum dots, contain cadmium, a carcinogen and neurotoxin, which is released upon photooxidation yielding epigenetic changes, apoptosis, and cell death. Toxicity has severely hindered the application of quantum dots in the clinic, even in a diagnostic capacity. Here, we examined toxicity behaviors of two alternative fluorescent nanoparticles: carbon dots and Mn-doped ZnSe quantum dots, comparing them to commercial quantum dots. Carbon dots were examined as bare, water soluble nanoparticles and in poly(lactide-co-glycolide) (PLGA) polymer nanocomposites, whereas Mn-doped ZnSe quantum dots were investigated with an organic surface coating, a water-soluble surface coating, and a polymer micelle surface coating. Four tests were used to assess cell health and possible pathways of cell death in human liver cells: an MTT assay for viability, a TUNEL assay for apoptosis, LIVE/DEAD screening for necrosis, and a test for production of reactive oxygen species (ROS). In addition, we examined particle sensitivity to photooxidation, a suspected cause of ion release and resulting toxicity. It was determined that carbon dots exhibit ~2-20X less toxicity than commercial quantum dots, and that coatings greatly enhance particle resistance to photooxidation.

GENOMIC-LEVEL EFFECTS OF CARBON NANOTUBES

Dimosthenis A. Sarigiannis[1], Graziella Cimino Reale[2], Angelo Collotta[3], Elisa Roda[4], Paolo Mustarelli[5], Teresa Coccini[6], Luigi Manzo[7]

[1]Aristotle University of Thessaloniki, Department of Chemical Engineering, Thessaloniki 54124, Greece, [2]Department of Experimental Oncology, Fondazione IRCCS Istituto Nazionale Tumori, Milan – Italy, [3]Institute for Health and Consumer Protection, European Commission - Joint Research Centre, Ispra, Varese, Italy, [4]Toxicology Division, Salvatore Maugeri Foundation IRCCS, Pavia, Pavia, Italy, [5]Department of Physical Chemistry, University of Pavia, Pavia, Italy, [6]Toxicology Division, Salvatore Maugeri Foundation IRCCS, Pavia, Pavia, Italy, [7]Department of Internal Medicine and Therapeutics, University of Pavia, Pavia, Italy

ABSTRACT

The widespread use of functionalized carbon nanotubes (CNTs) makes the understanding of potential harmful effects highly important. Two cell culture systems, human A549 pneumocytes and HaCaT keratinocytes, were used to assess the modulation of gene expression due to exposure to single and multi-walled CNTs. Moreover, CD-1 male mice were exposed to the CNTs tested by intra-tracheal instillation and lung samples were taken and analyzed after 1 day of exposure. Differentially functionalized CNTs (MW-COOH and MW-NH2) were tested in comparison with pristine multi-wall carbon nanotubes (MWCNTs) and single-wall carbon nanotubes (SWCNTs). Toxicogenomic analysis included whole genome micro-array analysis and quantitative PCR using micro-fluidic cards for inflammation genes. Comparison of gene expression between in vitro and in vivo exposure to CNTs revealed differences in the level of biological response induced towards oxidative stress, inflammation and apoptosis. Differential modulation in gene expression after in vivo exposure was observed as a function of single or multiple wall geometry and presence of specific functional groups. MW-COOH showed a very high degree of up-modulation of the genes coding for chemokine ligands clinically associated with the onset of lung fibrosis in humans. This effect was much less pronounced with MW-NH2 or SWCNT, whereas pristine MWVNT did not show any statistically significant modulation in gene expression. The main biological pathways induced by the tested CNTs were chemokine and cytokine induced inflammation and oxidative stress. This study indicates that CNT functionalization modulates the advent of early biological events affecting their health effects. Thus, it opens the way towards "intelligent" CNT functionalization that aims at reducing or eliminating potential health risks while delivering the added value with regard to applications such as enhanced drug delivery.

INTRODUCTION

Carbon nanotubes (CNTs) possess unique electrical, mechanical, and thermal properties, with potential wide applications in the electronics, computer, aerospace, and other industries. CNTs are also being developed for a range of biomedical applications including miniaturized biosensors, targeted drug delivery, and tissue engineering [1,2]. Wider application spectrum of CNTs in medicine can be expected due to recently developed functionalization schemes enabling the implementation of new functions that cannot otherwise be acquired by nonfunctionalized nanotubes [3]. Functionalized CNTs were shown to cross cell membrane and deliver attached cargos into cells, properties that could be exploited in applications such as gene delivery system, therapeutic siRNA silencing, and molecular imaging [4-7].

However, research findings have raised concern regarding safety and biocompatibility characteristics of CNTs [8,9]. Due to their small size and unique physicochemical properties, CNT particles may interfere with the normal biological processes when absorbed in the body. Human health risks have been especially related to inhalation exposure, based on reports indicating pulmonary inflammation, fibrosis, and asbestos-like responses induced by these fiber-shaped materials in experimental models [8,10]. Furthermore, CNTs were found to suppress systemic immune function [11], promote allergic responses [12] and cause platelet activation and prothrombotic effects in mice. While various applications are being proposed for CNTs in neurology and neurobiology [13,14], little information exists on biocompatibility of CNTs in the nervous system although neurocytological alterations have been described in some studies [15,16].

Numerous studies have been conducted in the last years on toxicological effects of CNTs. However, existing data are controversial and findings have been difficult to interpret in some cases. Toxicity and reactivity of CNTs were shown to vary depending on characteristics of the specific type of material tested, such as fiber length, fiber diameter, surface area, tendency to agglomerate, dispersibility in media, and the method used for synthesis, which can produce impurities and leave metal catalyst residues [3,8,9]. Effects of CNTs can also be modified by chemical functionalization [17]. In certain studies, a high degree of functionalization was shown to mitigate toxic effects [17,18]. For example, in human dermal fibroblasts, cytotoxicity of phenyl-SO_3H and phenyl-$(COOH)_2$ functionalized single-walled carbon nanotubes (SWCNTs) decreases as the functionalization of the tubes increased [17]. In principle, these findings may be of considerable practical importance, as they would support the notion that toxicity of CNTs can be attenuated with simple chemical changes. However, in other studies, functionalization increased pulmonary toxicity of SWCNTs in mice exposed in vivo [19]. Multi-walled carbon nanotubes (MWNTs) functionalized with either epsilon-caprolactam or L-alanine were more potent in altering the activation of the human serum complement system compared to unmodified MWNTs [20]. Experiments in bacterial systems also suggested a positive correlation between cytotoxicity and functionalization of CNTs [21].

In the present work, two cell culture systems, human A549 pneumocytes and HaCaT keratinocytes, were used to assess the modulation of gene expression due to exposure to single and multi-walled CNTs. Moreover, CD-1 male mice were exposed to the CNTs tested by intra-tracheal instillation and lung samples were taken and analyzed after 1 day of exposure. Differentially functionalized CNTs (MW-COOH and MW-NH_2) were tested in comparison with pristine multi-wall carbon nanotubes (MWCNTs) and single-wall carbon nanotubes (SWCNTs).

MATERIALS AND METHODS

Raw Materials

Commercial MWCNTs (95% purity) were obtained from Cheap tubes Inc. (Brattleboro, USA). MWCNTs with different functionalization (Figure 1) were prepared as described in [22]. The materials produced included carboxyl (COOH) functionalized MW-COOH, amine containing MWCNT-$CONH(CH_2CH_2O)_2$-$CH_2CH_2NH_2$.

MW-COOH

Five hundred mg of raw MWCNT were refluxed in 25 ml of a mixture of concentrated H_2SO_4/HNO_3 (3/1, v/v) (Figure 1 panel a). The resulting suspension was then diluted with 50 ml of icy water. The excess of acid and water was removed by filtration (Millipore, 0.5 μm). The black solid was washed thoroughly with NaOH 0.1 M and subsequently with HCl 0.1 M to remove oxidation debris. The solid was then washed with distilled water until the eluates were neutral. The resulting MW-COOH was dried under vacuum at room temperature. Approximately 400 mg of MW-COOH were obtained.

MW-$CONH(CH_2CH_2O)_2$-$CH_2CH_2NH_2$ (MW-NH_2)

A suspension of 450 mg of MW-COOH in 20 ml of oxalyl chloride was stirred at 62 °C for 24 h (Figure 1 panel b). The excess of oxalyl chloride was then eliminated in vacuo at 60 °C. The resulting solution of MWCNT-COCl and 20 ml of $NH_2(CH_2CH_2O)_2$-$CH_2CH_2NH_2$ in tetrahydrofuran was refluxed for 48 h. After cooling at room temperature, the derivatized CNT were filtered (Millipore, 0.5 m) and the filter washed several times with methanol. The resulting solid was air dried at room temperature. Approximately 400 mg of MW-NH_2 were obtained.

Physicochemical characterization of CNTs

IR spectroscopy was employed to verify that the functionalization procedure had successfully occurred. The IR spectra were obtained by means of a Jasco FTIR 410 spectrometer in diffuse reflectance mode. The degree of functionalization on the sidewalls of MWCNTs was determined by thermogravimetric analysis (TGA). TGA scans were recorded at 10 ºC/min under nitrogen flow with a 2950 TGA (TA Instruments), after an isotherm at 60 °C for 30 minutes in order to remove traces of the reaction solvents. The density of functionalization was reported as the ratio of MWCNTs carbon atoms to addends (amine or carboxyl groups).

Noncontact atomic force microscopy (NC-AFM) was used to image the MWCNTs at the nanometer level [22].

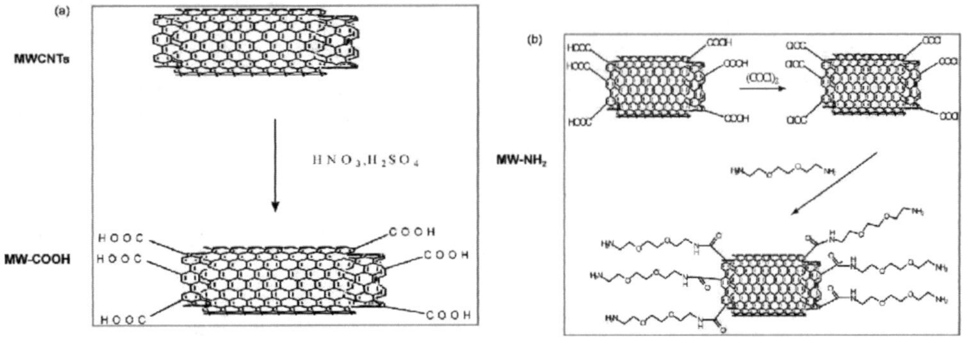

Figure 1: Scheme of the MWCNTs functionalization. Proposed resulting structures: MWCNTs, commercial MWCNTs (pristine); MW-COOH and MW-NH2, carboxyl- and amino-functionalized carbon nanotubes, respectively.

AFM images were collected with an Autoprobe CP Research Microscope (VEECO), operating in non-contact mode, by means of conical silicon tip (curvature radius, 10 nm) onto high resolution rectangular cantilevers (resonance frequency, 150 kHz; force constant, 5.5 N/m; NT-MDT Co). For the analysis, the samples were prepared by depositing 10 l of the solution/dispersion (0.1 mg/ml) of the functionalized nanotubes in the appropriate solvent onto a mica sheet (V-1 grade, SPI supplies). Several scans of different sizes were carried out for each sample with a scan rate of 1.0 Hz. A standard second-order flattening processing of the images was performed in order to correct the scanner nonlinearity.

Elemental impurities were measured by inductively coupled plasma optical emission spectrometry (ICP-OES, Perkin Elmer Optima 3300 DV) in 50 mg samples of material dissolved with nitric acid/hydrofluoric acid/hydrogen peroxide (2:0.5:1; 3.5 ml total volume) at 120 °C for 12 h.

Exposure conditions

Cells were seeded in 96-well plates at the density of 1x $104/cm^2$ in complete medium. After 24 h of cell attachment, plates were washed with 100 l/well of phosphate buffered saline (PBS). Cells were then exposed to suspensions (10 l) of the test materials (pristine MW-CNTs, MW-COOH, MWNH_2) at concentration between 1 and 100 μg/ml, for 24 or 48 h. No fetal bovine serum was used in these preparations as it was proven to interact with nanotubes [23].

Each experiment was repeated independently three times. Per experiment all treatments were repeated six times per 96-well plate. Identical treated cultures were taken as replicate measure for statistical tests. Significant effects ($p < 0.05$) were determined by one-way analyses of variance (ANOVA) (software package SPSS Inc. 1999)

Three groups of four CD-1 male mice were exposed to all tested types of CNTs (both pristine and functionalized) by intra-tracheal installation for one day. SW-CNTs were used at concentrations between 0.1 and 100 μg/ml, while all multiple wall CNTs (both pristine MW-CNTs and functionalized MW-COOH, MW-NH_2) were used at concentrations between 1 and 100 μg/ml. After the exposure period lung samples were taken from the exposed mice and the respective modulation in gene expression was analysed in the whole genome as described below.

Toxicogenomics

Gene expression profiling offers considerable potential for identifying chemical causation of effects induced in exposures to complex mixtures, and for understanding the mechanistic basis for their phenotypic effects. It entails a series of experimental steps, including RNA extraction and preparation, microarray hybridisation, chemiluminescent signal detection, image acquisition and image analysis of the microarrays, and microarray data analysis (in our case we used the Applied Biosystems Expression System software to extract assay signal and signal to noise values from the microarray images) [24]. Microarray data analysis identifies differentially expressed genes, applying different statistical filters to allow for only the statistically and biologically significant probes to be evaluated. Significant probe sets were evaluated for relevance to canonical pathway, molecular function and biological function using the Protein ANalysis THrough Evolutionary Relationships (PANTHER) Classification System (http://www.pantherdb.org). Statistically significant over- and under-represented annotation categories were determined by binomial statistics, using the observed number of expressed genes versus the numbers expected by chance within a certain annotation group. Categories with p-values $> 10^{-2}$ are usually rejected. The results of microarray analysis were validated using quantitative real-time polymerase chain reaction (PCR); the latter determines the expression level of messenger RNA (mRNA) in differentially expressed genes.

RESULTS

Physicochemical characterization

Some microstructural information of the test materials investigated in this study is summarized in Figures 2 and 3. MW-NH2 nanotubes were initially characterized by FTIR spectroscopy (Figure 2(left)). The spectrum showed the typical signals of the amide-group vibration and C-O and C-N stretching, at ~1660 cm^{-1} and 1100 cm^{-1} respectively, as an indicator of the amine reaction with the -COOH groups reflecting successful functionalization. The degree of functionalization of MW-COOH and MW-NH_2 was determined by thermogravimetric analysis from 60 to 800 °C under N_2 flow. Figure 2(right) shows the thermograms of the nonfunctionalized multi walled carbon nanotubes (MWCNTs), MW-COOH and MW-NH_2.

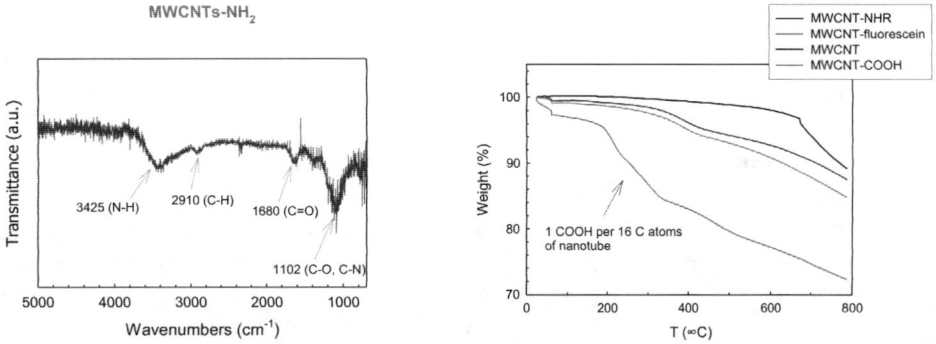

Figure 2: (a) FTIR spectrum of MW-NH2 collected in reflectance mode. (b) TGA thermograms obtained at 10 °C/min in N_2 purge of MW-COOH, MW-NH_2. The plot of the pristine MWCNTs is also reported for comparison.

The pristine MWCNTs were thermally stable up to about 700 °C. For the MW-COOH sample, about 20% loss occurred at the temperature range between 150 °C and 600 °C. Assuming that the weight loss was due to fragmentation and consequent pyrolysis of the sidewall organic groups, it can be estimated that one functional chain is anchored each ~16 carbon atoms of the CNTs surface [25]. Lower material loss (~10%) occurred in the MW-NH_2, which corresponds to one functional chain for about 100 carbon atoms of the CNT surface.

Figure 3 shows the noncontact atomic force microscopy (NC-AFM) images of MW-NH_2 (right panel). The NCAFM scan of nonfunctionalized MWCNTs is also reported for comparison (left panel). The pristine MWCNTs appear as unequally dimensioned ropes randomly arranged along the substrate, with some isolated bundles with a diameter 300 nm and length of about 1 μm, as indicated by the arrows in the image. The functionalization process apparently changed the morphology of pristine nanotubes. In the case of MW-NH_2 bundles shortened by the oxidative step and large spherical ropes of diameters ranging between 400 and 600 nm and of height ~ 70 nm were detectable.

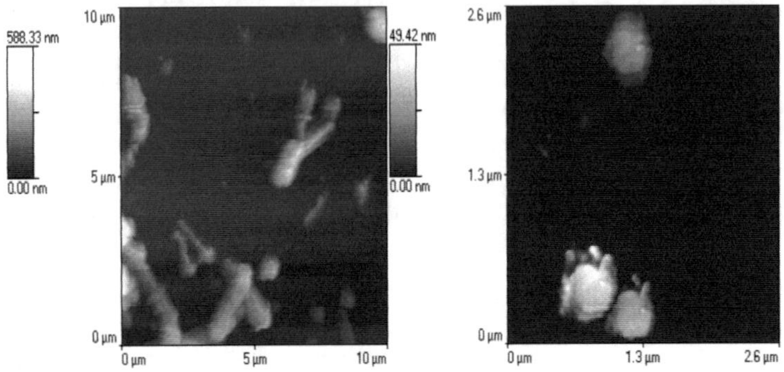

Figure 3: NC-AFM images of pristine MWCNT (left) and MW-NH_2 (right)

Agglomeration

Pristine MWCNTs, MW-COOH, and MW-NH_2 were either insoluble or little soluble in polar solvents and were difficult to evenly disperse in an aqueous matrix. They were sonicated to force their dispersion in the medium before application to cell cultures. In cells incubated with these materials significant agglomeration occurred. Nanoparticles were found to adhere to each other forming dense micron-sized assemblies completely covering the cell surface. Agglomeration increased with increasing doses, at both incubation time points (24 and 48 h) and occurred at similar extent in the HaCaT and A549 cell systems.

Impurities

The content of iron (Fe), cobalt (Co) and nickel (Ni) in the functionalized and nonfunctionalized MWCNTs was determined by inductively coupled plasma optical emission spectrometry (ICP-OES) analysis after mineralization of the samples by Teflon bomb digestion (50 mg, 140 °C overnight), 65% w/w HNO_3-70% w/w HF-30% w/w H_2O_2 (2 ml, 0.5 ml and 1 ml). After cooling, the solutions were filtered on 0.45 m Teflon membranes, taken to 25 ml volume with ultra-pure water and analyzed by ICP-OES. Data in pristine MWCNTs indicated the presence of Fe and Ni impurities at concentrations of 0.25 and 1.5%, respectively. Considerably lower metal concentrations were found in the functionalized nanotubes. In these materials, the content of Fe and Ni was 0.05-0.15% and 0.3-0.5%, respectively.

Gene expression in vitro

Transcriptomics analysis of the whole genome in the keratinocytes (HaCaT) selected as in vitro models in this study showed that gene expression significantly differs according to the duration of exposure. A large number of genes (between 1200 and 2200) changed their expression level when the cells were exposed for 6 hours to the different types of CNTs analysed, namely SWCNT, MWCNT, MW-NH_2 and MW-COOH. On the contrary, long-lasting exposure (48 hours) resulted in a significant reduction in the number of genes expression differentially compared to the control. Less than 500 genes were expressed differentially in the case of MWCNT, MW-NH2 and MW-COOH; only in the case of SWCNTs ca. 1800 genes showed differential expression after 48 hours of exposure.

Similar results were found when we analyzed the data isolating the 25 genes involved in the molecular pathway associated to oxidative stress. The results are summarized in the following table.

Table 1: Differential gene expression in the oxidative stress pathway

Duration (h)	SWCNT	MWCNT	MW-NH_2	MW-COOH
6	6	7	2	2
48	2	2	1	2

The picture changes in terms of modulated gene expression, when focusing on the genes involved in the pathway of inflammation mediated by chemokine and cytokine signalling. The following table summarized the data, which show that cells behaved differently when exposed to pristine and functionalized CNTs. SWCNTs and MWCNTs showed a reduction in the number of genes that had modulated their expression level after long-term exposure (48h); on the contrary, both types of functionalized CNTs showed practically no difference in the number of genes with modulated expression when comparing short and long-term exposure data.

Table 2: Differential gene expression in the chemokine/cytokine signalling-mediated inflammation pathway

Duration (h)	SWCNT	MWCNT	MW-NH_2	MW-COOH
6	9	17	4	7
48	5	3	4	6

The observed differences in the level of gene expression as a function of exposure duration may indicate that different cellular mechanisms are involved, namely that short-term response indicates the onset of defensive mechanisms encompassing several molecular pathways and, thus, a large number of genes. The observed effects after 48 h of exposure would seem to indicate the molecular processes that underlie actual cellular toxic response to CNT exposure. The persistence of the induction of genes driving chemokine and cytokine-signalling-mediated inflammation showed that further attention would have to be paid to the inflammatory effects of the tested CNTs, and in particular the functionalized ones. For this reason, we analyzed the gene expression of the genes involved in the inflammation pathway for the two types of in vitro models we used, i.e. HaCaT keratinocytes and A549 lung epithelial cells. Comparative analysis showed that the gene expression of A549 cells was much higher than the gene expression of the keratinocytes after 48 h of exposure (Table 3).

Table 3: Comparison of modulated gene expression in the chemokine/cytokine signalling-mediated inflammation pathway between HaCaT and A549 cells

Cell model	SWCNT	MWCNT	MW-NH2	MW-COOH
HaCaT	9	17	4	7
A549	21	22	20	14

Gene expression in vivo

The in vitro observations outlined above indicate that the main pathway to be analyzed in the case of in vivo experiments is inflammation in the lungs mediated by chemokine/cytokine signalling.

Identification of specific chemokines involved in inflammation in the lung would result in the discovery of specific biomarkers of effect that may be crucial for the early diagnosis of the toxic potency of CNTs, functionalized or not. The whole genome of lung tissue of CD-1 male mice has been analyzed after 1 day of exposure to CNTs via intra tracheal instillation (1 mg/kg bw). Table 4 gives an overview of the number of genes up- or down-modulated after exposure to the four types of CNTs tested in this study. Only genes with a more than 2-fold change in expression are reported in this table.

Table 4: Number of up-/down-modulated genes in CD-1 mice lung after 1 day of exposure to CNTs

# of modulated genes	SWCNT	MWCNT	MW-NH_2	MW-COOH
Up	258	258	147	372
Common	*6*	*6*	*6*	*6*
Down	200	278	133	330
Common	*9*	*9*	*9*	*9*

Similar numbers of genes modulate their expression level after exposure to pristine SW and MW CNTs, whereas significant differences exist between the two functionalized CNTs: MW-NH2 have a lesser effect than the non-functionalized MWCNTs; on the contrary, MW-COOH affect a much higher number of genes. Another interesting observation is that out of the 180-700 genes with significantly modulated expression only 15 are common across the four treatments.

The microarray results were validated with real-time PCR (polymerase chain reaction). Using bioinformatics analysis on the Integromics® software we made a comparative assessment of the main genes related to chemokine signalling-mediated inflammation. The results are given in table 5. Using a more stingent cut-off criterion for significance of modulation in gene expression (accepting as significant only the genes with more than 3-fold change in expression level) it was clear that MW-COOH had a very significant effect on the expression of all the genes of interest.

Table 5: Modulation of chemokines gene expression in CD-1 mice lung after exposure to CNTs (expressed in number of times gene expression is up-/down-modulated vs. the control)

Gene symbol	Gene name	SWCNT	MWCNT	MW-NH_2	MW-COOH
Ccl19	chemokine (C-C motif) ligand 19	1.2-1.5	1.6-1.8	1.2-1.7	**3.4-4.1**
Ccl2	chemokine (C-C motif) ligand 2	**4.1-4.8**	1.2-1.5	**2.9-3.9**	**27.9-31.8**
Cxcl10	chemokine (C-X-C motif) ligand 10	1.4-1.5	1.0-1.1	1.5-2.3	**17.6-19.0**
Cxcl11	chemokine (C-X-C motif) ligand 11	0.4-1.1	0.2-0.9	1.2-2.3	**6.2-49.6**

Using the PANTHER database coupled to the significance analysis using Integromics® we compared the induction of the inflammation mediated by chemokine and cytokine signalling pathway after exposure to MW-COOH vs. MWCNTs (figure 4), and after exposure to MW-NH2 vs. MWCNTs (figure 5). It was clear that the inflammation pathway was significantly induced by MW-COOH, and it was less so by MW-NH2 or pristine MWCNTs.

DISCUSSION

Carbon nanotubes are generally insoluble in any solvents, resulting in their poor processability for many proposed potential applications. Modification of the CNT surface by chemical functionalization, e.g. by adding carboxylic acid or amine groups, represents an effective method to increase water solubility and dispersibility [26]. Development of well-functionalized CNTs may also improve biological compatibility and pave the way to many useful medical applications of these materials [9].

However, there is still limited information on how chemical functionalization may affect CNT toxicity. Literature data are controversial showing either increased or reduced toxicity associated to functionalization. Recent studies [27] indicated good tolerability and no tissue accumulation of water-dispersible highly functionalized MWCNTs in mice treated by injection. The higher the degree of functionalization, the less was the retention of nanotubes in tissues. However, other studies in mice probed by Raman spectroscopy indicated slow excretion (about two months) of polyethylene glycol (PEG) functionalized SWCNTs given intravenously [28].

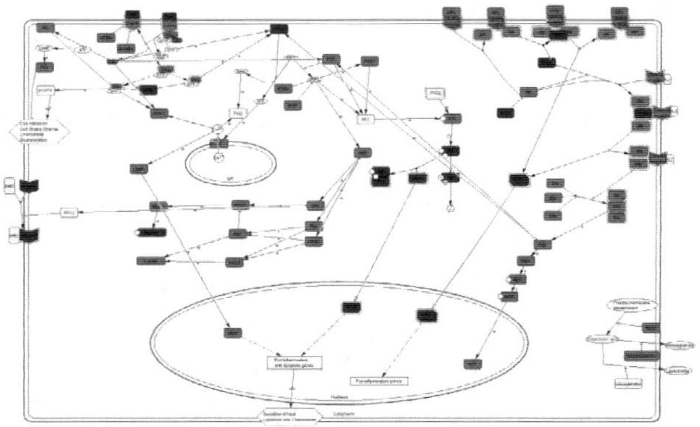

Figure 4: Inflammation mediated by chemokine and cytokine signalling pathway: comparison of results after exposure to MW-COOH (in blue) and MWCNT (in red); in yellow are marked the genes the expression of which is modulated after both treatments

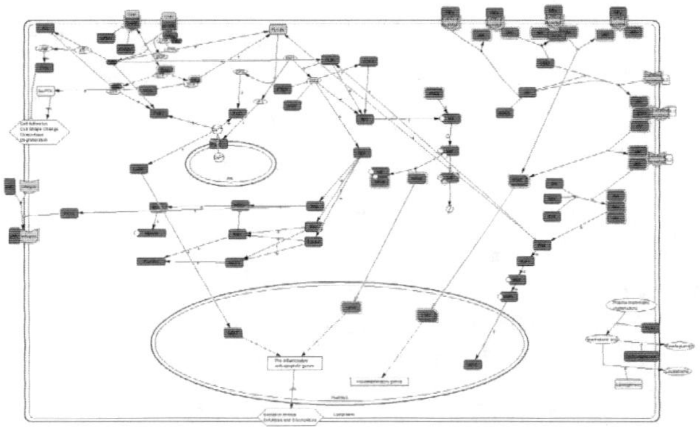

Figure 5: Inflammation mediated by chemokine and cytokine signalling pathway: comparison of results after exposure to MW-NH_2 (in green) and MWCNT (in red); in yellow are marked the genes the expression of which is modulated after both treatments

Analysis of the effects of different types of functionalized and pristine CNTs on gene expression showed that both the chemical family and the type of functionalization influence significantly the induction of key toxicity pathways such as inflammation mediated by chemokine and cytokine signalling. In particular, MW-COOH induced the over-expression of chemokine (C-C motif) ligand 2 (Ccl2) and ligand 19 (Ccl19). Both are cytokines involved in immunoregulatory and inflammatory processes. Ccl19 expression was induced significantly also after exposure to MW-NH2 and SWCNTs. Moreover, MW-COOH induced the over-expression of chemokine (C-X-C motif) ligand 10 (Cxcl10) and ligand 11 (Cxcl11), interferon gamma-induced proteins, involved in immunoregulation. More importantly, clinical observations have recently connected the abundant presence of these cytokines to hypersensitivity pneumonitis and, to a lesser extent, to lung fibrosis.

CONCLUSIONS

In summary, this study indicated that the biological effect of CNTs at the genomic level could be modified by chemical functionalization in association with changes in solubility and tendency of CNTs to form agglomerates. One of the goals of functionalizing CNTs is to increase their solubility in aqueous media, a feature that may make CNTs more compatible with physiological systems. Differential modulation in gene expression was observed as a function of single or multiple wall geometry and presence of specific functional groups. Comparison of gene expression between in vitro and in vivo exposure to CNTs revealed significant differences in the level of biological response induced towards endpoints such as oxidative stress or inflammation. In particular, MW-COOH induced the excessive coding for specific proteins associated clinically with lung fibrosis and other inflammatory processes. Toxicogenomics by Total Gene Expression Analysis is a powerful approach to identify early biological responses in in vitro and in vivo systems following exposure to carbon nanotubes.

The complexity of biological responses following exposure to engineered nanotubes was shown and the need for development of dedicated, comprehensive methodologies to approach the potential health effects of nanomaterials underlined. Before considering chemical functionalization as a general way to improve biocompatibility and safety characteristics of CNTs, a number of toxicological issues must be addressed and more extensive investigations on effects of functionalized CNTs in biological systems are warranted. This opens the way towards "intelligent" CNT functionalization that aims at reducing or eliminating potential health risks while delivering the added value of CNTs that are proposed for diagnostic or therapeutic applications bringing patients into direct contact with them [29]. While systematic approaches for evaluating potential adverse effects of nanomaterials are still to be implemented [30-33], the combination of cell culture systems [34,35] with in vivo experiments designed to minimize the use of laboratory animals in toxicity testing through the use of whole genome microarrays are expected to become a mainstay.

REFERENCES

[1] Kateb, B., Van Handel, M., Zhang, L., Bronikowski, M.J., Manohara, H., Badie, B. (2007). *Neuroimage.* 37, S9-S17.

[2] Veetil, J.V., Ye K. (2009). *Biotechnol. Prog.* 25, 709-721.

[3] Shvedova, A.A., Kisin E.R, Porter D., Sculte P, V.E. Kagan VE, Fadeel, B, Castranova V. (2009). *Pharmacol. Ther.* 121, 192-204.

[4] Bianco, A., Kostarelos, K., Partidos, C.D., Prato, M. (2005). *Chem. Commun.* (Cambridge, UK) 5, 571-577.

[5] Chen, J., Chen, S., Zhao, X, Kuznetsova, L.V., Wong, S.S., Ojima, I. (2008). *J. Am. Chem. Soc.* 130, 16778-16785.

[6] Cui, D. (2007). *J. Nanosci. Nanotechnol.* 7, 1298-314.

[7] Podestà, J.E., Al-Jamal, K.T., Herrero, M.A., Tian, B., Ali-Boucetta, H., Hegde, V., Bianco, A., Prato, M., Kostarelos, K. (2009). *Small* 5, 1176-1185.

[8] Donaldson, K., Aitken, R., Tran, L., Stone, V., Duffin, R., Forrest, G., Alexander, A., (2006). *Toxicol. Sci.* 92, 5-22.

[9] Simeonova, PP. (2009). *Nanomedicine* 4, 373-375.

[10] Lam, C.W., James, J.T., McCluskey, R., Arepalli, S., Hunter, R.L. (2006). *Crit. Rev. Toxicol.* 36, 189-217.

[11] Mitchell, L.A., Lauer, F.T., Burchiel, S.W., McDonald, J.D. (2009). *Nat. Nanotechnol.* 4, 451-456.

[12] Nygaard, U.C., Hansen, J.S., Samuelsen, M., Alberg, T., Marioara, C.D., Løvik, M., (2009). *Toxicol. Sci.* 109, 113-123.

[13] Cellot, G., Cilia, E., Cipolline, S., Rancic, V., Sucapane, A., Giordani, S., Gambizzi, L., Markram, H., Grandolfo, M., Scaini, D., Gelain, F., Casalis, L., Prato, M., Giugliano, M., Ballerini, L. (2009). *Nat. Nanotechnol.* 4, 126-133.

[14] Chao, T.I., Xiang, S., Chen, C.S., Chin, W.C., Nelson, A.J., Wang, C., Lu, J. (2009). *Biochem. Biophys. Res. Commun.* 384, 426-430.

[15] Belyanskaya, L., Weigel, S., Hirsch, C., Tobler, U., Krug, H.F., Wick, P. (2009). *Neurotoxicology* 30, 702-711.

[16] McKenzie, J.L., Waid, M.C., Shi, R., Webster, T.J. (2004). *Biomaterials* 25, 1309-1317.

[17] Sayes, C.M., Liang, F., Hudson, J.L., Mendez, J., Guo, W., Beach, J.M., Moore, V.C., Doyle, C.D., West, J. L., Billups, W. E., Ausman, K. D., Colvin, V.L. (2006). *Toxicol. Lett.* 161, 135-142.

[18] Zhang, L.W., Zeng, L., Barron, A.R., Monteiro-Riviere, N.A. (2007). *Int. J. Toxicol.* 26, 103-113.

[19] Tong, H., McGee, J.K., Saxena, R.K., Kodavanti, U.P., Devlin, R.B., Gilmour, M.I. (2009). *Toxicol. Appl. Pharmacol.* 239, 224-232.

[20] Salvador-Morales, C., Basiuk, E.V., Basiuk, V.A., Green, M.L., Sim, R.B. (2008). J. *Nanosci. Nanotechnol.* 8, 2347-2356.

[21] Kang, S., Mauter, M.S., Elimelech, M., (2008). *Environ. Sci. Technol.* 42, 7528-7534.

[22] Fagnoni, M., Profumo, A., Dondi, D., Merli, D., Mustarelli, P., Quartarone, E. (2009). *Adv. Mater.* 21, 1761-1765.

[23] Casey, A., Herzog, E., Davoren, M., Lyng, F.M., Byrne, H.J., Chambers, G. (2007). *Carbon* 45, 1425-1432.

[24] Sarigiannis, D., Gotti, A., Cimino Reale, G. Marafante, E. (2009). *Int. J. Risk Ass. Mgt.*, 13(3/4): 216–241.

[25] Cioffi, C., S. Campidelli, C. Sooambar, M. Marcaccio, G. Marcolongo, M. Meneghetti, D. Paolucci, F. Paolucci, C. Ehli, G.M.A. Rahman, V. Sgobba, D.M. Guidi, M. Prato (2007). *Journal of the American Chemical Society* 129, 3938-3945.

[26] Fu, K., Sun, Y.P., 2003. *J. Nanosci. Nanotechnol.* 3, 351-364.

[27] Lacerda, L., Ali-Boucetta, H., Herrero, M.A., Pastorin, G., Bianco, A., Prato, M., Kostarelos, K. (2008). *Nanomed.* 3, 149-161.

[28] Liu, Z., Davis, C., Cai, W., He, L., Chen, X., Dai, H. (2008). *Proc. Natl. Acad. Sci.* USA 105, 1410-1415.

[29] Dusinska, M. (2009). *Nanomedicine* 4, 605-607.

[30] Oberdörster, G., Maynard, A., Donaldson, K., Castranova, V., Fitzpatrick, J., Ausman, K., Carter, J., Karn, B., Kreyling, W., Lai, D., Olin, S., Monteiro-Riviere, N., Warheit, D., Yang, H.; ILSI Research Foundation/Risk Science Institute Nanomaterial Toxicity Screening Working Group. (2005). *Part. Fibre Toxicol.* 2, 1-35.

[31] Oberdörster, G., Oberdörster, E., Oberdörster, J. (2005). *Environ. Health Perspect.* 13:823-839.

[32] Stone, V., Johnston, H., Schins, R.P. (2009). *Crit. Rev. Toxicol.* 39, 613-626.

[33] Thomas, T., Bahadori, T., Savage, N., Thomas, K. (2009). *Nanomed. Nanobiotechnol.* 1, 426-433.

[34] Coccini, T., Roda, E., Sarigiannis, D.A., Mustarelli, P., Quartarone, E., Profumo, A., Manzo, L. (2010). *Toxicology*, 269(1): 41-53.

[35] Coccini, T., Roda, E., Sarigiannis, D.A., Mustarelli, P., Profumo, A., Manzo, L. (2009). *Toxicology Letters* 189 (S1): S183-S184.

(209e) Comprehensive Toxicity Evaluation of Carbon Nanoparticles

Yun Wu[1], Xi Zhao[2], Teh-hsun Chen[3], John J. Lannutti[4], Heather Powell[5], Sanjay Rajagopalan[6], Robert J. Lee[7] and Ly James Lee[2], (1)Nanoscale Science and Engineering Center, The Ohio State University, Columbus, OH, (2)William G. Lowrie Department of Chemical and Biomolecular Engineering, The Ohio State University, Columbus, OH, (3)CDC/NIOSH/HELD, Morgantown, WV, (4)Department of Material Science and Engineering, The Ohio State University, Columbus, OH, (5)Department of Materials Science and Engineering, The Ohio State University, columbus, OH, (6)Division of Cardiovascular Medicine, The Ohio State University, Columbus, OH, (7)College of Pharmacy, The Ohio State University, Columbus, OH

The burgeoning use of engineered nanomaterials (ENM) presents a rapidly evolving public health challenge as very little is known about their health effects. To address gaps in our current understanding, detailed studies on exposure assessment, physicochemical determinants of ENM toxicity, methods to measure and predict systemic health risk/toxicity and risk assessment of ENM at the cell, tissue and animal levels are warranted. Carbon nanoparticles such as buckyball, carbon nano tubes/nanofibers (CNTs/CNFs) represent by far the most widely studied ENM in industry and academia. The scientific community is most concerned with the potential toxicity of their exposure because of their increasing use in manufacturing, putative potential for cardiopulmonary effects, lack of monitoring tools, and limited understanding of physicochemical determinants of toxicity.

We have carried out a comprehensive evaluation of the toxicology of carbon nanoparticles from the single cell level *in vitro* to an animal model *in vivo* by combining a novel nanotechnology with conventional cellular toxicity and inhalational exposure study. At the cellular level, we used a nanochannel-electroporation (NEP) method that is capable of delivering nanoparticles with different size, geometry and surface chemistry, into individual cells with precise dose control. The basic element of NEP consists of two microchannels connected by a nanochannel. The cell to be transfected is positioned in one microchannel against the nanochannel and the other microchannel is filled with the agent to be delivered. A voltage pulse(s) lasting milliseconds (ms) is delivered between the two microchannels causing transfection. We achieve dose control by adjusting the duration and number of pulses. NEP provides a unique opportunity for quantitative studies of cytotoxicity of nanoparticles such as CNTs/CNFs at the single cell level. Cell viability, cell cycle, inflammation, oxidative stress, apoptosis, and CNTs/CNFs localization within the cell are compared with conventional cellular toxicity study of CNTs/CNFs using a large number of cells.

In vivo, a whole body exposure system is applied to carry out the nanoparticle delivery to mice. The particles are dispersed in air by an acoustical particle generator. The airborne particles are carried by air and entered the head/nose only exposure chamber for inhalation by a group of mice. Part of the aerosol is sampled through a Scanning Mobility Particle Sizer (SMPS, TSI Inc.) and an Ultrafine Water Condensation Particle Counter (UWCPC, TSI Inc.) to determine the particle size distribution and concentration. Cardiopulmonary health effects of particles are investigated under acute and possibly chronic *in vivo* exposures.

The combination of NEP, conventional cellular toxicity study and inhalational exposure study enable better understanding of toxicology of carbon nanoparticles. The methods developed may be applicable to other types of nanoparticles or toxic materials and in other exposure settings.

(209f) Life Cycle Ecotoxicity of Carbon Nanotubes

Matthew J. Eckelman, Civil & Environmental Engineering, Northeastern University, Boston, MA, Meagan Mauter, Harvard Kennedy School, Cambridge, MA, Jacqueline Isaacs, Northeastern University and Menachem Elimelech, Yale University, New Haven, CT

Carbon nanotubes (CNTs) comprise a promising and versatile class of materials with many potential environmental applications but with largely unknown environmental implications, with toxicity being the major concern driving research efforts in this area. Much current work tests the behavior of CNTs under various environmental conditions, particularly their end-of-life fate and transport in aquatic environments. Also important but less visible is work characterizing the impacts of CNT production using various synthesis methods, usually in terms of energy and material use. Taking a life cycle approach, this project quantifies and compares the inherent environmental impacts of CNTs released into the environment with those related to their synthesis. This was done using the single measure of aquatic ecotoxicity, which is the toxicity category most strongly supported by existing data.

As an emerging technology, CNTs are not included in any established life cycle assessment models. Complicating the picture is the fact that CNT toxicity varies along multiple parameters, including size distribution, chirality, and functionalization, making it very difficult to establish a single toxicity factor for CNTs as a class. Using published toxicity studies for single species from different trophic levels, we employed the USEtox model to develop a range of appropriate ecotoxicity characterization factors of approximately 1-10 mg/L (HC_{50}). USEtox is an integrated multi-media fate, transport, and toxicity model covering large classes of organic and inorganic substances. Two distinct release, fate and transport scenarios were considered: a highly conservative "worst case" scenario and a "realistic" scenario that draws from exiting literature on CNT fate, transport, and ecotoxicity. In terms of the production phase, a life cycle inventory of inputs and emissions was constructed by combining measured data for three different synthesis techniques (HiPco, CVD, and ablation) with the database of characterization factors provided with the USEtox model. Current manufacturing trends for reagent recovery and reuse and pollution controls were taken into account.

The results indicate that the ecotoxicity impacts of nanomaterial production processes are roughly equivalent to the ecotoxicity of CNT releases under the unrealistic worst case scenario, while exceeding the results of the realistic scenario by three orders of magnitude. Ecotoxicity from production processes is dominated by emissions of metals from electricity generation. Uncertainty exists for both production and release stages, and is modeled using a combination of Monte Carlo simulation and scenario analysis. The results of this analysis underscore the contributions of existing work on CNT fate and transport, as well as the importance of life cycle considerations in allocating time and resources toward research on mitigating the impacts of novel materials, in particular green chemistry efforts that reduce production ecotoxicity by using benign reagents or by increasing yields.

(221a) Surface-Dependent Biological Response of Human Monocytes to Iron Oxide Nanoparticles

Yaolin Xu[1], Jennifer Sherwood[2] and Yuping Bao[1], (1)Chemical and Biological Engineering, The University of Alabama, Tuscaloosa, AL, (2)Arizona State University, Metro Phoenix, AZ

Iron oxide nanoparticles (NPs) have been widely explored for targeted drug delivery, localized therapy, and as contrast agents for magnetic resonance imaging (MRI). For all these applications, iron oxide NPs need to be injected into the human body and directly interact with the patient immune system. Therefore, it is of great importance to study the immune response of monocyte cells to iron oxide NPs. Monocytes are essential in the body's primary response to pathogens, which make them an ideal candidate for immune response studies. In this study, the monocytes were treated with varying concentrations of polyacrylic acid (PAA)-, polyethylenimine (PEI)-, and polysorbate 80- coated iron oxide NPs. Concentration and time-dependent cytotoxicity and immune responses were conducted.

The 3-(4,5-Dimethylthiazol-2-yl)-2,5-diphenyltetrazolium bromide (MTT) assay confirmed that our iron oxide NPs can be practically used for biomedical applications due to the observed high cell viability (> 80%) for all the samples. Surprisingly, the cytotoxicity tests showed little dependence on the surface functionality based on the cell death trends from different types of NPs. The specific NP-cell interaction mechanisms will be investigated in the future. The effective internalization of the iron oxide NPs by the monocytes were clearly observed under transmission electron microscopy (TEM). Further, the cellular uptake efficiency of NPs as a function of surface chemistries was quantified using inductively coupled plasma-emission spectrometer (ICP). The immune response was primarily analyzed by monitoring the immune biomarkers: tumor necrosis factor (TNF)-α and toll-like receptors 2 (TLR-2). These studies will provide valuable information on the understanding of the surface-dependent cytotoxicity, cellular uptake, and immune response.

This work is funded by NFS-DMR 0907204 and EEC-1005191.

(221b) Development of Inhalable Iron Oxide Nanocomposites for Application in Lung Cancer Therapy

Nathanael Stocke[1], Samantha A. Meenach[1], Kimberly W. Anderson[1], Heidi M. Mansour[2] and James Z. Hilt[1], (1)Chemical and Materials Engineering, University of Kentucky, Lexington, KY, (2)Pharmaceutical Sciences - Drug Development Division, University of Kentucky, Lexington, KY

Chemotherapy results in many adverse side effects when administered systemically through oral or intravenous delivery, and the localization of this therapy can result in improved therapeutic outcomes and enhanced patient response. When treating lung cancer, targeted pulmonary administration of chemotherapy has shown enhanced therapeutic response and reduced side effects. Additionally, dual administration of chemotherapy and hyperthermia can result in a synergistic effect and the combination of hyperthermia with targeted pulmonary administration of chemotherapy poses a potential improvement on current cancer treatments. In this study, inhalable microparticle nanocomposite dry powders containing iron oxide nanoparticles for targeted pulmonary inhalation aerosol delivery as a dry powder inhaler (DPI) were rationally designed and successfully produced via dilute organic suspension advanced spray drying in closed-mode. An FDA-approved pharmaceutical excipient and non-reducing sugar alcohol with mucolytic properties, D-mannitol, was used in conjunction with various loadings of Fe_3O_4 nanoparticles and the physiochemical properties of the resulting powders were characterized. Scanning electron microscopy (SEM) and transmission electron microscopy (TEM) images showed size and spherical morphology of the particles. The size of the particles was also obtained by statistically analyzing the SEM images using SigmaScan® software. Energy dispersive spectroscopy (EDS) confirmed the presence of iron oxide in the composite microparticle powders. Thermal analysis by differential scanning calorimetry (DSC) and X-ray powder diffraction (XRPD) were used to determine the solid-state phase transitions and long range order, respectively. Thermogravimetric analysis (TGA) was used to determine the amount of Fe_3O_4 incorporated into the inhalable composite microparticle powder and the water content of the powder was quantified analytically via Karl Fisher titration. The emitted dose, respirable dose, fine particle fraction, and mass median aerodynamic diameter of the inhalable dry powders were determined with the Next Generation Impactor (NGI) and a DPI device approved for human use. The ability to formulate a dry powder magnetic composite could have a large impact on the scientific community due to the large interest in Fe_3O_4 for biomedical applications.

(221c) Engineering Magnetic Nanomaterial Production in Magnetotactic Bacteria Through Gene Regulation

Mary E. Wilson[1], Lina M. González[2], Philip R. LeDuc[1,2] and Warren C. Ruder[1], (1)Biomedical Engineering, Carnegie Mellon University, Pittsburgh, PA, (2)Mechanical Engineering, Carnegie Mellon University, Pittsburgh, PA

Magnetotactic bacteria possess the unique ability to synthesize intracellular magnetic particles, known as magnetosomes, which form magnetic chains and yield cells responsive to the Earth's ambient geomagnetic field. While several genomic regions have been identified that encode magnetosome-related genes, little is known about how these genes regulate magnetosome production. Determining magnetic field-induced signaling modules is a critical step for being able to eventually utilize magnetotactic bacteria as biomaterials-producing machines via controlling the production of magnetic nanoparticles by simply turning on or off a magnetic switch. Here, we explore the genetic response of *Magnetospirillum magneticum* strain AMB-1 to an applied electromagnetic field as a means to identify genes activated by magnetic stimulation. We first established the magnetic stimulation parameters for eliciting transcriptional regulation of specific genes. Optimal magnetic field strength and exposure time was determined via quantitative real time-PCR (qRT-PCR) utilizing in-house primer sets for genes related to the magnetosome island (MAI), magnetotactic islet, cytoskeleton and flagellar structures. AMB-1 cultures were subjected to a uniform, high magnetic field (10G) within a solenoid for 1 hour, 3 hours or 8 hours. After extracting total RNA and converting to cDNA, quantitative real time-PCR (qRT-PCR) was performed to measure relative transcription levels. After 1 hour of magnetic stimulation, gene expression was drastically down-regulated; whereas, gene expression was up-regulated after 3 hours. After 8 hours, gene expression was relatively unchanged, indicating that the AMB-1 cells may have adapted or equilibrated to the field. These results suggest that magnetic stimulation leads to altered AMB-1 signaling related to MAI activation/deactivation, as well as cytoskeletal and flagellar regulatory events, and show the challenging nature of magnetically-induced responses that are dynamically changing. We believe this research lays the groundwork for identifying endogenous gene network modules particularly active following application of concentrated magnetic field, which can be coupled with engineered gene circuits to create magnetically inducible synthetic devices. Ultimately, we aim to design synthetic devices that allow organisms to serve as machines that detect specified magnetic field gradients and react by performing a specific function – e.g., directed cell movement for engineered tissue morphogenesis or upregulated production of magnetic nanoparticles that could be exploited in biomedical applications such as cancer ablation therapy.

(221d) Fabrication of 'Theranostic' Magnetite and Antiretroviral Nanoparticles

Sharavanan Balasubramaniam[1], Sanem Kayandan[1], Tianyuzi Li[2], Xin-Ming Liu[2], JoEllyn M. McMillan[2], Michael D. Boska[2], Tatiana Bronich[2], Alexander V. Kabanov[2], Howard E. Gendelman[2], Judy S. Riffle[3,4] and Richey M. Davis[4,5], (1)Macromolecules and Interfaces Institute, Virginia Tech, Blacksburg, VA, (2)University of Nebraska Medical Center, Omaha, NE, (3)Chemistry, Virginia Tech, Blacksburg, VA, (4)Macromolecules and Interfaces Institute, Blacksburg, VA, (5)Chemical Engineering, Virginia Tech, Blacksburg, VA

In the past decade, multifunctional magnetite nanocarriers that integrate therapeutic agents into a single system have attracted considerable interest in theranostics integrating drug delivery and drug biodistribution. Using magnetic resonance imaging (MRI) to guide nanomaterial development can improve therapeutic endpoints. To this end nanoformulations of antiretroviral therapy (nanoART) previously developed for treatment of human immunodeficiency virus (HIV) infection were modified for theranostic testing. Herein, we describe the fabrication of polymer nanocarriers co-encapsulating the HIV protease inhibitor, ritonavir (RTV) within magnetite (Fe_3O_4) nanoparticles. Discrete ~ 8 nm nanoparticles (M_s = 65 A.m^2/kg) were synthesized by reducing iron (III) acetylacetonate in benzyl alcohol and their surfaces were subsequently modified to be hydrophobic by coating with oleic acid. A poly(ethylene glycol)-*b*-poly(D,L-lactide) block copolymer [mPEG(5 kDa)-*b*-PDLLA(10 kDa)] was synthesized by ring-opening polymerization of D,L-lactide in the presence of mPEG using stannous octoate as the catalyst. Drug-copolymer, magnetite-copolymer and drug-magnetite-copolymer 'theranostic' nanoparticles were fabricated by rapid nanoprecipitation in a multi-inlet vortex mixer.

The hydrophobic polyester core of the copolymer nanoparticles sequestered RTV and/or oleic acid-coated magnetite whereas the PEG chains extended into aqueous solutions to provide steric stability. Nanoparticles with hydrodynamic diameters in the range of 100-150 nm and polydispersity indices less than 0.2 were obtained, as measured by dynamic light scattering. For the drug-copolymer nanoparticles, the addition of a PLLA homopolymer in the nanoprecipitation step yielded monomodal size distributions. The drug and magnetite loadings determined by high-performance liquid chromatography (HPLC) and thermogravimetric analysis respectively indicated drug loading as high as 45 wt% and quantitative incorporation of magnetite. The transverse and longitudinal NMR relaxivities of the magnetite-loaded nanoparticles were ~ 200 s^{-1}.mM Fe^{-1} and ~ 1 s^{-1}.mM Fe^{-1} respectively. The transverse relaxivities enabled sensitive T_2-weighted MRI and the low longitudinal relaxivities are consistent with the hydrophobic nature of the cores of the nanoparticle clusters. The therapeutically-relevant drug loadings and high transverse relaxivities for magnetic resonance imaging render these nanoparticles valuable for development of HIV theranostics.

(221e) Injectable, Biodegradable Pnipam-Magnetite Nanoparticle Hydrogels

Scott B. Campbell and Todd R. Hoare, Chemical Engineering, McMaster University, Hamilton, ON, Canada

Injectable, biodegradable PNIPAM-magnetite nanoparticle hydrogels

Scott Campbell and Todd Hoare

Department of Chemical Engineering, McMaster University, Hamilton, ON, Canada

INTRODUCTION: The development of versatile hydrogel networks that elicit some change in response to one or more external stimuli could have a diverse range of potential applications. For example, composites with heat induction capabilities could be used for externally-activated but locally-induced hyperthermia treatment (i.e. exposing cancerous regions to elevated temperatures (~43°C) that will lyse the more temperature sensitive cancer tissues while being safe for normal tissues[1]) and/or as scaffolds for the delivery of therapeutic agents. Alternatively, thermosensitive drug delivering agents, such as micelles or microgels, could be incorporated into these scaffolds to allow for externally controlled drug release.[2] Such treatments may be combined into a single device to further increase the efficacy of treatment. Consequently, stategies involving heat inducting, thermosensitive composite materials utilizing magnetite nanoparticles in alternating magentic fields (AMF) have been widely pursued in recent years.

While many polymer-magnetite composite materials have been fabricated before, few are able to be directly injected to the desired site of interest with the hydrolytically-degradable crosslinks that are formed immediately after injection. The development of injectable materials that can provide similar release profiles to currently reported, macroscale-systems[3,4] would be highly beneficial to expand the potential applications and patient convenience of such devices. The further ability of these composite materials to degrade allows for the clearance of their comprising compontents from the body over time. Additionally, while most currently reported systems physically entrap the magnetite nanoparticles in a hydrogel matrix, by functionalizing the surface of the nanoparticles so that they also participate in the gelation process, the rheological properties of the final composites may be significantly altered, leading to exceptionally strong or elastomer-like hydrogels with unique properties and other potential applications (e.g. embolic or structural biomaterials). In response, this work describes the fabrication of novel thermosensitive, injectable, covalently-cross-linked, stimuli-responsive hydrogels in which one of the polymers used to form the hydrogel is physically adsorbed to magnetite nanoparticles. The hydrogel composites are characterized for their rheological and degradation properties, biocompatibility, and drug delivery properties.

EXPERIMENTAL: Magnetic, thermosensitive, *in situ*-injectable hydrogels were prepared by mixing hydrazide-functionalized poly(N-isopropylacrylamide) (PNIPAM) coated magnetite nanoparticles, in which the PNIPAM component is physically adsorbed to the magnetite nanoparticles, and aldehyde-functionalized dextran. The PNIPAM-based polymer is a copolymer of acrylic acid (20 wt.%) and NIPAM that is functionalized with hydrazide groups via the carbodiimide-mediated conjugation of adipic acid dihydrazide (ADH).[5] Adsoprtion of this polymer to the surface of iron oxide nanoparticles is performed directly after their synthesis. Here, in a similar procedure to that used by Hoare et al. (2009), $FeCl_3$ and $FeCl_2$ in a 2:1 molar ratio are coprecipated with ammonium hydroxide to form magnetite nanoparticles that are then heated to 70°C in the presence of hydrazide-functionalized PNIPAM to peptize the polymer to the surface of the nanopatrticles.[3] Oxidizing dextran with sodium periodate is used to synthesize the aldehyde-functionalized dextran.[7] The hydrogels are manufactured by placing PBS-based solutions of the hydrazide-functionalized NIPAM coated magnetite nanoparticles and aldehyde-functionalized dextran in opposite compartments of a double barrel syringe (with the desired drug if it is intended for drug delivery purposes). The hydrogels are formed when the solutions are co-extruded upon injection into pre-fabricated molds and the hydrazide and aldehyde groups of either polymer precursor condense to form hydrazone crosslinks (Figure 1).

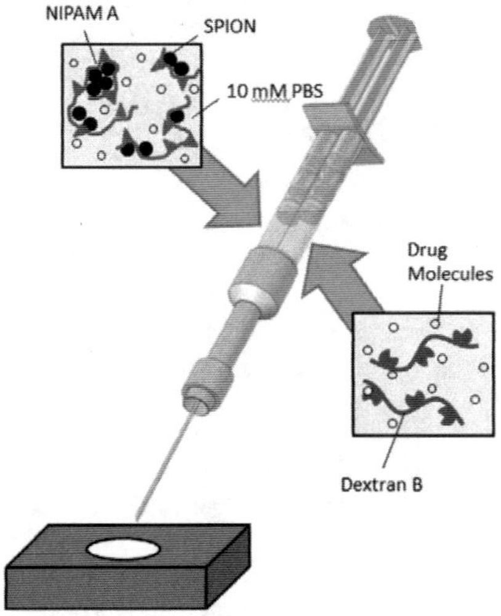

Figure 1: Hydrogel fabrication process. The double-barrel; syringe has two compartments: one with hydrazide functionalized NIPAM and the other with aldehyde functionalized dextran in PBS solutions. Upon injection, these two solutions mix and hydrazone bonds form in prefabricated moulds.

The hydrazone crosslinks that form will hydrolytically degrade over the course of several months under physiological conditions.[5] The PNIPAM component of the hydrogels is thermosensitive such that the hydrogels reversibly decrease in size when the temperature of their environment exceeds their lower critical solution temperature (LCST).[7] The magnetite nanoparticles can generate heat when placed in an AMF via hysteresis losses.[1] When the thermosensitive hydrogel and magnetite nanoparticles are combined in the same nanocomposite, heating of the nanoparticles induces a phase transition in the NIPAM component, which will cause the hydrogel to deswell and release a burst of pharmaceutical agents. Site-specificity is achieved by injecting the composite at the desired site, where it quickly gels *in vivo* using the rapid hydrazide-aldehyde chemistry. The thermosensitive nature of the hydrogel can be adjusted, along with numerous other parameters (for example, the degree of functionalization and the concentration of hydrogel precursor polymers, the phase transition temperature of the polymer precursors via copolymerization, etc.) to attempt to control the rate of drug release and the on-demand control of drug release under the presence of an AMF.

RESULTS: Stable, injectable, mechanically robust hydrogels were successfully formed with 2-10 wt% aldehyde-functionalized dextran content. The dried gels range from 70-75% iron oxide based by mass, relating to hydrated gels that have 10 - 22 wt.% iron oxide content. The hydrogels themselves have shear storage moduli orders of magnitude greater, and elasticities significantly greater, than conventional hydrogel materials, behaving on a bulk scale more like a conventional elastomer than a hydrogel (Figure 2).

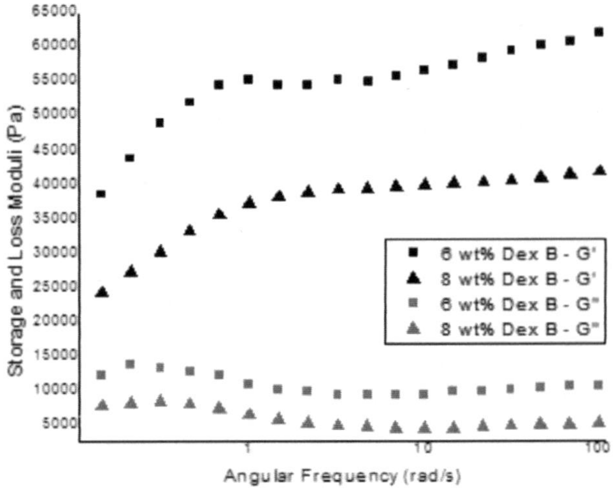

Figure 2: Storage (G') and loss (G") moduli of composites fabricated with varying dextran polymer (Dex B) contents (6 wt.% and 8 wt.%) in the hydrogel precursor solution.

Composites have been designed that are also able to be quickly broken apart under the application of an AMF, likely due to their lower crosslink density than conventional gels and the thermal motion of the magnetite nanoparticles under an AMF. These hydrogels could be used in alternate applications, such as a blood stopper in surgeries that can be broken apart after the operation is completed and as a quick supply iron source for the body. Drug-loaded hydrogels that are more stable under the presence of an AMF have also shown high-low release characteristics upon applications of short 5-10 minute pulses of an AMF (Figure 3). Increases in drug release when the AMF is applied may be partially attributed to the induction-enhanced degradation of the hydrogel itself. Additionally, the performance of these composites in cell viability assays will be discussed.

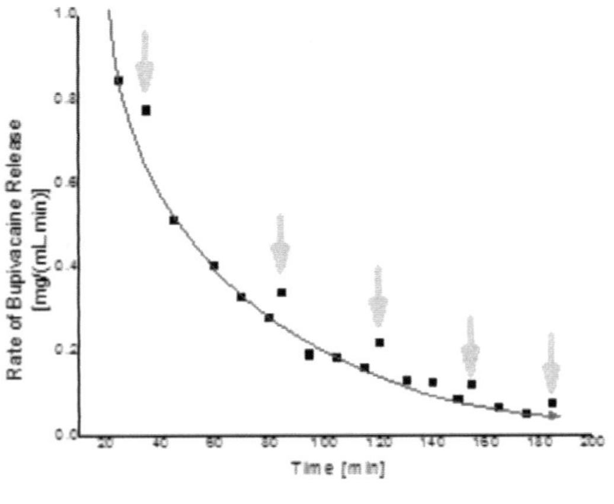

Figure 3: Rate of bupivacaine drug release under pulsed AMF conditions. The red line represents release without the application of an AMF, while the green lines indicate the enhanced rate of release after 5 minutes of applied AMF.

CONCLUSIONS: The combination of thermosensitive polymers with magnetite nanoparticles is a powerful tool in the fabrication of composites for therapeutic purposes. The injectable and externally-triggerable drug delivery composite material described herein offers significant advantages over many current therapies in terms of facilitating triggered changes in the composite properties using a patient-friendly and non-invasive triggering technology.

REFERENCES: (1) Laurent, S., Forge, D., Port, M., Roch, A., Robic, C., Elst, L.V., and Muller, R.N. (2008). Magnetic Iron Oxide Nanoparticles: Synthesis, Stabilization, vectorization, Physiochemical Characterizations, and Biological Applications. *Chemical Reviews, 108*, 2064-2110; (2) Sivakumaran, D., Maitland, D., and Hoare, T. (2011). Injectable microgel-Hydrogel Composites for Prolonged Small-Molecule Drug Delivery. *Biomacromolecules, 12(11)*, 4112-4120; (3) Hoare, T.; Santamaria, J.; Goya, G.F.; Irusta, S.; Lin, D.; Lau, S.; Padera, R.; Langer, R.; Kohane D.S. (2009). A Magnetically Triggered Composite Membrane for On-Demand Drug Delivery. *Nano Letters*, *9(10)*, 3651-3657; (4) Rovers, S.A., Hoogenboom, R., Kemmere, M.F., and Keurentjes, J.T.F. (2012). Repetitive On-Demand Drug Release by Magnetic Heating of Iron Oxide Containing Polymeric Implants. *Soft Matter, 8*, 1623-1627; (5) Patenaude, M., and Hoare, T. (2012). Injectable, Mixed Natural-Synthetic Polymer Hydrogels with Modular Properties. *Biomacromolecules, 13(2)*, 369-378;(6) Ito, T., Yeo, Y., Highley, C.B., Bellas, E., and Kohane, D.S. (2007). Dextran-based *in situ* Cross-Linked Injectable Hydrogels to Prevent Peritoneal Adhesions. *Biomaterials, 28*, 3418-3426; (7) Pelton, R. (2000). Temperature-sensitive aqueous microgels. *Advances in Colloid and Interface Science*, *85*, 1-33.

ACKNOWLEDGEMENTS: This research is funded by the Natural Sciences and Engineering Research Council of Canada (NSERC) and the J.P. Bickell Foundation (Medical Research Grant).

(221f) Protease Responsive and Biofunctional Peg Hydrogel Coated Magnetic Iron Oxide Nanoparticles for Targeted Delivery Into Tumor Cells

Caner Nazli, MATERIAL SCIENCE and Engineering PROGRAM, KOC UNIVERSITY, Istanbul, Turkey, Funda H. Yagci Acar, CHEMISTRY, KOC UNIVERSITY, ISTANBUL, Turkey and Seda Kizilel, Chemical and Biological Engineering, Koç University, Istanbul, Turkey

Specific targeting of tumor cells is one of the main limitations in most drug based therapies such as cancer treatment. However, side effects and resistance of cancer drugs prevent efficient therapy. In order to address these challenges, specific drug targeting, high accumulation of drug in target tissues, and controlled drug release strategies could be useful. Recently, we have shown that RGDS-functionalized poly (ethylene glycol) (PEG) hydrogel coated magnetic iron oxide nanoparticles (MIONPs) enhance accumulation and specific cellular uptake into cancer cells. In this study, we investigated coating of MIONP within covalently crosslinked PEG hydrogel functionalized with fibronectin derived RGDS and enzymatically degradable peptide sequences. MIONP coated within RGDS functional PEG hydrogel enhance specific internalization of the nanoparticles into cancer cells and maintain accumulation of nanoparticles around cancer tissue, and covalently bonded degradable sequence within PEG hydrogel structure allows for the release of therapeutic molecules within the coating towards their surrounding as a result of high expression of catalytic enzymes by cancer cells. This multifunctional property of coating may be useful for efficient therapy avoiding undesirable side effects and resistance of the drug. The approach developed here may also be useful for simultaneous imaging of cancer tissues and for targeted drug delivery into the tumor site.

(247a) Understanding the Mechanism of Single-Walled Carbon Nanotube Growth towards Designing a Chiral-Selective Catalytic System

Diego A. Gomez Gualdron, Chemical Engineering, Texas A&M University, College Station, TX and Perla B. Balbuena, Artie McFerrin Department of Chemical Engineering, Texas A&M University, TX

Due to their outstanding properties, single-walled carbon nanotubes (SWCNTs) possess the potential to revolutionize several fields through their use in the fabrication of power-efficient electronic devices, ultrasensitive and selective bio- and chemo-sensors, as well as in the design of new diagnosis and therapeutic strategies. A prerequisite for the exploitation of these technologies is the large-scale production of nanotubes with specific properties (e.g. metallic vs. semiconductor). Since such properties depend on the nanotube chirality, it is necessary to design a process where commercial quantities of predetermined nanotube chirality can be selectively synthesized.

A rationale design of such catalytic system must be guided by a thorough understanding of the nanotube growth mechanism, which is still subject to debate. Using molecular simulation techniques, we investigate the potential use of catalyst nanoparticles as templates to guide the nanotube growth toward a specific chirality during chemical vapor deposition (CVD) synthesis. Density functional theory (DFT) calculations show a correlation between nanotube and nanoparticle structures during the nucleation stage. It is found that, given the relative strengths of carbon-carbon, metal-carbon, and metal-metal bonds, an unsupported nanoparticle adapts to the nascent nanotube (inverse template effect), rather than imposing its structure to said nascent nanotube. To solve this issue we propose exploiting the interactions of supported nanoparticles with the support material to help control the nanoparticle structure during nanotube growth, thus achieving a direct template effect.

Reactive molecular dynamics simulations demonstrate that depending on the strength of the metal/support interaction and on the nanoparticle size, the nanoparticle structure and dynamics can be altered in such a way that may favor the growth of nanotubes of specific quality and chirality. As the nanotube nucleates and grows, it tries to maximize the occupation of hollow sites (by carbon atoms) on the nanoparticle surface. Thus the ever-evolving nanoparticle structure (at 1000 K) can act as template that influences the structure of the growing nanotube. The effectiveness of this template effect is positively affected by: 1) a good nanotube/nanoparticle contact, and 2) a high rate of defect annealing, whereas is negatively affected by: 3) a heterogeneous nanoparticle surface, and 4) excessive nanoparticle mobility. Since there are some opposing trends on how these factors are affected by nanoparticle size and interaction with the support, a compromise must be made in optimizing the conditions that favor templated nanotube growth. Based on this we suggest a range of conditions that can potentially lead to templated growth in experimental nanotube synthesis.

In order to take advantage of templated growth it is necessary to find conditions that promote nanoparticle structure stability, but also growth of high quality SWCNTs. But to correctly predict these conditions, it is first necessary to elucidate key aspects of the nanotube growth mechanism. We use reactive molecular dynamics to investigate the carbon transport mechanism and carbide formation. We aim to demonstrate that some of the nanoscale synthesis conditions established by the widely known vapor-liquid-solid (VLS) mechanism are not mandatory for successful nanotube growth, thus in principle widening the range of experimental conditions within which nanotube synthesis can be optimized.

Effect of the Oxygen Content of Precursors on the Structural Properties of Nitrogen doped Multiwall Carbon Nanotubes

Ali Qajar, Maryam Peer, Ramakrishnan Rajagopalan, Henry C. Foley

Abstract:

Nitrogen doped carbon nanotubes are synthesized by using liquid phase precursors. Acetonitrile is used as carbon and nitrogen source and ferrocene is employed as the catalyst. Also, toluene and THF are used as auxiliary carbon sources. All reactions are performed at 900 °C and constant flow rates of 1000 sccm for Ar, the carrier gas and liquid injection rate of 10 ml/hr for the reactant solution. The structural properties of the nanotubes are characterized by using SEM and TEM imaging, Raman spectroscopy, XRD and XPS. Our study shows despite pervious results, addition of THF does not significantly increase the nitrogen content of nanotubes. These outcomes are consistent with newer results, suggesting that order of 0.1% or less nitrogen doping can significantly change the structural properties of nanotubes.

(247c) Catalyst-Support Interactions and Their Influence in Water-Assisted Carbon Nanotube Carpet Growth

Placidus B. Amama[1], Cary L. Pint[2], Francesca Mirri[3], Matteo Pasquali[4], Robert H. Hauge[2] and Benji Maruyama[5], (1)Materials & Manufacturing Directorate, RXBT, Air Force Research Laboratory/UDRI, WPAFB, OH, (2)Richard E. Smalley Institute for Nanoscale Science and Technology, Rice University, Houston, TX, (3)Department of Chemical & Biomolecular Engineering, Rice University, Houston, TX, (4)Chemical and Biomolecular Engineering, Rice University, Houston, TX, (5)Materials & Manufacturing Directorate, RXBN, Wright Patterson Air Force Research Laboratory, Dayton, OH

The growth and catalyst morphological changes occurring for thin Fe layers deposited on Al_2O_3, MgO, TiN, and ZrO_2 using water-assisted chemical vapor deposition are compared. The growth behaviors of the catalyst substrates were evidently different, with Al_2O_3/Fe supporting high-quality carbon nanotube (CNT) carpet growth and showing the highest activity and longest lifetime. The TiN/Fe catalyst also supported CNT carpet growth, albeit with much lower activity, shorter lifetime, and lower CNT quality while MgO/Fe and ZrO_2/Fe did not support CNT carpet growth under standard growth conditions. Studies using a combination of atomic force microscopy and X-ray photoelectron spectroscopy revealed a general correlation between the catalyst behavior (activity and lifetime) and the evolution of the catalyst (Ostwald ripening and subsurface diffusion rates). Analysis of MgO/Fe and ZrO_2/Fe catalyst systems raise interesting questions related to additional chemical interactions between the substrate and catalyst that could influence nucleation and CNT growth. This work provides a step toward understanding the challenges that arise in engineering efficient CNT growth processes on a desired substrate.

(247d) Epitaxial Matching As Route to Grow Chirally Selective Single-Walled Carbon Nanotubes

Debosruti Dutta, Department of Chemical and Biomedical Engineering, University of South Florida, Tampa, FL, R. Mohan Sankaran, Chemical Engineering, Case Western Reserve University, Cleveland, OH and Venkat R. Bhethanabotla, Department of Chemical Engineering, University of South Florida, Tampa, FL

One of the major goals in nanoscience involves the synthesis of single-walled carbon nanotubes (SWCNTs) with a predefined chirality because of its potential impact on a wide range of emerging technologies including nanoelectronics, renewable energy, nanocomposites, sensor and medical. SWCNTs are typically grown using the chemical vapor deposition (CVD) method by passing acetylene gas over small metallic nanoparticles. Recently, we showed that changing the composition of Ni_xFe_{1-x} bimetallic nanoparticles via a novel direct-current atmospheric-pressure microplasma synthesis route resulted in chiral-selective growth of specific nanotube chiralities. In this work, an epitaxial nucleation model for single-walled carbon nanotube (SWCNT) growth on bimetallic catalysts surfaces is reported in support of experimental observations of chiral enrichment. Tuning composition will perturb the crystal structure, and affect the lattice matching of the nanoparticle with certain chiralities of the SWCNT, promoting chiral-selective growth. We model the bimetallic catalyst surfaces as a 2D (111) surface consisting of Ni or a combination of Ni and Fe atoms, with varying average bond length between nearest neighbor atoms which corresponds to the crystal structure of the alloys. The energies associated with nanotube cap formation on these various surfaces are calculated using density functional theory (DFT). We find that certain cap chiralities, such as (8, 4), are more stably bound to a surface that resembles a Ni0.27Fe0.73 bimetallic catalyst, whereas other chiralities, such as (9, 4), are more stable on a pure Ni surface. These results help explain the predominance of certain chiralities on specific bimetallic catalysts and provide a potential route to controlling the chirality of as-grown SWCNTs.

(247e) Role of Dissolved Carbon On the Nickel-Catalyzed Growth of Carbon Nanotubes

Ali Rinaldi[1], **Jean-Philippe Tessonnier**[1], Manfred E. Schuster[1], Raoul Blume[1], Frank Girgsdies[1], Qiang Zhang[1], Timo Jacob[2], Sharifah Bee Abd Hamid[3], Dangsheng Su[1] and Robert Schlögl[1], (1)Inorganic Chemistry, Fritz Haber Institute of the Max Planck Society, Berlin, Germany, (2)Institut für Elektrochemie, Universität Ulm, Ulm, Germany, (3)Combicat, University of Malaya, Kuala Lumpur, Malaysia

In catalysis, the support is known to act on heat and mass transfer, but also on the active phase through strong metal-support interactions (the SMSI effect). In the case of black carbon, the structural order (graphitic character) adds one level of complexity but also one degree of freedom for logical catalyst design. It is of crucial importance to study how the graphitic character influences the anchoring of the nanoparticles and eventually modifies the catalytic activity of the metal.

In the present work, we studied how the interaction between Ni nanoparticles and carbon supports with different graphitic characters influences the Ni-catalyzed growth of carbon nanotubes (CNTs). By combining *in situ*techniques (XPS and XRD) with aberration-corrected HRTEM and theoretical calculations, we showed that the carbon support plays a dramatic role on the structure of the CNTs grown by catalytic chemical vapor deposition (CCVD) of ethylene. We observed that during the activation of the catalyst under hydrogen and further ramping to reaction conditions, carbon atoms from defective supports diffuse inside the Ni nanoparticles to form a non-stoichiometric metastable carbide. The insertion of carbon sub-surface generated strain and caused the reconstruction of the Ni nanoparticles. The particles became strongly faceted, which led to the growth of highly disordered carbon nanofibers during the CCVD step. In contrast, highly ordered multiwalled carbon nanotubes were grown under the same reaction conditions on Ni particles of same size and shape but supported on graphitic carbon.

Our results are relevant for the synthesis of carbon nanotubes and of hierarchical carbon-carbon composites, but we anticipate that they can also be extended to the growth of graphene. This offers also new possibilities to control the catalytic activity of metal particles and might open new routes for the rational design of heterogeneous catalysts.

(247f) Selective Synthesis of (9,8) Single Walled Carbon Nanotubes Using CoSO4/SiO2 Catalyst

Yuan Chen, Hong Wang and Li Wei, School of Chemical and Biomedical Engineering, Nanyang Technological University, Singapore, Singapore

A new chirality selective catalyst ($CoSO_4/SiO_2$) has been developed for single walled carbon nanotube (SWCNT) synthesis. The $CoSO_4/SiO_2$ catalyst is prepared by dispersing cobalt (II) sulphate heptahydrate on high surface area porous silica using an incipient wetness impregnation method. Photoluminescence, absorption and Raman spectroscopy results show the $CoSO_4/SiO_2$ catalyst has high selectivity toward (9,8) nanotubes. The chirality selectivity of the catalysis can be tuned by varying catalyst reduction conditions. At an optimal reduction condition, when the catalyst is reduced under H_2 up to 540 ^{o}C, photoluminescence results show the (9,8) nanotubes count for 51.7 % of all semiconducting SWCNTs. Theoretical study shows that the diameter of structurally stable metal nanoparticles is scattered, other than continual. The Co_{147} cluster is the most stable cluster in its diameter range. The selectivity toward the (9,8) nanotube is credited to large Co nanoparticles with diameter comparable with that of (9,8) nanotubes. Forming such large Co nanoparticles is because of the lower reduction temperature (470 ^{o}C) of Co species in the $CoSO_4/SiO_2$ catalyst. We also found that sulfur compounds resulting from reducing SO_4^{2-} ions play a critical role in stabilize such large Co nanoparticles. This research opens up routes toward highly selective growth of SWCNTs.

(247g) "Zero Dimensional" Single Walled Carbon Nanotubes

Kaladhar Kamalasan[1,2,3,4], **Riccardo Gottardi**[1,4,5,6], Susheng Tan[7], Yanan Chen[8], Bhaskar Godugu[8], Sam N. Rothstein[1], Anna C. Balazs[1], Alexander Star[8] and Steven R. Little[1,2,3,4], (1)Department of Chemical and Petroleum Engineering, University of Pittsburgh, Pittsburgh, PA, (2)Department of Immunology, University of Pittsburgh, Pittsburgh, PA, (3)Department of Bioengineering, University of Pittsburgh, Pittsburgh, PA, (4)McGowan Institute for Regenerative Medicine, University of Pittsburgh, Pittsburgh, PA, (5)Department of Orthopaedic Surgery, University of Pittsburgh, Pittsburgh, PA, (6)Fondazione RiMED, Palermo, Italy, (7)Department of Electrical and Computer Engineering, University of Pittsburgh, Pittsburgh, PA, (8)Department of Chemistry, University of Pittsburgh, Pittsburgh, PA

Single-walled carbon nanotubes (SWNTs) are unique, one dimensional, tubular carbon allotropes with exceptional properties. Individual and dispersible SWNTs are, however, rarely available, since their side wall hydrophobicity and disproportionate length drive them to form long bundles. These factors represent an obstacle for wet nanoscale applications of SWNTs. One approach to overcome these issues is to dramatically reduce the tube length so as to make the tube side-wall hydrophobicity and dispersibility more manageable. Herein, we report the first synthesis and characterization of SWNTs in the 2nm-5nm length range, the shortest so far achieved to our knowledge. In effect, we have reduced a long (and essentially) one dimensional material to tubes whose lengths are on the same order of their diameter, thus effectively rendering them "zero dimensional" (or "0dSWNTs"). We observe that 0dSWNT maintain the structural properties of nanotubes but possess increased dispersibility and complete water-solubilization (>3 months) following appropriate chemical (hydroxylation) functionalization. Finally, we examined 0dSWNTs by mass spectrometry, which is the first instance, to our knowledge, of positive identification of the appropriate signature of a carbon nanotube using MALDI-TOF-MS.

The 0dSWNTs may represent unique building blocks for a bottom-up approach to the exploitations of nanotubes as their production process from SWNTs can be scaled up and automatized. Furthermore, their low dimensional anisotropy renders them well dispersible in water. Furthermore, additional functionalization is also possible, such as hydroxylation (OH-0dSWNTs) rendering these tubes water soluble for long periods of time and allowing them to fly in a MALDI-TOF mass spectrometer. Alternate functionalization could also be used to direct, their self assembly, depending on the final structure required for electronics, nanofiltration, energy, materials science, or biomedical applications. In addition, as 0dSWNTs have the dimensions typical of transmembrane proteins and the alternate hydrophilic-hydrophobic pattern of a lipid layer, they could potentially insert themselves in a lipid membrane acting as an artificial channel to shuttle ions, nutrients, peptides. Such structure would be capable of performing some of the complex behaviors typical of biological cells, opening the way to new "synthetic cellular therapies". Proof-of-principle preliminary data will be presented that suggest that this is now a possibility.

(261a) Invited: Biological Consequences of Engineered Nanoparticle Interactions with Environmental Microbial Communities

Stacy Wirth[1], Teresa L. Kirschling[1], Gregory V. Lowry[2] and **Robert D. Tilton**[3], (1)Department of Chemical Engineering, Carnegie Mellon University, Pittsburgh, PA, (2)Department of Civil and Environmental Engineering, Carnegie Mellon University, Pittsburgh, PA, (3)Department of Chemical Engineering and Department of Biomedical Engineering, Carnegie Mellon University, Pittsburgh, PA

One of the many drivers for nanotechnology development is the potential for suitably designed nanomaterials to interact with and guide biological systems. By virtue of their size, their often unique physical and chemical properties and their relative ease of functionalization for specific biomolecular recognition, nanoparticles hold great promise in medicine and biotechnology. This potential for strong biological interaction, combined with considerable realized progress in biofunctional nanosystem development and significant commercial deployment of nanomaterials in a variety of markets, has prompted research efforts aimed at providing the scientific basis for nanomaterial risk assessment. While this research has not surprisingly lagged behind nanomaterial and nanotechnology research and development, the new fields of "nanotoxicology" and "nanoecotoxicology" are growing rapidly. One source for potential concern is inadvertent release of biologically potent nanoparticles that were engineered for biomedical applications. There is also a more general concern that inadvertent release into the environment of nanoparticles that were engineered for other purposes may have deleterious ecological or biological consequences. One subject of considerable contemporary research attention concerns the implications of the fairly common use of silver nanoparticles in clothing and other consumer products, because of their broad spectrum biocidal activity. Pathways have been identified by which these nanoparticles may migrate via municipal wastewater treatment plants from households to farm fields and surface waters, where they might exert as yet unknown ecological effects. Similarly, zero valent iron nanoparticles ("nanoscale zero valent iron" or NZVI) have been engineered for in situ remediation of groundwater contamination by chlorinated solvents or toxic metals. Although they are still in the research and development stage and not yet widely deployed, these nanoparticles are highly reactive and designed to be injected into groundwaters in high concentrations. As with silver nanoparticles, their actual or expected ecological effects are unknown.

Much of the research into potential environmental impacts of engineered nanoparticles concerns their effects on microbes, especially bacteria. This is because bacteria reside at the base of ecological food nets, and they play a critical role in molecular and elemental cycling in the environment. Significant disruption of microbial communities could therefore lead to significant ecosystem disruption. This presentation will focus mostly on the effects of exposing environmental bacterial communities to engineered nanoparticles, particularly NZVI and silver nanoparticles. Both NZVI and silver nanoparticle exposure have been demonstrated in laboratory culture experiments to be toxic to bacteria. The majority of these culture experiments consider planktonic, or freely suspended, bacterial cultures. In the environment, bacteria in the planktonic state are relatively rare compared to the vast majority of bacteria that reside in biofilm communities. These are adhered to solids and are protected by a secreted extracellular matrix. As of yet, there is little understanding of how these environmentally relevant community structures influence bacterial susceptibility to engineered nanoparticle exposure. Most planktonic culture experiments minimize the complexity of the bacterial surroundings. Real ecological samples have complex compositions, and their effects on nanoparticle/microbe interactions are as yet just beginning to be understood. The environment presents nanoparticles with many surfaces on which to adhere and macromolecules that may adsorb to their surfaces. These surfaces could function as sinks and adsorbed macromolecules could alter nanoparticle colloidal stability, transport and the surface forces they experience at biological cell walls.

The impact of NZVI on bacterial communities present in trichloroethylene-contaminated aquifers was studied in microcosms, where nanoparticles were added to slurries of the aquifer solids and their associated bacterial communities. These microcosm studies indicated no overall biocidal effect of NZVI, but the water chemistry changes caused by NZVI redox reactions did alter the microbial species balance. Communities shifted in favor of sulfate reducing and methanogenic organisms. Depending on the type of macromolecules used to disperse NZVI, it was found in some cases to have an overall biostimulatory effect.

The effects of the biofilm growth mode on bacterial susceptibility to silver nanoparticles were studied. Silver nanoparticles present more than one potentially toxic mode of interaction with bacteria, since dissolved silver ions are toxic. The relative degree to which toxicity is caused by direct nanoparticle action on cells or by dissolved silver ion effects can differ from one type of organism to another. Single species biofilms of *Pseudomonas fluorescens* were cultivated and exposed to poly(vinylpyrrolidone)-stabilized silver nanoparticles. Biofilm viability was decreased by silver nanoparticle exposure in a concentration-dependent manner, but to a far smaller degree than was observed for planktonic cultures of this bacterium. Nanoparticle concentrations that eliminated all detectable viability in planktonic culture left a significant fraction of viable bacteria in biofilms. Evidence for a nanoparticle-specific toxic effect was observed and found to correlate with the colloidal stability of the nanoparticles. Biofilm susceptibility to silver nanoparticle exposure was partially mitigated by the presence of humic acid, a representative form of natural organic matter common to most environments. This suggests a possibly important protective role for natural organic matter in the environment.

While most of this presentation will focus on nanoparticle effects on bacterial communities, the presentation will conclude with evidence that bacterial communities present in the environment are capable of acting on the nanoparticles themselves, altering them in a way that would affect their transport in the environment. In particular, evidence will be shown that bacterial communities can biodegrade polymeric stabilizers of the sort that are frequently attached to nanoparticle surfaces to control their colloidal stability. Stabilizer biodegradation was found to drive nanoparticle aggregation. It is apparent that nanoparticle/bacterial interactions "go both ways." This will likely drive a significant amount of future research using systems engineering approaches to support environmental nanoparticle risk assessment.

(261b) Assemblies of Plasmonic Nanoparticles and Their Biomedical Applications

Nicholas A. Kotov, Chemical Engineering, University of Michigan, Ann Arbor, MI

Assemblies of Plasmonic Nanoparticles and Their Biomedical Applications

Prof. Nicholas Kotov

Department of Chemical Engineering, University of Michigan, Ann Arbor, MI, 48198, USA

Plasmons of gold nanoparticles (NPs) represent one of the best examples of collective properties of nanoscale systems. Engineering of nanoscale assemblies of Au NPs opens great prospects for biomedical applications. The presentation will include the state-of-the-art account of preparation and properties of both extended and discrete assemblies of plasmonic NPs. The first type of assemblies includes supercrystals of gold spheres and nanorods (NRs) and layered composites from the same building blocks made by the layer-by-layer assembly. The second type of assemblies covers the small superstructures of 3-30 NPs or NRs. Methods to control the structure of both assemblies and how these properties affect the application-defining optical, electrical, and mechanical properties will be discussed.

The application space to be covered in this presentation includes cancer/inflammation detection using photoacoustic imaging, drug delivery, photodynamic cancer treatment, continuous monitoring of cellular metabolism, and neuroprosthetic dev ices. The latter represent particularly unexplored area of research for gold nanostructures and will greatly benefit from the unique combination of optical, mechanical, and biological properties of plasmonic NP assemblies. New directions of research in this area will also be mapped out and include (1) chiral plasmonic nanoassemblies; (2) artificial viruses from inorganic NPs; (3) NPs from non-metal plasmonic materials; and (4) biodegradable inorganic NPs and assemblies.

Special attention will be made on new methods of evaluation of toxicology of nanoparticles involving three-dimension cell cultures in spheroids of liver cells which displayed unexpected transport characteristics of NPs in the tissue.

Selected References:

J. Lee *et al*, Exciton–plasmon interactions in molecular spring assemblies of nanowires and wavelength-based protein detection, *Nature Materials,* **2007**, 6(4), 291-295

A. Agarwal *et al,* Targeted gold nanorod contrast agent for prostate cancer detection by photoacoustic Imaging, *Journal of Applied Physics*, **2007**, 102, 064701.

W. Chen, *et al,* Multifunctional Magnetoplasmonic Nanoparticle Assemblies for Cancer Therapy and Diagnostics (Theranostics), *Macromol. Rapid Commun.*, **2009**,30,228.

J. Lee, *et al,* **In vitro Toxicity Testing of Nanoparticles in 3D Cell Culture,** *Small,* **2009**, 5 (10), 1213-1221.

Podsiadlo, P.; *et al,* Gold Nanoparticles Enhance the Anti-Leukemia Action of a 6-Mercaptopurine Chemotherapeutic Agent. *Langmuir,* **2008**, 24(2), 568-574.

R. A. Alvarez-Puebla, *et al*, Real-Time Detection of Scrambled Prions on 3D Gold Nanorod Supercrystals, *Proceedings of the National Academy of Sciences,* **2011**, 108 (20) 8157-8161.

Yang, M. *et al,* SERS-Active Gold Lace Nanoshells with Built-in Hotspots. *Nano Letters* **2010**, 10(10), 4013-4019.

S. Srivastava, *et al,* Light-Controlled Self-Assembly of Semiconductor Nanoparticles into Twisted Ribbons, *Science*,

2010, 327, 1355, 1355-1359.

Y. Xia, *et al,* Self-assembly of virus-like self-limited inorganic supraparticles from nanoparticles, *Nature Nanotechnology,* **2011**, 6, 580-587.

L. Xu, *et al,* Regiospecific Plasmonic Assemblies for *in-situ* Raman Spectroscopy in Live Cells, *J. Am. Chem. Soc.,* **2012**, in press.

H. Zhang, *et al,*, Gold Nanoparticles for Neural Prosthetics Devices, *Nano Letters,* **2012**, in press.

(261c) Droplet Deposition From Consumer Products

Patrick T. Spicer and Marco Caggioni, Microstructured Fluids Group, Procter and Gamble Co., West Chester, OH

The last decade has seen amazing innovations in the design, synthesis, and fabrication of monodisperse colloids with complex shapes in relatively significant quantities. However, many of the breakthrough examples in colloidal design are synthesized from certain materials that might not be usable in all applications. In liquid consumer products many benefits are delivered via deposition of dispersed droplets onto hair, skin, and fabric and many breakthrough colloidal materials made from solids can not be used. In this talk I will discuss our recent work on developing hybrid droplet materials that enable us to control shape, surface, and other properties. Physical models of stability and deposition properties of the shaped droplets are used to design and produce droplets with a range of applications. Measurements of these new materials' impact on relevant performance metrics, like adhesion, as well as the potential for droplet shape to impact product performance, will also be discussed.

(276a) Monodisperse Carbon Nanotubes Produced Using Iterative Orthogonal Density Gradient Ultracentrifugation

Alexander A. Green, Wyss Institute for Biologically Inspired Engineering, Harvard University, Boston, MA and Mark C. Hersam, Northwestern University, Evanston, IL

The variation in physical and electronic properties of as-synthesized carbon nanotubes can hinder their application as optical, electronic, and energy harvesting materials. Here, we describe the use of iterative orthogonal density gradient ultracentrifugation (DGU) to isolate carbon nanotubes that are monodisperse in physical structure and electronic properties. This approach achieves concurrent separation by electronic type (semiconducting or metallic) and diameter, thus enabling the production of highly pure semiconducting single-walled carbon nanotubes (SWNTs) in which 99% of the nanotubes have diameters within 0.01 nm of each other. Applying this strategy to a polydisperse mixture of multi-walled carbon nanotubes, we demonstrate the enrichment of double-walled carbon nanotubes (DWNTs) having outer walls of either semiconducting or metallic electronic type. The semiconducting character of these sorted carbon nanotubes is verified through direct electrical measurements in carbon nanotube network thin film transistors (TFTs). TFTs fabricated from semiconducting outer wall DWNTs provide on/off ratios two orders of magnitude higher than comparable metallic DWNT devices. Furthermore, semiconducting SWNT devices exhibit on/off ratios approaching 10^5 and mobilities up to 37 $cm^2V^{-1}s^{-1}$, while maintaining on-state width-normalized conductances of 0.19 S/m. These figures of merit suggest future application of these monodipserse SWNTs in integrated circuits and near-infrared optoelectronic light emitters and photodetectors.

(276b) Single-Walled Carbon Nanotube Dynamics in Rock-Like Porous Media

Shannon L. Eichmann, Department of Chemical and Biomolecular Engineering, Rice University, Houston, TX and Matteo Pasquali, Chemical and Biomolecular Engineering, Rice University, Houston, TX

Single-Walled Carbon Nanotubes (SWNTs) have unique electronic and mechanical properties that have led to an increased interest for their use in a wide range of applications (i.e., composite materials, drug delivery, and electronics). Because of their extremely high aspect ratio and slenderness (~1 nm diameter), SWNTs can increase the electrical conductivity of fluids more effectively and access much tighter spaces than other particles of equivalent volumes. Moreover, under appropriate conditions, SWNTs precipitate into highly conductive chains that can span macroscopic length scales.[1] Therefore, SWNTs can be useful in oil reservoirs as tools for sensing, targeting and as contrast agents. Recently, we used video microscopy to observe individual SWNTs by tagging with fluorescent dyes[2] and individual semiconducting SWNTs by near-infrared (NIR) fluorescence.[3,4] These techniques allowed us to directly measure SWNT diffusion, bending dynamics and stiffness in aqueous media. In addition, we studied SWNT dynamics in agarose gels, a special case of porous media mimicking biological tissues and cells, and showed how media pore size and SWNT stiffness affect diffusion.[5] We now study diffusion and bending dynamics of individual SWNTs within a rock-like porous media as a model for oil reservoirs. Porous media is produced by packing colloidal silica particles between clean glass surfaces, where the packed layer is a few particles thick, allowing for optical imaging without index matching. We image both SWNT motion within rock-like porous media by intrinsic NIR fluorescence microscopy and the packed silica structure using transmitted light microscopy. Results are presented comparing SWNT translational and rotational diffusivity within packed silica to that previously observed in agarose gels.[5] We find that the lower porosity and large, fixed obstacles of the silica pack versus that of agarose gels lead to highly path dependent dynamics and show how pore structure (i.e. diameter and tortuosity) affects not only SWNT diffusion but also bending mode dynamics. Ongoing work is focused on comparing measured SWNT mobility to models for anomalous diffusion based on bending energy and mean passage time.[6,7] Future work will focus on investigating the effects of surface functionalization toward controllable localization and formation of conductive SWNT networks for signaling, as well as, specific targeting of oil-water interfaces within porous media.

1. Kamat, P. V., *et. al.,* (2004) *JACS* **126,** 10757

2. Duggal, R. & Pasquali, M. (2006) *PRL* **96,** 246101

3. Weisman, R. B., *et. al.,* (2004) *App Phys A* **78,** 1111

4. Fakhri, N., *et. al.,* (2009) *PNAS* **106,** 14219

5. Fakhri, N., *et. al.,* (2010) *Science* **330,** 1804

6. Odijk, T. (1993) *Macromolecules* **26,** 6897

7. Saxton, M. J. (1994) *Biophys J* **66,** 394

(276c) Diagnostic Sodium Ion Sensor for the Real-Time Screening and Diagnosis of Cystic Fibrosis

Evan K. Wujcik, Daniel Trowbridge and Chelsea N. Monty, Department of Chemical and Biomolecular Engineering, The University of Akron, Akron, OH

We present the conception of a single atomic layer of carbon - functionalized electrochemical sodium ion sensor for the real-time screening and early diagnosis of cystic fibrosis (CF), a genetic disease that attacks the lungs, pancreas, liver, and intestines. CF is characterized by the abnormal transport of sodium and chloride ions across the epithelium and is therefore most commonly diagnosed by sweat testing. In this work, a flexible electronic sensor has been developed to quantify the amount of sodium ions in sweat (in real-time), alleviating the wait time, large sample size, possible contamination, and analytical equipment currently associated with sweat testing. This sensor is composed of a multi-wall carbon nanotube (MWCNT) or graphene coated Nylon-6 mat, for a flexible yet conductive sensor. The atomic layer of carbon is then functionalized with a cyclo-oligomeric calix[4]arene, which has been shown to selectively form a supramolecular complex with sodium ions. Upon complex formation with the sodium analyte, the charge carriers are drawn away from the carbon layer, hence the current is impeded, and the sodium ion detection is prevalent at levels appropriate for accurate diagnosis of CF. In this presentation, the optimization of carbon- and calix[4]arene- functionalization, nylon conditions, as well as the selectivity and sensitivity of the sensor will be discussed.

(276d) In Vitro Selection of DNA Aptamers for the Single-Wall Carbon Nanotubes

Oxana Selivanova and Ming Zheng, Polymers Division, NIST, Gaithersburg, MD

In Vitro Selection of DNA Aptamers for the Single-Wall Carbon Nanotubes

Oxana Selivanova and Ming Zheng

Polymers Division, National Institute of Standards and Technology

100 Bureau Dr., Gaithersburg, MD 20899

Aptamers are nucleic acid oligonucleotides selected from random DNA sequence libraries. They exhibit high affinity and selectivity to various targets thus serving as the molecular recognition elements. DNA aptamers are typically selected via a Systematic Evolution of Ligands by Exponential Enrichment (SELEX) procedure (1). In this work we use the SELEX method to identify DNA aptamers that strongly bind and recognize the single-wall carbon nanotubes (SWCNT). DNA is known to bind to SWCNTs, which has been used for the SWCNT separation into individual chirality and uniform lengths (2, 3). SELEX will allow identification of the strongest binding sequences that will aid in the design of the SWCNT separation technology and nano-medicine applications. In this work we perform several selection rounds via SWCNT-DNA binding followed by the asymmetric PCR (aPCR). After several rounds only the strongest SWCNT binding DNA sequences remain, which are identified via cloning followed by sequencing. The sequence pattern revealed will be discussed.

1. Tuerk, C., Gold, L. 1990. Systematic evolution of ligands by exponential enrichment: RNA ligands to bacteriophage T4 DNA polymerase. Science 249: 505-510.

2. Tu, X., Manohar, S., Jagota, A., Zheng, M. 2009. DNA sequence motifs for structure-specific recognition and separation of carbon nanotubes. Nature Letters 460: 250-253.

3. Tu, X., Zheng, M. 2008. A DNA-based approach to the carbon nanotube sorting problem. Nano Res. 1: 185-194.

(276e) Detection of Carbon Nanotubes in Biological Samples Through Microwave-Induced Heating

Fahmida Irin[1], Babina Shrestha[2], Mohammad Saed[3], Jaclyn Canas[2] and **Micah Green**[1], (1)Chemical Engineering, Texas Tech University, Lubbock, TX, (2)Environmental Toxicology, Texas Tech University, Lubbock, TX, (3)Electrical Engineering, Texas Tech University, Lubbock, TX

We demonstrate a novel technique for quantitative detection of carbon nanotubes (CNT) in biological samples by utilizing the thermal response of CNT under microwave irradiation. In particular, rapid heating of CNT due to microwave absorption is employed to quantify CNT uptake in agricultural samples with excellent sensitivity. We inject alfalfa (Medicago sativa) roots with a known quantity of CNT (single walled and multi walled) and expose the samples to a microwave field (30-50 W) to generate standard temperature-CNT concentration relationships; this detection method is then used to accurately determine CNT uptake by alfalfa plant roots grown in CNT-laden soil. The threshold for detectable CNT concentration is much lower (<0.1 μg) than common analytical methods such as electron microscopy and Raman spectroscopy. Considering the lack of effective detection methods for CNT uptake in plants, our concept is not only unique but also practical, as it addresses a major problem in the field of nanomaterial characterization and nanotoxicology risk assessment.

(276f) Characterization of Lysozyme/Single-Walled Carbon Nanotube/ Polyvinyl Alcohol Films

A. Gloria Nyankima[1], Daniel W. Horn[2] and Virginia A. Davis[2], (1)Chemical Engineering, Auburn University, Auburn University, AL, (2)Department of Chemical Engineering, Auburn University, Auburn, AL

Through the integration of lysozyme (LSZ) and single-walled carbon nanotubes (SWNT), a robust film coating was developed that could potentially be placed in public environments and reduce the transfer of communicable illness. LSZ, an enzyme found in egg whites, tears, and saliva, possesses antibacterial characteristics but lacks mechanical strength. Previous research has shown that SWNT can provide strength to lysozyme films and coatings without hindering antibacterial properties. Coatings made by layer-by-layer assembly without any polymer binder possessed high mechanical strength but limited wear and solvent resistance. To increase mechanical robustness, polyvinyl alcohol (PVA) was incorporated with the LSZ-SWNT dispersions. The resulting films were characterized for their electrical, mechanical, optical, thermal properties and antibacterial activity. Films were produced from both concentrated bulk mixtures and supernatant dispersion. Both types of films lacked electrical conductivity, but maintained antibacterial activity. Mechanical testing showed that the supernatant films were stronger; this is likely the result of improved dispersion. To evaluate the films' durability in various environments, solvent rub testing involving different acidic and basic solutions were conducted. These results demonstrated the versatility of the films' potential applications.

(276g) Physical and Electrical Characterization of Single and Multi-Walled Carbon Nanotube Bulk Networks

Christopher M. Schauerman[1], Timothy P. Maher[1], Paul R. Jarosz[1], Jamie E. Rossi[1], Thomas L. Mastrangelo[1] and Brian J. Landi[2], (1)NanoPower Research Laboratories, Rochester Institute of Technology, Rochester, NY, (2)Chemical & Biomedical Engineering, Rochester Institute of Technology, Rochester, NY

Carbon nanotubes (CNTs) are a promising material to enhance the next generation of energy storage and transmission systems. When assembled into bulk structures, CNTs can be engineered into many different form factors (e.g. CNT papers, ribbons, yarns, ropes, wires, and cables) to meet the design requirements of a particular application. Conductors composed of CNT materials can benefit from their unique nanoscale properties, and CNTs offer improvements in weight savings, flexibility, and corrosion resistance. Advancements are still needed, however, in material processing (e.g. purification, chemical doping, and nanoscale network modification) to realize the electrical properties predicted from theory and measured at the individual nanotube level in these bulk systems. The susceptibility to doping and densification depends heavily on the CNT diameter and CNT nearest neighbor interactions which may result from close-packed bundling, which is prevalent for single-wall carbon nanotubes (SWCNTs). In comparison, double walled carbon nanotubes (DWCNTs) and multi walled carbon nanotubes (MWCNTs) have network packing that is more rigid and rod-like. These differences in the nanoscale morphology translate into variations in the way the bulk network can be ordered and the resulting physical properties. Several strategies to understand and control these differences, in the present work, have been employed to measure the bulk properties of CNT materials including the combined effects of CNT type purity, mechanical densification, chemical doping, and nanoscale alignment. A constructed sample set of SWCNTs and MWCNTs mixed at varying weight loadings (0 % - 100 % w/w SWCNTs in MWCNTs at 20 % intervals) was used to measure differences in the way the two materials behave at the bulk scale. Four-point probe measurements of the electrical conductivity (S/m) were taken on Van der Pauw electrode squares of the SWCNT-MWCNT materials. Differences in the mixed CNT network's susceptibility to known chemical dopants (e. g. $KAuBr_4$ and I_2 gas) were measured and show a correlation between CNT diameter and relative extent of SWCNTs present on conductivity. Additionally, the material was characterized using Raman spectroscopy, scanning electron microscopy (SEM), and thermal gravimetric analysis. Differences in the D and G bands of the Raman spectra before and after exposure to the chemical dopants indicate differences in SWCNT and MWCNT responses to chemical treatment. SEM analysis of the surface morphology shows reduced void volume in the SWCNT samples in comparison to the higher MWCNTs weight loadings. The impact of these measurements will be discussed in the context of CNT technologies and applications, for both energy storage and transmission systems.

(276h) Expanding the Application Space of Graphene-Based Materials

Mainak Majumder, Mechanical and Aerospace Engineering, Nanoscale Science and Engineering Laboratory (NSEL), Monash University, Victoria, Australia

Graphene – a 2D arrangement of C-atoms is touted as a wonder material that can transform technologies in the electronics and energy-storage industry. However, some of its attributes such as the inherent high surface area and the conformal nature of these ultra-thin films can realize applications in sectors such as water purification and protection of corrosion of metallic components. In this talk, I will discuss our experiences in transforming 'sand' – a main-stay filter material to 'super-sand' - graphene-oxide coated sand which improves filtration efficiencies by five to six times compared to 'sand' filtration. This new material may enable cost-efficient water purification systems for developing economies, given the combined benefits of activated-carbon and sand filtration in this novel composite filtration medium processed through soft-chemistry principles [1]. Further, I will elaborate on our recent observations of improvement in corrosion-resistance by one and half orders of magnitude from conformal ultra-thin films of graphene over Cu-substrates formed by Chemical Vapor Deposition [2]. In summary, it is hoped that enabling technologies such as these may help 'graphene-research' bridge the esoteric world of academia and urgent humanitarian problems.

References:

1. Wei Gao, Mainak Majumder*, Lawrence B. Alemany, Tharangattu N. Narayanan, Miguel A. Ibarra, Bhabendra K. Pradhan , and Pulickel M. Ajayan*ACS Appl. Mater. Interfaces, 2011, 3 (6), pp 1821–1826
2. R.K. Singh Raman, P. Chakraborty, Banerje, Derrek E. Lobo, Hemtej Gullapalli, Madusha Sumandasa, Ashwin Kumar, Lokesh Choudhary, Rachel Tkacz, Pulickel M. Ajayan, Mainak Majumder* "Protecting Copper from Electrochemical Degradation by Graphene Coating", Carbon, just accepted, (http://dx.doi.org/10.1016/j.carbon.2012.04.048)

(286a) Separating Magnetically Labeled and Unlabeled Biological Cells within Microfluidic Channels Using Magnetic Nanodisks

Atul Bharde[1], Tom Byvank[1], Aaron Chen[2], Greg Vieira[2], Brandon L. Miller[3], Jessica O. Winter[4], **Jeffrey J. Chalmers**[5] and R. Sooryakumar[6], (1)Physics, The Ohio State University, Columbus, (2)Department of Physics, Ohio State University, Columbus, OH, (3)Chemical and Biomolecular Engineering, The Ohio State University, Columbus, OH, (4)William G. Lowrie Department of Chemical and Biomolecular Engineering, the Ohio State University, Columbus, OH, (5)William G. Lowrie Department of Chemical and Biomolecular Engineering, The Ohio State University, Columbus, OH, (6)Department of Physics, The Ohio State University, Columbus, OH

The transport of microscopic objects that rely on magnetic forces has numerous advantages including flexibility of controlling many design parameters and the long range magnetic interactions generally do not adversely affect biological or chemical interactions. We present results on the use of magnetic micro-arrays imprinted within microfluidic channels that benefit from of these features and the ability to rapidly reprogram the magnetic energy landscape for cell manipulation and sorting applications. A central enabling feature is the very large, tunable, magnetic field gradients ($> 10^5$ T/m), designed within the microfluidic architecture. These gradients are created with circular permalloy disks and linear wires (40 nm thick) which are patterned on silicon surfaces using lithographic technology.[1,2] Through use of magnetic microsphere tethers to label biological cells, results on the transport and sorting of heterogeneous cell populations will be presented. The role of micro-array design parameters, competing magnetic forces and hydrodynamic drag forces as well as cell-labeling efficiency in the effectiveness of cell separation will be discussed.

1. Henighan, T., Chen, A., Vieira, G., Hauser, A.J., Yang, F.Y., Chalmers, J.J., Sooryakumar, R. Transporting Biological Cells via Programmable Mobile Magnetic Traps. Biophys J.. 98(3):412-417. 2010. PMID: 20141754
2. Vieira, G., Heninghan, T., Chen, A., Hauser, A.J., Yang, Y., Chalmers, J.J., Sooryakumar, R. Magnetic wire traps and programmable manipulation of biological cells. *Physical Review Letters,* 103:128101, 2009.

(286b) Internalization Pattern of Functionalized Magnetic Nanoparticles and the Prospects of Intracellular Hyperthermia

Robert J. Wydra[1], Younsoo Bae[2], Kimberly W. Anderson[1] and J. Zach Hilt[1], (1)Chemical and Materials Engineering, University of Kentucky, Lexington, KY, (2)Pharmaceutical Sciences, University of Kentucky, Lexington, KY

Internalization pattern of functionalized magnetic nanoparticles and the prospects of intracellular hyperthermia

Robert J. Wydra[1], Younsoo Bae[2], Kimberly W. Anderson[1], J. Zach Hilt[1]

[1]Department of Chemical and Materials Engineering, University of Kentucky

[2]Department of Pharmaceutical Sciences, University of Kentucky

Magnetic nanoparticles are being studied for a wide range of biomedical applications such as imaging, targeted delivery, and thermal therapy of cancer. In most cases, the nanoparticles rely on passive targeting to systemically circulate and accumulate in tumors via the phenomenon known as the enhanced permeation and retention effect. To increase specific interactions with cells, nanoparticles can be functionalized with appropriate ligands. It has recently been demonstrated by Rinaldi and others that internalized nanoparticles can induce cellular death when exposed to an alternating magnetic field without a measurable temperature rise. In this study, four nanoparticle systems – uncoated iron oxide, citric acid coated iron oxide, (3-aminopropyl)trimethoxysilane (APTMS), and poly(ethylene glycol) (PEG) coated iron oxide – were incubated with cancer cells and assessed for their ability to be internalized and to be used for intracellular hyperthermia. The iron oxide core nanoparticles, which are selected for their ability to remotely actuate in an AMF, were prepared utilizing the facile co-precipitation technique. The citric acid stabilizer was added in a one-step method and provided a negative surface charge in physiological conditions. APTMS was attached to the particle surface through a ligand exchange and provided a posited charged surface. The PEG coating was selected to increase circulation time and avoid reticuloendothelial system clearance. This coating was prepared with a surface initiated polymerization, atom transfer radical polymerization, where-by an initiator is initially attached to the particle surface and the polymerization extends from the surface. The particles were characterized using Fourier transform infrared spectroscopy to verify surface functionalization; thermal gravimetric analysis to quantify mass percent of coating; dynamic light scattering to determine particle size; and UV-Vis spectroscopy to determine particle stability in a variety of media. Nanoparticles were incubated with A549 lung adenocarcinoma cells and PC3 prostate cancer cells and the internalization pattern was determined by visualization using fluorescent microscopy. Nanoparticle uptake by cancer cells was quantified with the 1,10-phenanthroline colorimetric assay. Cells with internalized nanoparticles were exposed to an AMF to determine the potential for intracellular hyperthermia.

ADDIN EN.REFLIST

(286c) Magnetic Block Ionomer Complexes for Imaging and Therapeutics

Richey M. Davis[1], Nipon Pothayee[2], Nikorn Pothayee[2], Neeta Jain[3], Lindsay M. Johnson[2], Sharavanan Balasubramaniam[2], Nammalwar Sriranganathan[3] and Judy S. Riffle[4], (1)Chemical Engineering, Virginia Tech, Blacksburg, VA, (2)Macromolecules and Interfaces Institute, Virginia Tech, Blacksburg, VA, (3)VA-MD Regional School of Veterinary Medicine, Virginia Tech, Blacksburg, VA, (4)Chemistry, Virginia Tech, Blacksburg, VA

Multifunctional nanocarriers comprising a combination of therapeutic and imaging agents are of great interest for delivering drugs and tracking their biodistribution. With an aim to develop nanocarriers with high drug loadings that integrate magnetic nanoparticles for imaging into one system, we report on 2 types of antibiotic-laden magnetic block ionomer complexes based on assembly of block ionomers with nanomagnetite and cationic drugs. One type of complex - magnetic block ionomer complex (MBICs) - consisted of aggregates of superparamagnetic magnetite nanoparticles held together by noncovalent forces. The second type of complex - magnetic block ionomer clusters (MBIClusters) - were held together by covalent crosslinks between polymer chains. In both cases, the magnetite core was synthesized via reduction of an Fe(III) organometallic precursor. Nonionic-ionic block copolymers were synthesized through controlled free-radical polymerization (ATRP). The polymer was bound to the magnetic nanoparticle surfaces via ligand adsorption of the PAA, thereby creating a double corona structure with a nonionic PEO shell and an ionic layer of PAA. The portion of carboxylates that were not attached to the magnetite provided binding sites for drug loading via ionic complexation. In the case of the MBIClusters, amine groups at the tips of the H_2N-PEO corona chains were subsequently crosslinked through reaction with a poly(ethylene glycol) diacrylate oligomer to yield clusters. The nanoscale average sizes of the clusters were controlled by adjusting the concentration of the reactants in the crosslinking step.

For both classes of complexes, gentamicin (a multi-cationic drug) was employed as model for encapsulation. Results suggest that the MBICs are sufficiently stable in physiological conditions to be suitable as drug carriers. Dispersions of the MBICs were stable in PBS containing 0.14 M NaCl for up to 10 days, and were stable in PBS containing fetal bovine serum (FBS). Physicochemical properties of the MBICs were significantly altered upon gentamicin encapsulation (31 wt % drug). A near zero-order release of gentamicin (pH 7.4 in PBS) that reached ~35 wt% of the initial gentamicin within 10 h was observed, and this was followed by slower release of another 7% by 18 h. The unloaded complexes (no drug) had a transverse relaxivity of 82 $s^{-1}(mM\ Fe)^{-1}$, which is ~ 2X higher than that reported for commercial T_2 contrast agents at 37 °C and 1.5 T. The increase in size of the complexes upon gentamicin encapsulation was accompanied by an increase in r_2 to 161 $s^{-1}(mM\ Fe)^{-1}$. The increase in r_2 with formation of small aggregates is consistent with recent findings in our group that controlled aggregation leads to considerably shorter T_2 relaxation times or higher r_2 relaxivities. Gentamicin loading in the MBIClusters was as high as 38 wt%, corresponding to a drug loading efficiency of ~ 95%. The MBIClusters exhibited markedly high transverse relaxivities (r_2) ranging from 190 to 604 s^{-1} $mMFe^{-1}$. The effect of cluster size on relaxivity will be discussed.

(286d) The Influence of the Stabilizing Polymer Brush On Transverse Relaxation Rates in Magnetic Resonance Imaging: Implications of Interparticle Interactions

O. Thompson Mefford, Materials Science and Engineering, Clemson University, Clemson, SC and Steven Saville, Clemson University, Clemson, SC

There have been significant advances recently in the use of magnetic nanoparticles for various biomedical applications, with one potential use as contrast enhancement agents in magnetic resonance imaging (MRI).[1] In order to study the effect chain formation on MRI contrast enhancement and its relationship with the stabilizing brush length, a matrix of 22nm particles with varied ligand lengths was synthesized using iron oxide particles with poly(ethylene glycol) ligands. These systems were characterized with dynamic light scattering, transmission electron microscopy, dark-field scattering, and proton transverse relaxation measurements. The dark field scattering experiments and transverse relaxation measurements were done in a similar magnetic field under the same time scale to correlate the reduction of the transverse relativity with the formation of linear magnetic chains. Our results suggest that varying the ligand length has a direct effect on the transverse relaxation mechanism due to the contribution of ligand length to the colloidal stability of the system, including differences in chain formation rate and size. With increasing ligand length, interparticle interactions are limited, which results in slower chain formation and shorter widespread chains. This data indicates that understanding the colloidal arrangement of these systems is paramount, and that both particle stability and time dependence both play a key role in determining the effect of iron oxide nanoparticles on surrounding water protons.

In this talk we will discuss the synthesis of unique heterobifunctional polymers, magnetite particle synthesis and functionalization, as well as calculations of interparticle interaction based on a modified DLVO theory.

1. Sun, C.; Lee, J. S. H.; Zhang, M. Q., Magnetic nanoparticles in MR imaging and drug delivery. *Advanced Drug Delivery Reviews* **2008,** 60, (11), 1252-1265.

2. Carroll, M. R. J.; Huffstetler, P. P.; Miles, W. C.; Goff, J. D.; Davis, R. M.; Riffle, J. S.; House, M. J.; Woodward, R. C.; St Pierre, T. G., The effect of polymer coatings on proton transverse relaxivities of aqueous suspensions of magnetic nanoparticles. *Nanotechnology* **2011,** 22, (32).

3. Bertoni, G.; Torre, B.; Falqui, A.; Fragouli, D.; Athanassiou, A.; Cingolani, R., Nanochains Formation of Superparamagnetic Nanoparticles. *The Journal of Physical Chemistry C* **2011,** 115, (15), 7249-7254.

(286e) Controlled Magnetite Nanoparticle Morphology for Better Cellular Uptake: Experiments and Models

Chettiannan Ravikumar and Rajdip Bandyopadhyaya, Chemical Engg., IIT Bombay, Mumbai, India

Introduction

Aqueous dispersions containing water-insoluble magnetite (Fe_3O_4) nanoparticles are widely used in various biomedical applications, like hyperthermia, magnetic labeling of cells, enhanced contrast in MRI of tissues etc. In these applications, the final diameter and morphology of the nanoparticles (either isolated primary particles or aggregates of primary particles) are the most important parameters determining the efficacy of cellular uptake of these particles. Several experimental reports suggest that Fe_3O_4 nanoparticles synthesized in the aqueous medium can be either of isolated or an aggregated morphology. It is of interest to explain the particle formation mechanism, so that one can optimize the experimental conditions to achieve nanoparticles of a specific diameter and morphology, which would determine colloidal stability and shelf-life of these particles or their dispersions, for most effective cellular uptake of these formulations. To this end, we develop the mechanism of Fe_3O_4 nanoparticle formation (a water-insoluble material) in an aqueous medium, the resultant structure of the particulate-dispersion and its uptake by human cells.

Methodology and results: Time-scales, mechanism, morphologyy and uptake

We have conducted both chemical coprecipitation and thermal decomposition methods for nanoparticle synthesis. Different biocompatible coating agents such as carboxymethyl cellulose (CMC), dextran, citric acid or poly (acrylic) acid (PAA) were used to coat the particles, in order to disperse them and achieve a stable aqueous dispersion. The crystal structure and diameter of Fe_3O_4 nanoparticles was confirmed by X-Ray Diffractometer (XRD). Fourier transform infra red (FTIR) spectroscopy of particles showed that the coating agents were chemically adsorbed on the particle surface. Transmission electron microscopy (TEM), atomic force microscopy (AFM) and dynamic light scattering (DLS) of coated particles together gave both core particle diameter and the hydrodynamic diameter in the dispersion. Now, in absence of a coating agent, particles synthesized by the coprecipitation route always resulted in an aggregated morphology, having core, primary particle diameter in the range of 6 - 9 nm. However, with citric acid coating; nearly completely dispersed, coated particles were achieved by first treating the uncoated particles with tetra methyl ammonium hydroxide (TMAOH), followed by citric acid coating of the treated particles. In contrast, PAA coated particle were still aggregated. On the other hand, in thermal decomposition route, completely isolated and monodisperse PAA coated particles having average core particle diameter of 6 nm with a standard deviation of 1.1 nm were obtained.

Therefore, to understand the aforesaid contrasting morphology of PAA coated particles obtained via the two routes, a time-scale based mechanisms of particle formation for each route was developed. The relative rates (or in other words the characteristic time scales) of each of the individual steps leading to particle formation, namely, adsorption of coating agent on a nanoparticle, diffusion of particle on a polymer chain, Brownian collision and coagulation of particles etc., were estimated a-priori. We find that the coagulation time scale in thermal decomposition route is much smaller than the experimental aging time, resulting in complete coagulation and formation of isolated PAA coated particles. In contrast, during coprecipitation, absence of coagulation during the experimental aging time resulted in formation of aggregates of particles. Thus from our mechanistic study, we conclude that the high temperature synthesis of Fe_3O_4 in thermal decomposition leads to faster coagulation rate and results in isolated nanoparticle formation.

Coated particles in a dispersion require a long shelf-life for *in-vivo* use. Stability against sedimentation depends on the number of primary particles in an aggregate (single primary particle, two particles forming a dimer, and higher aggregates). We calculated the aggregate number density distribution of our synthesized particles in the aqueous dispersion using Monte Carlo simulation. The simulation was performed by calculating the total interaction potential between two nanoparticles as a function of their interparticle distance, and applying a criterion for the two particles to aggregate; the criterion being that the minimum depth of the secondary minimum in the total interaction potential must be at least equal to thermal energy, k_BT. Total interaction potential curve for citric acid coated particles showed a secondary minimum (formation of reversible aggregates) and dextran coated particles showed a primary minimum (formation of permanent aggregates). The former is due to low shell thickness possessed by citric acid coated particles and latter is due to bridging attraction of dextran coated particles. PAA coated particles did not show either primary or secondary minimum, because of its larger shell thickness, thereby justifying the completely isolated particles obtained in experiments. Number density distribution predicted by simulation showed that citric acid coated particles will be in the form of moderate aggregates, dextran in complete aggregate and PAA in isolated form. Simulation predictions were compared with experimental results and they showed good agreement with each other. This new approach of determining the states of particles in dispersions provides a-*priori* information in choosing an appropriate coating agent for magnetic nanoparticles, in order to achieve isolated particles in dispersions.

Finally, dispersions containing different morphology of coated particles were incubated (*in-vitro*) with human hepatoma cell lines (HepG2), in order to understand the effect of morphology and surface charge of the coated nanoparticles on cellular uptake. Better cell viability was observed with citric acid and dextran coated nanoparticles, compared to PAA coated particles. We find that the neutral dextran coated particles in the form of aggregates were not taken up by the cells, whereas negatively charged and nearly isolated citric acid coated particles showed very good uptake. Rate of uptake of citric acid coated particles in cells was fitted to a mathematical model based on a two stage mechanism of particle adsorption and internalization. Based on this model, the average mass of Fe (in the form of nanoparticles) internalized by a single HepG2 cell was predicted. From this analysis, we conclude that the coated particles having isolated morphology and a surface charge can lead to high cellular uptake.

Conclusions

Individual (primary) nanoparticle diameter, charge and morphology of an ensemble of these nanoparticles - either in an isolated or an aggregated form - are key parameters dictating colloidal stability and shelf-life of aqueous particulate dispersions aimed towards long-term usage of these dispersions for in-vivo cellular uptake. Our time-scale based nanoparticle formation mechanism models primary particle diameter, and interparticle potential-driven Monte Carlo simulation predicts their equilibrium state of aggregation and dispersion stability. This in conjunction with a model of cell-particle interaction explains experimental data on coated magnetite nanoparticle uptake by HepG2 cells, correctly predicting contrasting behavior of different coating agents on extent of uptake. We find that citric acid coated, negatively charged particles, having a nearly isolated morphology, leads to a high cellular uptake; this compared to dextran coated particles, which are large aggregates of uncharged particles. The experimentally validated model is therefore useful in designing synthesis conditions to achieve controlled size, shape and aggregation of primary nanoparticles, aimed towards maximum uptake of solid particles, or extending further, in elucidating diffusion and accessibility issues of guest molecules in porous nanoparticles. The simulation results can also help in screening of a proper coating agent to achieve any desirable state of aggregation of particles in the dispersion, aimed towards optimum features in cellular uptake of nanoparticles.

(313a) Hierarchical Structure of Graphene Oxide-Polymer Nanocomposites

Karl Putz[1], Marc Palmeri[2], Charles Wood[1], Zhi An[3], SonBinh Nguyen[3] and L. Catherine Brinson[4], (1)Mechanical Engineering, Northwestern University, Evanston, IL, (2)Materials Science and Engineering, Northwestern University, Evanston, IL, (3)Chemistry, Northwestern University, Evanston, IL, (4)Department of Mechanical Engineering and Department of Materials Science and Engineering, Northwestern University, Evanston, IL

Biological organisms utilize hierarchical structures to create materials with outstanding properties. Synthetic composites typically utilize inorganic reinforcement with one length scale of structure, resulting in materials that involve tradeoffs between properties such as strength and stiffness. Synthesis of biomimetic composites requires materials with multiple length scales of structure. Graphene oxide papers are simple to fabricate, yet elegantly demonstrate the power of hierarchical structures. Work on graphene oxide papers has typically focused on the nanometer, sheet-to-sheet spacing because this is the simplest to measure through techniques such as x-ray diffraction. However, recent investigations have shown the existence of longer length scale structures that have a profound impact on the properties of the material. In this presentation, the hierarchical structure of graphene oxide papers, and their associated composites will be shown on the nanometer (sheet-to-sheet spacing), 100-nm (lamellae), and micron (whole paper) length scales. Through modification of these structures with small molecules and polymers we can begin to understand the structure-property relationships in these hierarchical composites.

(313b) Controlled Orientation and Actuation of High Performance Polyimide Nanocomposites Using Magnetic Nanoparticles Tethered Graphene

Mitra Yoonessi, Graftech International, Parma, OH, Daniel Scheiman, ASRC, Cleveland, OH, John Peck, University of Akron, Akron, OH, James Gaier, NASA Glenn Research Center, Cleveland, OH, Cleveland,, OH and Michael A. Meador, Structure and Materials Division, NASA Glenn Research Center, Cleveland, OH

Many new technologies require control of thermal and electrical transport direction in lightweight advanced composites. Polyimide is a high performance polymer with superior properties such as high glass transition temperature, excellent thermal stability, high modulus, low dielectric constant, low coefficient of thermal expansion, and radiation resistance. Hybrid nickel graphene nanoparticles with magnetic and high electrical, thermal and mechanical properties were prepared using the thermal decomposition method. The nickel nanoparticles were covered with an amorphous carbon layer to protect from oxidation. Ultra-small angle X-ray scattering (USAXS) of the Ni-graphene polyimide nanocomposites revealed a dual size distribution of nickel nanoparticles when estimated as spheres, which was in agreement with HR-TEM and detailed magnetic hysteresis measurements. The Ni-graphene nanoparticles were dispersed in the polyimide in the presence of magnetic field yielding a high degree of nanoparticle orientation in the polyimide matrix. This orientation resulted in high in-plane electrical and mechanical properties of polyimide nanocomposites. The nanoparticle orientation was confirmed by changes to saturation magnetization as the sample was rotated in the magnetic field, as well as by transmission electron microscopy. The saturation magnetization of Ni-graphene polyimide nanocomposites containing randomly oriented nanoparticles increased with increasing magnetic nanoparticle content. The nanocomposites exhibited actuation in the presence of a magnetic field.

(313c) Electrospun Poly(vinyl alcohol)/α-Zirconium Phosphate Nanocomposite Fibers

Monira Lizu[1], Xi Zhang[1], Jayanthi Sampathi[1], Luyi Sun[2], Matthew F. Milner[2] and Suying Wei[1], (1)Chemistry and Biochemistry, Lamar University, Beaumont, TX, (2)Department of Chemistry and Biochemistry, Texas State University-San Marcos, San Marcos, TX

Poly (vinyl alcohol) (PVA)/α-zirconium phosphate (ZrP) nanocomposite fibers were successfully prepared via the simple and low-cost electrospinning method. In order to obtain the defect-free, uniform polymer nanocomposite fibers, a number of parameters were investigated namely concentration of polymers, feed rate, applied voltage and working distance between needle tip and the fiber collecting substrate. Morphological properties of the nanofibers were characterized by scanning electron microscopy, and smooth and nonwoven mat of fibers were revealed. Strong intermolecular interactions between the PVA matrix and the included ZrP nanofillers were indicated by the attenuated total reflectance Fourier transform infrared spectroscopy (ATR-FTIR). Incorporation of the ZrP nanofillers also drastically improved the thermal stability of the polymer nanocomposite fibers, which demonstrated clear dependence on both the aspect ratio and loading level of the nanofiller. Furthermore, the viscoelastic properties of the polymer nanocomposite solutions were studied by rheological analysis; the higher loading of the nanofiller resulted in higher viscosity, and showed different fluid nature as compared to the lower loading level.

(313d) Polypropylene (PP)-Elastomer Nanocomposites: Effect of PP and Elastomer MFI On Toughness and Thermal Expansion Behavior

Rajkiran R. Tiwari and Donald R. Paul, Department of Chemical Engineering, The University of Texas at Austin, Austin, TX

Extruded PP/PP-g-MA/MMT/elastomer nanocomposites were prepared at fixed 30 wt% elastomer and 0-7 wt% MMT. The ratio of maleic anhydride grafted polypropylene, PP-g-MA and organoclay was fixed at 1.0. Elastomer particle morphology in the presence of MMT was evaluated for three different molecular weight grades of polypropylene (PP) (H= high, M= medium and L= low) and five different ethylene-co-octene elastomers (EOR) with different melt flow index (MFI). The MMT particles are located exclusively in the PP phase in the PP/PP-g-MA/MMT/EOR nanocomposites as seen from TEM images. PP/PP-g-MA/MMT/elastomer nanocomposites with toughness in range from 450 J/m to 800 J/m (super toughness) are obtained based on the molecular weight of the PP, elastomer type and MMT content. Elastomers having a melt flow index (MFI) in the range of 0.5-1.0 showed significant improvement in the impact strength of L-PP based nanocomposites while most H-PP/EOR blends gave 'super-tough' materials without MMT and maintain this toughness in the presence of MMT. For the similar elastomer particle sizes in nanocomposites, the impact strength varies as H-PP> M-PP> L-PP. The ductile-brittle (D-B) transition temperature decreases with increased molecular weight of the PP and the addition of MMT. Elastomers having a melt flow index (MFI) in the range of 0.5-1.0 showed a significant decrease in the D-B transition temperature for both L-PP and H-PP nanocomposites.

The effect of PP-g-MA/organoclay ratios varying from 0 to 1.5 was also studied on the toughness of L-PP/PP-g-MA/MMT/EOR nanocomposites prepared from elastomer having MFI of 0.5. The elastomer particle size reduced significantly and toughness is observed at lower MMT content as the PP-g-MA/organoclay ratio is increased. The PP-g-MA facilitates dispersion of MMT in the PP matrix which acts as a barrier to coalescence of elastomer particles leading to much smaller particle sizes at lower MMT content. The thermal expansion coefficient is significantly reduced in the presence of MMT and as the PP-g-MA/orgnoclay ratio is increased. The tensile modulus and yield stress improved with the MMT content and PP-g-MA/organoclay ratio. Overall, extruder-made TPO nanocomposite provides a well balance of stiffness and toughness and offers the advantage of using different molecular weight grades of PP and elastomer, and MMT to formulate TPOs in terms of particle size control and properties based on end use applications.

(313e) Comparative Study of Thermal Behavior of Hdpe/Clay Nanocomposites Prepared with Modified Montmorillonite with Two Functional Organic Agents

María Verónica Carranza Oropeza[1,2], Renato Godoy[2], Eduardo Figueiredo[2], Henrique Perez[3], Ademar Lugao[3] and Reinaldo Giudici[4], (1)Chemical Engineering, São Paulo University, São Paulo, Brazil, (2)Orbys High Performance in materials, São Paulo, Brazil, (3)IPEN, São Paulo , Brazil, (4)Chemical Engineering, Polytechnic School of the University of São Paulo, São Paulo, Brazil

Comparative study of thermal behavior of HDPE/clay nanocomposites prepared with modified montmorillonite with two functional organic agents

Verónica Carranza[1,3], Renato Godoy[1], Eduardo Figueiredo[1], Henrique Perez[2], Lugão[2], Reinaldo Giudici[3]*

[1]*Orbys High Performance in Materials - São Paulo, Brazil*

[2]*Institute of Nuclear Researches - São Paulo, Brazil*

[3]*Chemical Engineering Department of São Paulo University, São Paulo, Brazil*

**Email: mvco@usp.br*

Abstract -

Nanocomposites are the focus for materials engineers to overcome the limitations of traditional micrometer-scale polymer composites. In particular, the field of nanocomposites based on organically modified clays. Thus, the addition of clay, especially montmorillonite in a polymer matrix can have significant influences on the properties of this matrix. For example, increasing of thermal stability of polymer/clay nanocomposites, strength and heat resistance, decreasing gas permeability and increasing biodegradability of biodegradable polymers. The potential of these property improvements usually depend on the degree of delamination and dispersion of the clay in the polymer matrix, which depends on a combination of the proper chemical treatment and optimized processing. Polymer/clay nanocomposites have been successfully used in several kinds of polar polymers. However, for polymers with low polarity, such as polyolefins, the current results are not satisfactory due to the low compatibility between the clay and the polyolefins . In this way, the focus of this study was to evaluate the effect of two kind of modified montmorillonite on the structure and thermal properties of new nanocomposites based in high density polyethylene (HDPE). The two HDPE/modified montmorillonite nanocomposites were prepared by direct melt-mixing procedure using a twin-screw extruder. The modified montmorillonite with a quaternary ammonium salt was used with 5, 7,5 and 20 wt% to prepare the first nanocomposite. The second one was prepared with 7,5 wt% of modified montmorillonite with amino functional alkoxysilane. Changes in the surface of montmorillonite and the dispersion of modified clays in the polymer matrix were evaluated using X-ray diffraction (XRD). The thermal behavior of the clay and nanocomposites were analysed by differential scanning calorimetry (DSC) and thermogravimetric analysis (TG). The XRD results show that almost complete exfoliation of the silicate layers took place and nanocomposite structure was obtained for all new materials. The thermal behavior of the PE/clay nanocomposites can be improved in the case of use 7,5wt% of modified montmorillonite with amino functional alkoxysilane.

(313f) Flexible Multifunctional Superparamagnetic Nanocomposite Films

Georgios A. Sotiriou, Department of Mechanical and Process Engineering, Particle Technology Laboratory, ETH Zurich, Zurich, Switzerland, Christoph O. Blattmann, Particle Technology Laboratory, ETH Zurich, Zurich, Switzerland and Sotiris E. Pratsinis, Particle Technology Laboratory, Institute of Process Engineering, Department of Mechanical and Process Engineering, ETH Zurich, Zurich, Switzerland

The synthesis of multi-layer polymer nanocomposite films with multifunctional filler particles was investigated here. The resulted films exhibit the characteristic properties of each component, such as magnetic, luminescent and plasmonic [1]. The filler particles are homogeneously dispersed within the polymer films, allowing for uniform properties along the whole range of the film. Such structures are advantageous for several applications that require multiple functionalities such optical filters, biosensors, luminescent materials. Here, the plasmonic- and luminescent-superparamagnetic nanocomposite are made and performance of these polymer composites is demonstrated. Furthermore, free-standing microstructures can be made by depositing the polymer nanocomposite films on a sacrificial layer. The filler film stays intact after the addition of the polymer, resulting in high-filler-fraction polymer nanocomposites with uniquely homogeneous distribution. Such structures ideal for their integration in micro- and nano-fabrication techniques, for example fr the formation of superparamagnetic cantilevers [2].

References

[1] Sotiriou, G. A., Hirt, A. M., Lozach, P. Y., Teleki, A., Krumeich, F. & Pratsinis, S. E. Hybrid, silica-coated, Janus-like plasmonic-magnetic nanoparticles. *Chem. Mater.* **23**, 1985-1992 (2011).

[2] Suter, M., Ergeneman, O., Zürcher, J., Moitzi, C., Pané, S., Rudin, T., Pratsinis, S. E., Nelson, B. J. & Hierold, C. A photopatternable superparamagnetic nanocomposite: Material characterization and fabrication of microstructures. *Sens. Act. B-Chem.* **156**, 433-443 (2011).

(326a) Award Submission: A Novel Anionic Nanoparticle Delivery System for Microrna-29b Targets FLT3 and KIT Receptor Tyrosine Kinase Expression in Acute Myeloid Leukemia

Xiaomeng Huang[1], Sebastian Schwind[2], Bo Yu[1], Shujun Liu[2], Jiuxia Pang[2], Ramasamy Santhanam[2], Yue-Zhong Wu[2], Kenneth K. Chan[3], William Blum[2], Clara D. Bloomfield[2], Danillo Perroti[2], Ramiro Garzon[2], John C. Byrd[2], Natarajan Muthusamy[2], Robert J. Lee[3], Guido I. Marcucci[2] and Ly James Lee[4], (1)Nanoscale Science and Engineering Center, The Ohio State University, Columbus, OH, (2)Comprehensive Cancer Center, The Ohio State University, Columbus, OH, (3)College of Pharmacy, The Ohio State University, Columbus, OH, (4)William G. Lowrie Department of Chemical and Biomolecular Engineering, The Ohio State University, Columbus, OH

MicroRNAs, small non-coding RNA molecules, contribute to the pathogenesis of various human cancers, including acute myeloid leukemia (AML). One of these microRNAs - *miR-29b* - is frequently downregulated in AML, has antileukemic activity and was shown to target genes involved in leukemogenesis.[1] Furthermore, higher pretreatment levels of *miR-29b* were associated with improved clinical response to treatment with the anti-leukemic agent decitabine.[2] Recently we demonstrated that *miR-29b* also downregulates the expression of the receptor tyrosine kinases (RTKs) KIT and FLT3 in AML.[3,4] The aberrant activation of the RTKs KIT and FLT3, by mutations or overexpressed is frequently found in AML and much effort has been undertaken to target these RTKs. In AML, due to development of early resistance, most inhibitors of RTKs have not led to substantial clinical benefit. However, instead of targeting the activity of RTKs, decreasing their expression may be a valuable therapeutic alternative. Thus the artificial increase of *miR-29b* levels in AML blasts to downregulate the RTKs KIT and FLT3 may be of high clinical benefit. Although some miR delivery methods have been developed for solid tumors, these methods have limitations that prevent their use in leukemia therapy. Cationic lipid/polymer formulations frequently used in solid tumors have the tendency to accumulate in lungs, kidney and liver and may cause a non-specific immune response, for which the consequences in AML have not been fully explored yet. We developed a transferrin-conjugated anionic lipid-based nanoparticle (NP) system to deliver synthesized *miR-29b* mimic molecules to leukemia cells. Low-molecular-weight polyethylenimine (PEI) was used to provide a positive charge, which can easily capture the negatively charged miR molecules achieving high entrapment efficiency of more than 90%. The lipid-based carrier was made of 1,2-Dioleoyl-sn-Glycero-3-Phosphoethanolamine (DOPE), Linoleic acid and 1,2-Dimyristoyl-sn-glycerol, methoxypolyethylene Glycol (DMG-PEG). We chose linoleic acid because of its low binding affinity to small RNAs, which implies easy dissociation of miRs from the lipopolyplex core. Moreover, the NPs were protected from reticuloendothelial system clearance by DMG-PEG, which allowed long *in vivo* circulation times. Since AML blasts overexpress the transferrin receptor, we conjugated the NPs with transferrin (Tf) using a post-insertion method to increase the uptake by targeted delivery. The average size and zeta potential of empty NP and miR-loaded NP were 129.6 ± 1.02nm (± Standard Deviation [SD]) and -9.82 ± 1.53mV (± SD) and 137.6 ± 0.96nm and 22.48 ± 1.44mV, respectively. After the Tf-conjugation of the miR-loaded NP the size increased to 147.3 ± 4.74nm and the zeta potential was 5.82 ±1.87mV. To evaluate the efficiency of NP uptake we encapsulated FAM-fluorescent labeled miR and treated two AML cell lines (i.e. Kasumi-1 and MV4-11). Tf-conjugated NP (Tf-NP) treated cells showed the highest uptake by confocal microscopy and flow-cytometry (mean fluorescence intensity (MFI): 21.0 and 22.9 for Kasumi-1 and MV4-11 respectively) compared to non-Tf-conjugated NPs (MFI: 8.3 and 13.0). Next, we evaluated the efficiency of *miR-29b* mimics delivery to AML cell lines and primary patient blasts with the Tf-NP system. After treatment with *miR-29b* loaded Tf-NPs (Tf-NP-*miR-29b*) we observed an increase of intracellular levels of *miR-29b,* 420 folds increase in Kasumi-1 cells, 220 folds increase in MV4-11 cells and an average 1000 folds increase in patient blasts compared to negative control-loaded Tf-NP (Tf-NP-nc) treatment after 24 hours. The expression levels of an unrelated miR, i.e. *miR-140,* remained unchanged. To assess whether the increased *miR-29b* levels would lead to a decrease of KIT and FLT3 expression we performed western blot analysis 48 hours after treatment. Tf-NP-*miR-29b* treatment led to 3.4 and 2 folds decrease of FLT3 and 1.5 and 1.4 folds decrease of KIT in Kasumi-1 and MV4-11 cells respectively and on average 1.5 folds decrease of FLT3 and 2.5 folds decrease of KIT in cells derived from three AML patients compared to Tf-NP-nc treated cells. The Tf-NP-*miR-29b*-mediated downregulation of these targets had antileukemic effects on AML cells. Tf-NP-*miR-29b* treatment reduced growth (per day) from 32.2% to 25.3% (P=0.01) in Kasumi-1 cells and from 53.0% to 43.9% (P=0.007) in MV4-11 cells, compared to Tf-NP-nc treatment. We also observed a reduction of numbers of colony formed by 55-66% after Tf-NP-*miR-29b* treatment after 2 weeks. The average numbers of colonies formed were 143±9(±SD) and 65±6 (P<0.001) for Kasumi-1 and 213±7 and 80±5 (P<0.001) for MV4-11 of Tf-NP-nc and Tf-NP-*miR-29b* treatment, respectively. Since we recently demonstrated that higher pretreatment levels of *miR-29b* were associated with improved clinical response to decitabine treatment,[4] we also tested whether priming AML cells with Tf-NP-*miR-29b* prior to decitabine treatment would have additional therapeutic effect. Tf-NP-*miR-29b* treatment (for 48h) followed by 0.5 μM decitabine (for 48h) decreased the cell viability compared to Tf-NP-nc treatment or unprimed cells by 40% (P=0.001) and 30% (P=0.01), respectively in Kasumi-1 and by 16% (P=0.003) and 22% (P=0.001), respectively in MV4-11 cells. We confirmed these findings in three AML patient blasts, Tf-NP-*miR-29b* priming, followed by 5 μM decitabine treatment decreased cell viability for patient 1 by 20% (P=0.01), for patient 2 by 40% (P=0.003) and for patient 3 by 22% (P=0.002), respectively compared to Tf-NP-nc priming. In conclusion, we report here a novel anionic NP-based delivery system for miRs in AML. Our delivery system for *miR-29b* showed targeting activity against the important RTKs KIT and FLT3 that are associated with leukemogenesis, antileukemic effects and improved response to decitabine treatment in AML cells. The modulation of *miR-29b* levels with the here proposed delivery system as a single agent or in combination with other anti-cancer drugs - i.e. decitabine - may represent a promising new targeted treatment option for AML patients in the near future.

1. Garzon R, Liu S, Fabbri M, et al. MicroRNA-29b induces global DNA hypomethylation and tumor suppressor gene reexpression in acute myeloid leukemia by targeting directly DNMT3A and 3B and indirectly DNMT1. *Blood*. 2009;113(25):6411–6418.
2. Blum W, Garzon R, Klisovic RB, et al. Clinical response and miR-29b predictive significance in older AML patients treated with a 10-day schedule of decitabine. *PNAS*. 2010;107(16):7473–7478.
3. Liu S, Wu L, Pang J, et al. Sp1/NFκB/HDAC/miR-29b Regulatory Network in KIT-Driven Myeloid Leukemia. *Cancer Cell.* 2010; 17(4): 333-347
4. Blum W, Schwind S, Tarighat SS, et al. Clinical and Pharmacodynamic Activity of the Combination Bortezomib and Decitabine: a Phase I Trial in Patients with Acute Myeloid Leukemia. *Blood.* 2012; published ahead of print as doi:10.1182/blood-2012-03-413898.

(326b) Award Submission: Microrheological Measurements of the Effects of Cholesterol On the Mechanical Properties of Lung Surfactant Monolayers

Kyuhan Kim, Chemical Engineering, Univerisity of California, Santa Barbara, Santa Barbara, CA, Todd M. Squires, Chemical Engineering, University of California, Santa Barbara, Santa Barbara, CA and Joseph A. Zasadzinski, Department of Chemical Engineering and Materials Science, University of Minnesota, Minneapolis, MN

Cholesterol plays a significant role in biological interfaces, especially in the organization of membrane rafts in cell membranes and in clinical lung surfactants used to treat neonatal respiratory distress. We are particularly interested in lung surfactant (LS) monolayers, where the role of cholesterol remains unclear. Different clinical surfactants used for treatment of premature infants have widely different cholesterol contents; one of the main difficulties is that even though a small mole fraction of cholesterol (5-10%) has been reported in extracts of mammalian LS, there is still a question whether cholesterol originates from the LS or from cell debris. We show, using a novel, micron-scale surface rheometer, that small fractions of cholesterol dramatically change the elastic and viscous properties of LS monolayers. As little as 2 mol% cholesterol decreases the viscosity of a LS monolayer by two orders of magnitude, while likely enhances the spreading of LS as the alveolar area changes during respiration. 5 mol% cholesterol induces an elastic response in the monolayer, which may help retain LS in the deep lung by resisting the Marangoni flow caused by the surface tension gradient between the deep lung and bronchi. Using a combination of atomic force microscopy and optical fluorescence microscopy, we show the change in monolayer morphology induced by cholesterol provides important clues to the dynamic processes. We conclude that a cholesterol content of about 4 mol% is a critical concentration around which rheological properties can vary dramatically. Subtle changes in composition around this concentration can enable both a low viscosity to improve spreading during respiration, as well as a high elasticity to minimize Marangoni flow out of the lungs.

(326c) Award Submission - Poly(amino ether)-Gold Nanorod Assemblies for shRNA Delivery

James Ramos, Biomedical Engineering, Arizona State University, Tempe, AZ and Kaushal Rege, Chemical Engineering, Biomedical Engineering, Arizona State University, Tempe, AZ

Poly(amino ether)-Gold Nanorod Assemblies for shRNA Delivery

James Ramos, Kaushal Rege

As the field of nanotechnology continues to grow, many novel nanomaterials are showing strong potential for use in the biomedical field. Gold nanorods (GNRs) have emerged as one such nanomaterial with the capacity for use in biosensing, imaging, photothermal therapy and gene delivery as a cancer therapy. Using polymers from a novel poly(amino ether) library synthesized in our laboratory, we have synthesized poly(amino ether) coated GNRs (PAE-GNRs) by depositing polyelectrolyte multilayers on cetyltrimethyl ammonium bromide (CTAB)-based gold nanorods. These PAE-GNRs exhibit an increase in the nanoparticles stability as well as the capacity to electrostatically bind to nucleic acids. We have previously shown the PAE-GNRs ability to deliver plasmid DNA to two different prostate cancer cell lines for enhanced transgene delivery when compared to GNRs deposited with polyethylenimine, a current standard for polymer mediated gene delivery. Here we show the capability for the PAE-GNRs to delivery shRNA for gene knockdown. By loading the PAE-GNRs with shRNA targeted against the luciferase reporter gene and using them to treat cancer cell lines that constitutively express the luciferase, we demonstrate the ability of the PAE-GNRs to deliver the shRNA for gene silencing. This is confirmed by the observed reduction in luciferase expression levels compared to no treatment and delivery of non-silencing nucleic acids. By optimizing the shRNA loading amount, treatment times, and PAE-GNR concentrations we have determined the ideal treatment conditions for maximum gene silencing. These conditions can then be translated for delivery of shRNA targeted against a therapeutically relevant target as a gene therapy to combat disease. Additionally, due to the unique optical properties of the GNRs, we also demonstrate the ability to image the nanoassemblies after delivery *in vitro* using two-photon induced luminescence of the GNRs. This can be translated as a labeling/detection application for PAE-GNR assemblies modified with moieties targeted to a specific cell type. Our results indicate that the engineering of the PE-GNRs leads to a gold nanoassembly that combines high stability with low cytotoxicity and the capacity to mediate shRNA delivery for gene silencing, as well as be used for TPL imaging. These PE-GNRs also have the potential to be used for photothermal ablation, photothermal enhanced delivery, and carrying of other small anionic molecules for delivery making them a powerfully potential single platform theranostic nanomaterial.

(326d) Award Submission: Surface-Dependent Biological Response of Human Monocytes to Iron Oxide Nanoparticles

Yaolin Xu[1], Jennifer Sherwood[2] and Yuping Bao[1], (1)Chemical and Biological Engineering, The University of Alabama, Tuscaloosa, AL, (2)Arizona State University, Metro Phoenix, AZ

Iron oxide nanoparticles (NPs) have been widely explored for targeted drug delivery, localized therapy, and as contrast agents for magnetic resonance imaging (MRI). For all these applications, iron oxide NPs need to be injected into the human body and directly interact with the patient immune system. Therefore, it is of great importance to study the immune response of monocyte cells to iron oxide NPs. Monocytes are essential in the body's primary response to pathogens, which make them an ideal candidate for immune response studies. In this study, the monocytes were treated with varying concentrations of polyacrylic acid (PAA)-, polyethylenimine (PEI)-, and polysorbate 80- coated iron oxide NPs. Concentration and time-dependent cytotoxicity and immune responses were conducted.

The 3-(4,5-Dimethylthiazol-2-yl)-2,5-diphenyltetrazolium bromide (MTT) assay confirmed that our iron oxide NPs can be practically used for biomedical applications due to the observed high cell viability (> 80%) for all the samples. Surprisingly, the cytotoxicity tests showed little dependence on the surface functionality based on the cell death trends from different types of NPs. The specific NP-cell interaction mechanisms will be investigated in the future. The effective internalization of the iron oxide NPs by the monocytes were clearly observed under transmission electron microscopy (TEM). Further, the cellular uptake efficiency of NPs as a function of surface chemistries was quantified using inductively coupled plasma-emission spectrometer (ICP). The immune response was primarily analyzed by monitoring the immune biomarkers: tumor necrosis factor (TNF)-α and toll-like receptors 2 (TLR-2). These studies will provide valuable information on the understanding of the surface-dependent cytotoxicity, cellular uptake, and immune response.

This work is funded by NFS-DMR 0907204 and EEC-1005191.

(326e) Award Submission: Spatially-Patterned Collagen-GAG Scaffolds for Regulating MSC Fate

Steven R. Caliari[1], Daniel W. Weisgerber[2], Douglas O. Kelkhoff[2], Manuel A. Ramirez[3] and Brendan A. Harley[1,4], (1)Chemical & Biomolecular Engineering, University of Illinois at Urbana-Champaign, Urbana, IL, (2)Materials Science and Engineering, University of Illinois at Urbana-Champaign, Urbana, IL, (3)Chemical and Biomolecular Engineering, University of Illinois at Urbana-Champaign, Urbana, IL, (4)Institute for Genomic Biology, University of Illinois at Urbana-Champaign, Urbana, IL

The functional regeneration of multi-tissue structures such as the tendon-bone junction (TBJ) is a central challenge in modern tissue engineering that requires new technologies permitting simultaneous control of microstructural, mechanical, and biochemical properties in a spatially-defined manner. The TBJ is one of the most mechanically severe interfaces in the body, is a common site for injury, and current clinical strategies that rely on mechanical fixation rather than biological re-integration are inadequate. Collagen-glycosaminoglycan (CG) scaffolds are regulatory compliant extracellular matrix (ECM) analogs that have been applied to a variety of regenerative medicine challenges, most recently tendon tissue engineering [1]. Here we present an approach to spatially pattern microstructural (pore size and shape), mechanical, and biochemical (mineral content, biomolecule supplementation) properties within a single CG scaffold to drive spatially-defined multi-lineage mesenchymal stem cell (MSC) specification for TBJ regenerative applications.

TBJ scaffolds were fabricated by combining a directional solidification lyophilization technique [1] with a previously described liquid-phase co-synthesis method [2] to form scaffolds with a distinct zonal structure containing spatially-graded mineral content and pore anisotropy mimicking the native TBJ. Scaffolds supported long-term tenocyte (TC) and MSC viability for up to 8 weeks. Anisotropic 3D scaffolds with larger pores and higher relative density were more mechanically competent and able to maintain aligned contact guidance cues, resulting in long term TC phenotypic stability with increased expression of scleraxis (15-fold) compared to 2D culture. MSC-seeded mineralized CG scaffolds displayed up-regulation of bone markers osteocalcin and bone sialoprotein as well as depressed expression of chondrogenic markers (collagen II, SOX9, aggrecan) compared to non-mineralized scaffolds. Soluble factor supplementation was shown to further influence cell bioactivity in a dose and cell-type dependent manner with BMP-2 supplementation eliciting significant increases in alkaline phosphatase expression. Efficient (~50%) factor immobilization to the scaffolds via carbodiimide chemistry enabled long-term factor efficacy with minimal effects on scaffold mechanical and microstructural properties. These factors remained bioactive throughout extended (> 7 days) culture periods and elicited similar phenotypic changes compared to equivalent soluble doses, providing a pathway to create biomolecularly patterned scaffolds to drive multi-lineage MSC differentiation.

We describe a CG system where scaffold anisotropy, mineralization, and biomolecule supplementation can be tailored in a spatially-defined manner for TBJ engineering. Ongoing work is integrating mechanical stimulation with biomolecule-immobilized CG scaffolds to more efficiently drive multi-lineage MSC differentiation and long-term phenotypic maintenance across the graded scaffold.

References: 1) Caliari SR et al., Biomaterials, 2011; 2) Harley BA et al., J Biomed Res A, 2010

(326f) Award Submission:Detecting Toxicity of Carbon Nanotubes Using Nanoindentation

Chenbo Dong[1], Michael L. Kashon[2], David H. Lowry[2], Linda M. Sargent[2] and Cerasela Zoica Dinu[1], (1)Chemical Engineering, West Virginia University, Morgantown, WV, (2)National Institute for Occupation Safety and Health, Morgantown, WV

Carbon nanotubes are being explored for broadest applications including electrical circuits, hydrogen storage, fiber optics, and conductive plastics. In recent years, their biocompatibility with small molecules, such as proteins and nucleic acids, opened up exciting biomedical applications. Thus, concerns about their potential cytotoxicity on both human and environmental have risen. It is necessary to develop significant and sensitive analysis technologies for carbon nanotubes cytotoxicity investigation. Our research aims at elucidating the mechanisms of interactions of carbon nanotubes with biological systems and evaluating the overall effects on the cell fate. To unravel the molecular mechanisms of such interaction, we need to use the latest technological advances in which both structure and function of the cells upon exposure are being analyzed. Here we employed Atomic Force Microscopy (AFM) to carry on system analysis on cellular topography and mechanical properties of immobilized cells treated with carbon nanotubes with high spatial resolution and increased sensitivity. Our results indicated that cellular exposure to carbon nanotubes lead to significant changes in cell biophysical and biomechanical properties. Moreover, such analysis provided the possibility to develop biomechanical based methods to probe the structure-function relationship in live cells and cells incubated with different nanomaterials.

(326g) Award Submission: Effect of Uptake Pathway On Non-Viral Gene Delivery in Vitro and in Vivo

Victor Wing Tat Shum, Chemical and Biomolecular Engineering, University of Illinois at Urbana-Champaign, Urbana, IL and Daniel W. Pack, Chemical & Biomolecular Engineering, University of Illinois Urbana-Champaign, Urbana, IL

Efficient non-viral gene delivery often involves the conjugation of a cell-specific ligand to the vector, which directs the vector to its intended target. However, little is known in terms of how the process of endocytosis affects the performance of the delivery vehicle. Previously, we have demonstrated that caveolin-mediated endocytosis is important to *in vitro* polyethylenimine (PEI)-mediated gene delivery in HeLa cells through the use of small molecule drugs and small-interfering RNA. The goal of this project is to elucidate the effects of cellular uptake mechanism on non-viral gene delivery *in vitro* and *in vivo* utilizing small hairpin RNA (shRNA) Tet-on system. Specifically, we are investigating the efficacy of PEI-mediated gene delivery upon internalization via clathrin-dependent endocytosis or caveolin-dependent endocytosis in HeLa and MDA-MB-231 cell cultures and xenograft tumors. We attached transferrin and folate ligands to PEI to direct internalization by clathrin- and caveolin-mediated mechanisms, respectively, and transfected HeLa and MDA-MB-231 cells with targeted and untargeted PEI in the presence of small molecule drugs that inhibit clathrin- or caveolin-mediated endocytosis. We have demonstrated that caveolin-dependent endocytosis is important to successful PEI-mediated gene delivery in both HeLa and MDA-MB-231 cells. In order to further understand the impact of endocytic pathways on gene delivery efficiency *in vitro*, we have modified HeLa and MDA-MB-231 cells to express small hairpin RNA (shRNA) that target clathrin, caveolin, or lamin A/C (negative control) protein expression (HeLa-CLTC, HeLa-CAV, HeLa-LAM, 231-CLTC, 231-CAV, and 231-LAM) in the presence of tetracycline. Using the shRNA Tet-on system, we are able to demonstrate ~90% clathrin knockdown and ~60% caveolin knockdown in both modified HeLa and MDA-MB-231 cells. Using this shRNA knockdown mechanism, we have confirmed that caveolin-dependent endocytosis is more important than clathrin-dependent endocytosis in PEI-mediated gene delivery, which corroborates the results observed in unmodified cell lines. We will also report the effect of clathrin-dependent and caveolin-dependent endocytosis on *in vivo* PEI-mediated gene delivery in a xenograft murine tumor model. Such studies of intracellular trafficking and its effect on PEI gene delivery efficacy in cancer cell lines and tumors enable researchers to design more efficient non-viral gene delivery vectors for cancer gene therapy, by synergistically combining the design of cell targeting ligands and the understanding of cellular uptake mechanism for polymeric carriers.

(326h) Award Submission: High Throughput Sequencing of Antibody Heavy and Light Chain Pairings

Brandon DeKosky[1], Ryan P. Deschner[1], Brandon Rawlings[1], Gregory Ippolito[1], C. Grant Willson[2], Andrew D. Ellington[3] and George Georgiou[4], (1)Department of Chemical Engineering, The University of Texas at Austin, Austin, TX, (2)Departments of Chemical Engineering and Chemistry, The University of Texas at Austin, Austin, TX, (3)Department of Biochemistry, University of Texas at Austin, Austin, TX, (4)Departments of Chemical Engineering and Biomedical Engineering, The University of Texas at Austin, Austin, TX

Next generation DNA sequencing (NGS) is transforming the field of immunology by giving researchers the potential to sequence immune repertoires at unprecedented depth. However, the utility of NGS for repertoire analysis is severely curtailed by its inability to resolve the pairing between antibody heavy and light chains and T-cell receptor α and β strands. To directly address this issue, we have developed an emulsion-based method to sequence natively paired antibody heavy and light chains at high-throughput. V_H-V_L pairings have been validated using mixtures of various immortalized B-cell clones with known antibody sequences to confirm accuracy. High-throughput sequencing of the heavy and light chain pairing is a significant step toward greater understanding of the immune repertoire, and potential applications include immune response analysis and antibody discovery.

This work was supported by the University of Texas at Austin Cockrell School of Engineering, the Harrington Foundation, the National Science Foundation, and The Hertz Foundation.

(346a) Measured Structures of Adsorbed Surfactant Shells On Doubly Sorted Nanotubes

Jeffrey A. Fagan[1], Carlos Silvera-Batista[2], Constantine Khripin[3], Ming Zheng[3] and Angela Hight Walker[4], (1)Polymers Division, National Institute of Standards and Technology, Gaithersburg, MD, (2)Polymers Division, National Institute Of Standards And Technology, Gaithersburg, MD, (3)Polymer Division, National Institute of Standards and Technology, Gaithersburg, MD, (4)National Institute Of Standards And Technology, Gaithersburg, MD

One key that determines the behavior of single-wall carbon nanotubes in solution, whether for their processing or optical properties, is the interfacial layer of bound surfactant or macromolecules. This layer helps control these properties by controlling access to the nanotube surface, affecting the local dielectric environment, having desorption/adsorption kinetics that affect colloidal aggregation, and by displaying different functional groups to the environment. In this talk I will present results for the characterization of interfacial layer on SWCNTs that have been doubly or triply sorted for properties including water-filling, their length distribution, and the electronic type / species. These finely separated populations exhibit dramatically narrow sedimentation coefficients reminiscent of biological macromolecules when characterized in an analytical ultracentrifuge. Taking advantage of the high resolution, we explore the changes in the bound layer while varying the surfactant molecule for multiple populations. The implications of the measured dependencies for processing control and improved separations will be discussed.

(346b) Dispersions of Non-Covalently Functionalized Graphene with Minimal Stabilizer

Dorsa Parviz[1], Tanvir Ahmed[2], Sriya Das[1], Fahmida Irin[1] and **Micah Green**[1], (1)Chemical Engineering, Texas Tech University, Lubbock, TX, (2)Mechanical Engineering, Texas Tech University, Lubbock, TX

The dispersion of graphene in solvents often requires excessive quantities of stabilizers (such as surfactants or polymers). We demonstrate that functionalized pyrene derivatives are quite promising for stabilization of single- and few-layer graphene flakes in aqueous dispersions. The graphene/stabilizer ratio obtained by this method is exceptionally high relative to conventional nanomaterial stabilizers. The mechanism of stabilization by pyrene derivatives were investigated by studying parameters such as stabilizer concentration, initial graphite concentration, the type and number of functional groups, counterions, and the electronic interaction between the functional groups and the basal plane. Zeta potential measurements were used to quantify the stability of the dispersions as a function of pH. The dispersions were also used to prepare graphene/epoxy composites. These composites displayed mechanical and electrical properties comparable to polymer-stabilized graphene/epoxy composites while containing significantly less impurities. These dispersions also show promise for applications to organic solar cells, conductive films, and inkjet-printed electronic devices.

(346c) Characterization of Pristine and Functionalized Multi-Walled Carbon Nanotubes in Unsaturated Polyester Resin

Esteban Urena-Benavides, Chemical Engineering, Auburn University, Auburn, Matthew J. Kayatin, Chemical Engineering, Auburn University, Auburn, AL and **Virginia A. Davis**, Department of Chemical Engineering, Auburn University, Auburn, AL

Most carbon nanotube – thermoset nanocomposite studies have focused in epoxy based resins, but the global unsaturated polyester resin market comprises more than 80 % of all thermoset resins. Therefore, it is important to develop an understanding of the dispersion of carbon nanotubes in UPR and characterization of the resulting systems. UPR research has however been hindered due to styrene evaporation and developing functionalization schemes which are compatible with the resin. In this research, dispersions of functionalized and pristine MWNT in UPR were prepared using high shear mixing for three consecutive days. The MWNTs had a length of ~ 5000 nm and a diameter of 8.5 nm. The structure of the suspensions was studied using optical microscopy and rheological analysis. The experimental setup allowed prolonged rheological measurements with minimal solvent evaporation during testing. A mechanical percolation threshold of 0.097 vol % was obtained for unmodified MWNTs, which is very close to the theoretical value of 0.085 vol % calculated for individually dispersed MWNTs. Data previously reported by us for SWNTs in UPR yielded an experimental threshold of 0.098 %; close to the theoretical value of 0.1 vol % for individually dispersed SWNTs with L/D = 500. The fractal structure of the dispersions was described based on plots of shear storage modulus (G') and critical strain for linear viscoelasticity (γ_c) vs. vol. %. A fractal dimension $d_f = 1.28$ was obtained for MWNTs, while for SWNTs $d_f = 1.88$; the higher d_f for SWNTs reflect a denser network consistent with a lower aspect ratio. The rigidity of the flocs was determined by the parameter m, which has values of 0.23 – 0.29 for rigid flocs while for soft flocs $0.4 < m < 0.5$. For the MWNTs, m was 0.34, while for the SWNTs it was 0.31 indicating a higher rigidity for SWNTs than for MWNTs.

Chemical modification of the MWNTs was achieved by simultaneous polymerization and grafting of polystyrene to the sidewall of the nanotubes using a modified Billips reaction. The polymerization was confirmed by Fourier Transform Infrared (FTIR) spectroscopy. Thermogravimetric analysis was used quantify the weight percentage of polystyrene on the surface, and Raman spectroscopy confirmed the covalent attachment of the polymeric chains to the nanotubes. Testing the rheological properties of the modified MWNTs in UPR yielded a lower G' than the unmodified nanotubes, which may be an indication of better compatibility with the resin. The weaker attractive interactions of the modified nanotubes result in a weaker network yielding a lower G'.

The results of this research demonstrate the dispersion of MWNTs in UPR by high shear mixing and the effects of sidewall functionalization on the dispersion state. They provide a foundation for incorporating CNT into systems where solvent evaporation and surface chemistry

(346d) Oxygen Effect On Elastic Modulus of Graphene Oxide

Ken-Hsuan Liao, Chemical Engineering and Materials Science, University of Minnesota in Twin Cities, Minneapolis, MN, Jong Seok Jeong, Chemical Engineering & Materials Science, University of Minnesota, Minneapolis, MN, Greg D. Haugstad, Characterization Facility, University of Minnesota, Minneapolis, MN, K. Andre Mkhoyan, Materials Science and Engineering, University of Minnesota, Minneapolis, MN and **Christopher W. Macosko**, Department of Chemical Engineering and Materials Science, University of Minnesota, Minneapolis, MN

Oxygen Effect on Elastic Modulus of Graphene Oxide

Ken-Hsuan Liao, Jong-Seok Jeong, Greg Haugstad, K. Andre Mkhoyan, Chris Macosko

Chemical Engineering & Materials Sciences, University of Minnesota, Twin Cities

Characterization Facilities of University of Minnesota, Twin Cities

Graphene/polymer nanocomposites are one of the most attractive applications for graphene. Incorporation of graphene has been reported to effectively improve mechanical and electrical properties of polymers. Chemically reduced graphene from graphite oxide has been mainly used since it appears to be the most economic production process. Different chemically reduction processes lead to various oxygen contents of the produced graphene. However, according to reported results, there is still no conclusion how oxygen content effect the elastic modulus of graphene. With this motivation, we have studied the elastic modulus of graphene oxide with various oxygen content. We used *in situ* AFM nano-indentation to measure the influence of oxygen on the elastic modulus of graphene oxide with various carbon/oxygen (C/O) ratios. The results show that chemical reduction (lower oxygen contents) will decrease the elastic modulus of graphene oxide. We speculate that chemical reduction of oxygen atoms of epoxy groups on graphene oxide surface will remove the bridging effect between carbon atoms, which leads to more flexible sheets.

(346e) Ultra-Highly Conductive, Strong, Lightweight Fibers of Carbon Nanotubes by Scalable Processing

Natnael Behabtu[1], Colin C. Young[2], Dmitri Tsentalovich[3], Anson Ma[4] and **Matteo Pasquali**[2], (1)Department of Chemical and Biomolecular Engineering, Rice University, Houston, TX, (2)Chemical and Biomolecular Engineering, Rice University, Houston, TX, (3)Rice University, Houston, TX, (4)Institute of Materials Science, Department of Chemical, Materials, and Biomolecular Engineering, University of Connecticut, Storrs, CT

Since their discovery 20 years ago, carbon nanotubes (CNTs) have held great, yet unfulfilled, promise for lightweight multifunctional materials. A long held tenet is that useful mechanical strength and electrical conductivity can be attained only with very long CNTs (~mm). Such requirement limits severely potential applications, because of poor scalability of growth and processing of long CNTs. Here we report high-performance fibers of short, bulk-grown CNTs (below 10 µm) produced by industrially scalable wet spinning. The spinning process is similar of that of PPTA, i.e., Kevlar and Twaron. The CNTs are dissolved in a strong acid to form a liquid crystalline dope; the liquid crystalline dope is spun into a coagulant bath to form a solid fiber, which is wound on a rotating drum. These fibers combine the specific strength and stiffness of carbon fibers to the specific electrical conductivity of metals. The specific thermal conductivity of these CNT fibers exceeds that of metals and rivals the specific thermal conductivity of the best pitch-based carbon fibers. This combination of properties and processing defines a new area of material design.

(346f) Dispersion of Functionalized Graphene Sheets in Nonpolar Solvents

Jean-Philippe Tessonnier and Mark A. Barteau, Chemical and Biomolecular Engineering, University of Delaware, Newark, DE

The dispersion of reduced graphene oxide (RGO) as single- and few-layer sheets in aqueous and in polar organic solvents is well-established. Successful approaches for example include the use of surfactants or the careful selection of the solvent depending on its cohesive energy density (compatible Hansen Solubility Parameters – HSP values – between the solvent and the sheets). The dispersion of RGO in nonpolar organic solvents such as toluene or alkanes is more challenging and the sheets typically form large flocs within a few minutes. We developed a 2-steps one-pot technique to covalently attach alkyl chains to RGO using the residual oxygen-containing functional groups present of the surface of the sheets as anchoring points. The synthesized alkyl-chain-grafted RGO readily dispersed in a variety of polar and nonpolar organic solvents. By studying the stability of the suspensions, we found that the grafted alkyl groups increased the range of compatible HSP values, meaning that the synthesized material disperse in a broader variety of organic solvents, including methanol, ethanol, toluene and heptane. ATR-FTIR, UV-Vis and Raman spectroscopy showed that the grafting reaction did not significantly alter the structure and defect density of the sheets. Furthermore, TG-MS data indicated that most groups were stable below 300 °C. This technique is particularly interesting to design nanostructured heterogeneous catalysts for liquid phase reactions. In addition, we believe it could be particularly useful for the further processing of defect-engineered graphene sheets, to achieve particular electronic or optical properties.

(346g) High Resolution Length Fractionation of Surfactant Dispersed Single-Wall Carbon Nanotubes

Constantine Khripin[1], Xiaomin Tu[2], Jeffrey A. Fagan[3], John Howarter[4] and Ming Zheng[1], (1)Polymer Division, National Institute of Standards and Technology, Gaithersburg, MD, (2)Polymers Division, NIST, Gaithersburg, MD, (3)Polymers Division, National Institute of Standards and Technology, Gaithersburg, MD, (4)Purdue University

Length fractionation of colloidal carbon nanotube dispersions is required for many studies. Size-exclusion chromatography (SEC) has been reported as a reliable method for high resolution length fractionation of DNA-dispersed single-wall carbon nanotubes (SWCNTs). However, no high resolution length fractionation method is currently available for the commonly used bile salt dispersed SWCNTs. Here we describe extension of the SEC method to high resolution length fractionation of bile salt dispersed SWCNTs using porous silica-based beads as the stationary phase, and bile salt solution as the mobile phase. The optical absorption spectra of the resulting fractions vary with fraction length, with diminished excitonic transition peaks for short tubes. We measure the length dependence of the extinction coefficient to enable quantification of the concentration of length-sorted fractions. We present evidence that the spectral variation is due to a large concentration of defective SWCNTs in the short fractions. Finally, the new SEC method is used to length-fractionate bile salt dispersed SWCNTs pre-purified by density gradient centrifugation methods. Unlike length-fractionated samples obtained from the raw material, and all length-fractionated samples previously reported, these double-purified samples show little dependence of optical absorbance spectra on length. These results support the notion that the optical absorbance spectrum of the *defect-free* SWCNT is approximately length-independent.

(346h) Reduction of Freely-Suspended Crumpled Graphene Oxide Nanopaper

Xiaofei Ma[1], Michael Zachariah[1] and Christopher Zangmeister[2], (1)Department of Mechanical Engineering and Chemistry, University of Maryland, College Park, MD, (2)National Institute of Standards and Technology, Gaithersburg, MD

The reduction of freely suspended graphene oxide (GO) paper was investigated in a flow reactor configuration as a method for fast reduction of GO to graphene-like materials. Aerosolization of GO aqueous solution followed by rapid solvent evaporation produces stand-alone crumpled nanopaper-like sheets of about 150nm in size. Reduction of GO was conducted in a flow reactor with residence time of ~ 5 seconds in both air and argon environments. Reduced GO nanopaper was evaluated using X-ray photoemission spectroscopies (XPS) for chemical analysis, photoacoustic spectrometer for light absorption measurement, and cavity ring down method for optical extinction measurement. Online size selection and aerosol mass analysis was used to measure the mass loss of individual crumpled GO paper during reduction. The reduction kinetics from mass measurement was correlated with optical property characterizations. TEM images of reduced GO thin film indicated that the pre-crumpling of GO sheet contributes to the reduction in activation energy.

(393a) Enantioselective Properties of Chiral Surface Modified Gold Nanoparticles and Their Variation with Temperature, Wavelength, and Particle Size

Nathan Khosla, Carnegie Mellon University, Pittsburgh, PA and Patrick Downey, Colorado School of Mines, Golden, CO

Surfaces of chemically synthesized gold nanoparticles were modified with both L and D enantiomers of glutathione and cysteine to make them chiral. The chirality of these nanoparticles was examined by using various chiral probe molecules such as 2-Butanol, glucose, tartaric acid and alanine. We have used simple polarimetery to test the chirality at 25 degrees celsius and 436 nm wavelength. Our work shows that L-cysteine is better ligand for adsorbtion of the probe molecules than L-glutathione. L-cysteine coated Au nanoparticles show larger enantioselective separation in comparison to L-glutathione coated Au nanoparticles. In addition, glucose has highest enantioselective separation in comparison to 2-butanol, tartaric acid, and alanine. Thus we see that this method for separating chiral molecules on the surface of chirally modified gold nanopartices works with different efficacy for a range of proble molecules. Furthermore, these effects are altered depending on which ligand was used to modfy the gold nanoparticle surface.

(393aa) Development of an Integrated Computational and Experimental Framework for Understanding and Controlling Nanoparticle Interactions

Chris S. Ewing[1], J. Karl Johnson[2], Joseph J. McCarthy[1] and Götz Veser[3], (1)Chemical and Petroleum Engineering, University of Pittsburgh, Pittsburgh, PA, (2)Dept. of Chemical and Petroleum Engineering, University of Pittsburgh, Pittsburgh, PA, (3)Department of Chemical and Petroleum Engineering, University of Pittsburgh, Pittsburgh, PA

Nanomaterials hold vast potential for novel, more efficient, and "greener" processes and products due to unique properties that are fundamentally different from their bulk counterparts. These properties, in combination with their extremely small size, make nanoparticles of great interest for use as building blocks in the assembly of complex functional materials to be used in a multitude of applications from electronic devices to new construction materials with truly unique properties. Nanoparticles are also already used in a variety of industrial products and processes, from which they will be released into the environment where their "nanoproperties" may have negative impacts on human health and the environment. Thus, a better understanding of the fundamental interaction forces between nanoparticles would not only provide significant aid in the development and manufacture of complex functional nanomaterials, but may more importantly provide a means for predicting or even controlling the toxicity of nanomaterials. Controlling the interactions between nanoparticles could have vast implications concerning human welfare, as well as environmental health and safety, in view of the ever-increasing use of nanomaterials.

While the interaction forces governing both the macroscopic and molecular size regimes are well understood, our understanding of the interactions that govern the nanoscale regime (i.e., 1 - 100 nm) is only poorly developed to date. Continuum theories, which allow description of the macroscopic regime, break down as particle sizes approach the nanoscale regime. At the same time, atomistic simulations, which can describe interactions on an atomic scale, become impractical due to the large number of atoms in systems composed of collections of nanoparticles.

Using silica as a base model, we are developing a bottom-up framework based on material specific properties, for accurately modeling both static and dynamic nanoparticle systems. First an interparticle potential is constructed by contributions from van der Waals and electrostatic forces, both of which are implemented via a form of coarse-grained molecular dynamics. Van der Waals forces are accounted for by packing small silica clusters, represented as single points, into the particle geometry. Electrostatics are modeled by calculating the sum of electrostatic interactions between silica nanoparticles for which atomic charges were determined with a large-scale DFT implementation and the resulting electron density gradient fitted to a function dependent on particle size. Validation of the model is completed with the colloidal probe technique in atomic force microscopy (AFM). An accurate interparticle potential can then be implemented in dissipative particle dynamics (DPD) to simulate large nanoparticle systems.

The proposed model, and the insight it would provide, may allow for predicting or even controlling the toxicity of current nanomaterials and guide the development of novel functional materials constructed via self-assembly.

(393ab) Facilitating Seed Germination by a Novel Method Using Nanoparticle Electrospray

Gang Wu[1], Li Huang[2], Jennifer Head[2], Da-Ren Chen[3] and Yinjie J. Tang[3], (1)Enery, Environmental and Chemical Engineering Department, Washington University in St. Louis, Saint Louis, MO, (2)Washington University in St. Louis, St. Louis, MO, (3)Department of Energy, Environmental and Chemical Engineering, Washington University, St Louis, MO

Attributed to the rapid development of nanotechnology, its applications in many fields such as materials engineering and biomedicine have been extensively studied during past years. However, less research concerning about nanoparticle applications in agriculture has been reported. Here, we first time demonstrated that plant seed germination can be promoted by a novel nanoparticle electrospray technique. Germination rate of some plant seeds (e.g., lettuce seed) could be significantly improved by TiO_2 nanoparticle electrospray. TEM and SEM-EDX images of seed cross section indicated the presence of mono-dispersed nanoparticle inside seed after electrospray treatment. Therefore, mechanisms contributed to seed germination enhancement have been speculated as that electronically dispersed nanoparticles can penetrate through seed coat and create "nano-holes" on seed coats, resulting in improved germination conditions. To verify its practicability, the shelf life test was also performed and significant enhancement effect was still observed on sprayed seeds even one month after the treatment, implying a promising application of this method in agriculture.

(393ac) Electronic and Mechanical Properties of Superlattices of Crystalline Domains Embedded in Twisted Bilayer Graphene

Andre R. Muniz, Department of Chemical Engineering, Federal University of Rio Grande do Sul, Porto Alegre, Brazil and Dimitrios Maroudas, Department of Chemical Engineering, University of Massachusetts Amherst, Amherst, MA

We report results of a systematic computational study on mechanical and electronic properties of carbon nanostructures generated upon covalent interlayer C-C bonding in twisted bilayer graphene (TBG); such interlayer bonding is the result of chemical functionalization, namely, hydrogenation of the graphene layers in TBG by controlling the hydrogen coverage and hydrogenation pattern. The structural features of these configurations are determined by the twist angle, i.e., the angle of rotation of the graphene planes with respect to each other. Interlayer covalent bonding generates superlattices of diamond-like nanocrystalline domains and of caged fullerene-like configurations embedded within the graphene layers for small twist angles (near 0 degrees) and for large twist angles (near 30 degrees), respectively. The analysis is based on a combination of first-principles density functional theory (DFT) calculations, density-functional tight-binding (DFTB+) computations, and classical molecular-dynamics simulations based on a reliable reactive bond-order potential (AIREBO).

We have calculated the electronic band structure of these superlattice structures and we show that it depends strongly on the hydrogenation pattern and density of interlayer bonds; the predicted band gaps span a broad spectrum from ~0 eV, i.e., in the band-gap limit of single-layer graphene, to ~3 eV, i.e., in the band-gap limit of bulk diamond phases. We have also carried out MD simulations of the dynamic deformation of the superlattice structures at constant strain rate and temperature. We have obtained stress-strain curves and determined the Young modulus, tensile strength, and fracture strain of all of the generated nanostructures. These physical properties are determined over the entire range of twist angles and density of interlayer C-C bonds; this allows for a fundamental understanding of how these structural parameters affect material properties and for the development of criteria for choosing such parameters optimally to fine tune electrical and mechanical response, aiming at specific technological applications.

(393ad) Nest-Like Carbon/TiO2-B Nanocomposites As Anode Materials for Lithium Ion Batteries

Cunyu Zhao and Ying Li, University of Wisconsin-Milwaukee, Milwaukee, WI

TiO_2 has been investigated as lithium ion battery (LIB) anode materials due to its advantages of cost, safety and rate capacity. In particular, TiO_2-B(bronze) has shown a favorable channel structure for lithium mobility, which results in fast charge-discharge capacity of lithium cell [1]. However, poor electronic conduction of TiO_2 is one of the major drawbacks that hinders the applications of this material in LIB. In this study, a novel nest-like composite material of carbon coated TiO_2(B) nanowires was fabricated via a combined hydrothermal and ultrasonic spray pyrolysis processes. The nest structure of nanowire aggregates resulted in a high surface area, a good packing density, and a conductive network. The electrochemical performance of the C/TiO_2-B nanocomposites was tested in a lithium half cell, which demonstrated higher rate capacity and cycle life compared with randomly oriented TiO_2-B nanowires.

References:

[1] Hansan Liu, Zhonghe Bi, Xiao-Guang Sun, Raymond R. Unocic, M. Parans Paranthaman, Sheng Dai, and Gilbert M. Brown. Adv. Mater. **2011**, 23, 3450–3454

EFFECT OF SURFACTANT ADSORPTION ON MULTI-WALLED CARBON NANOTUBE DISPERSION

Wipawan Mattavakul [a] *and Boonyarach Kitiyanan* [a,b]
[a] *The Petroleum and Petrochemical College, Chulalongkorn University, Bangkok, Thailand*
[b] *Center of Excellence on Petrochemical and Materials Technology, Bangkok, Thailand*

Abstract

Carbon nanotubes usually agglomerate due to strong van der Waals interactions. One of the approaches to disperse them is non-covalent modification by using surfactant adsorption. The purpose of this work was to relate the adsorption behaviour of surfactants on multi-walled carbon nanotubes (MWCNTs) and MWCNTs dispersion. In this study, sodium dodecyl benzenesulfonate (SDBS), cetylpyridinium chloride (CPC) and polyoxyethylene octyl phenyl ether ($OP(EO)_{10}$) were used as anionic, cationic and non-ionic surfactants, respectively. The effect of surfactant concentration and the molecular structure of surfactants on the MWCNTs in solution were investigated. The adsorbed amount of surfactants on MWCNTs showed an order of SDBS > $OP(EO)_{10}$ > CPC. Moreover, the dispersion of MWCNTs at different concentrations was strongly affected by the charge of surfactant and alkyl chain length.

Introduction

The discovery of carbon nanotubes (CNTs) has inspired many researchers to innovate advanced materials for new applications since CNTs exhibit outstanding mechanical, electrical, and thermal characteristics. However, CNTs usually agglomerate due to strong van der Waals force, surface area, and hydrophobic interactions; these entanglements are able to cause many defects in composite materials. On the other hand, CNTs suspension is a significant factor to improve the properties of the original medium.

To disperse CNTs, a number of methods are suggested. Some are based on covalent modification of the tubes, which inevitably leads to undesired deterioration of the CNT properties. Alternatively, non-covalent modification through amphiphilic self-assembly can be used to stabilize the dispersion of CNTs in aqueous solution, by the electrostatic repulsion or steric hindrance of the micelles formed around them, without changing their chemical and electrical property. While, surfactants are employed in CNT dispersions, surfactant molecules work by adsorption at the sold-liquid interface and self-accumulation into supramolecular structures. Therefore, the CNTs structures are maintained after the modification.

The purposes of this work are to study the adsorption of surfactants on MWCNTs and to investigate multi-walled carbon nanotubes (MWCNTs) dispersion by using surfactants solution. Sodium dodecyl benzenesulfonate (SDBS), cetylpyridinium chloride (CPC) and polyoxyethylene octyl phenyl ether ($OP(EO)_{10}$) are used as anionic, cationic and non-ionic surfactants, respectively. The impact of surfactant concentration and the molecular structure of surfactants on the MWCNTs in solution are also investigated.

Experimental

A. Adsorption Experiment

The adsorptions of surfactants from aqueous solution onto the carbons were carried out at 25 ^{o}C. The surfactant solutions were prepared and diluted by distilled water. Surfactant solutions (20 mL) with different concentrations (0-10,000 μmol/L) were added to 25 mL glass vials with Teflon screw cap containing 2.5 mg of carbons. The vials were shaken in shaker bath for 5 days. After the adsorption reached the equilibrium, the supernatant was separated from the adsorbents by using 0.45 μm Nylon syringe filters. These solutions (CPC, SDBS and $OP(EO)_{10}$) were analysed by using the photometric mode of Ultraviolet -Visible (UV-Vis) spectrophotometer (Shimadzu Lamda 10) at wavelength of 259, 224 and 275 nm, respectively. The supernatant surfactant concentration was determined by fitting the absorbance to the calibration curve. The adsorbed amounts of surfactant on the adsorbent (Γ, μmol/m^2) were calculated on the difference between the initial concentration and the equilibrium concentration and the following equation was applied:

$$\Gamma = \frac{(C_o - C_e) * V}{W * A_s} \qquad (1)$$

Where C_o and C_e are the initial and equilibrium concentration of surfactant solution (μmol/L), respectively, V is the batch liquid volume (mL) and M is the adsorbent mass (mg). All experiments were duplicated.

B. Dispersion Experiment

Dispersion experiments were performed by using the aqueous solutions of various surfactants in distilled water at concentrations: 0.1 CMC, CMC and 10 CMC. Surfactant solutions (20 mL) were added to 25 mL glass vials with Teflon screw cap containing 0.1 mg of MWCNTs. Each vial was sonicated for an hour. Then, the equilibrium concentrations of dispersed MWCNTs were measured by using the kinetic mode of UV-Vis spectrophotometer (Shimadzu Lamda 10) at wavelength 650 nm, which was used to examine the surfactant-adsorbed MWCNTs. Essentially, the spectra collected between 400 and 800 nm do not contain signals by surfactants molecules, besides, the same concentration of surfactant aqueous solution was used as reference to eliminate the absorbance of surfactant in suspensions.

C. Determination of Specific Surface Area

The specific surface areas of the adsorbents were determined by nitrogen adsorption BET measurement by a surface area analyser (Quanta Chrome, Autosorb-1).

Results and Discussion

A. Adsorption Isotherm of CPC, SDBS and $OP(EO)_{10}$ on MWCNTs

The pH_{pzc} values defined for MWCNTs adsorbents was 7.3 (Carabinero *et al.,* 2011). For pH values lower than pH_{pzc}, the adsorbent presents a positive surface charge. On the other hand, the adsorbent acts as a negative surface when the pH value is higher than pH_{pzc}. The dissolved SDBS is negatively charged in water solution; because it possesses SO_3^-. The adsorption of SDBS preformed when the adsorbent presented a positive surface charge. For MWCNTs, the electrostatic interaction occurs for pH < 7.3. Thus, this behaviour explained the high adsorption capacity for MWCNTs for SDBS solution. Whereas, the $OP(EO)_{10}$ and CPC solutions were neutrally and positively charged, respectively. The adsorption still occurred on a positive surface charge, since the pH values of CPC and $OP(EO)_{10}$ solution were 5.4 and 6.1, respectively. Therefore, the adsorption of CPC and $OP(EO)_{10}$ were lower compared to SDBS. Among the remaining two, the pH value of CPC was lower than of $OP(EO)_{10}$. The lower the pH values from the pH_{pzc}, the more positive the surface of the adsorbent. From the high positive surface charge of MWCNTs in CPC solution, the repulsive interaction between CPC surfactant and the surface of MWCNTs effected to the adsorption decreased. Thus, the order of surfactant adsorption on MWCNTs was SDBS > $OP(EO)_{10}$ > CPC.

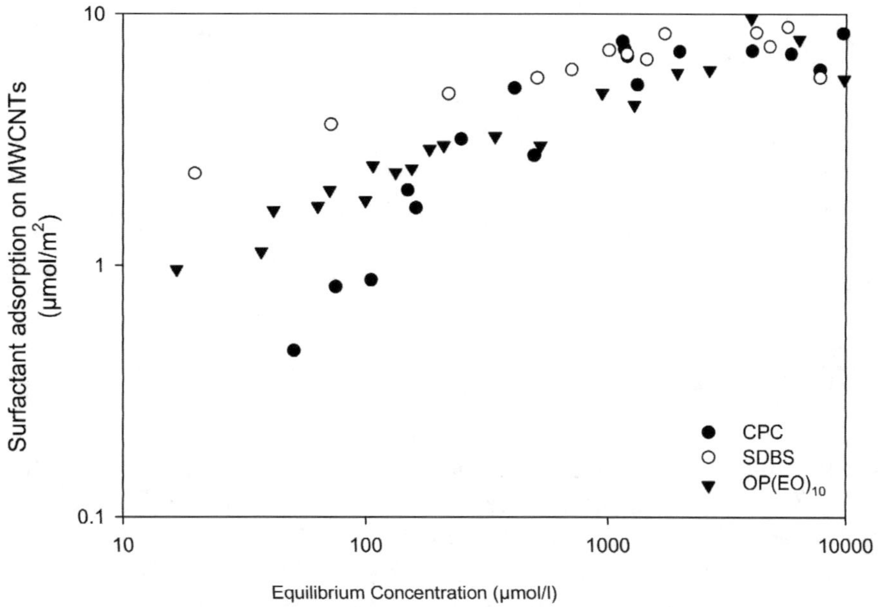

Figure 1 Adsorption isotherms of CPC (•), SDBS (○) and $OP(EO)_{10}$ (▼) on multi-walled carbon nanotubes 25 °C.

B. Dispersability of MWCNTs in Aqueous Surfactant Solutions

The dispersion stability of MWCNTs in the presence of the different surfactant concentrations are illustrated by the UV-Vis measurement of the suspension as a function of time as exhibited in Figure 2 to 4. The results showed that the dispersability of MWCNTs without surfactant solution was not different from the system with the 0.1 CMC of surfactant. It should be considered that all surfactants were not able to stabilize the MWCNTs dispersion at 0.1 CMC as shown in Figure 2. This is ascribed to the insufficient surfactant molecules to adsorb and disperse the agglomerate MWCNTs. Therefore, the higher surfactant equilibrium concentrations were required to reach the maximum dispersion of MWCNTs. The UV-Vis absorbances of the dispersions obviously enlarged at CMC as a result of the increased micelle formation or surfactant aggregations to disperse the MWCNTs.

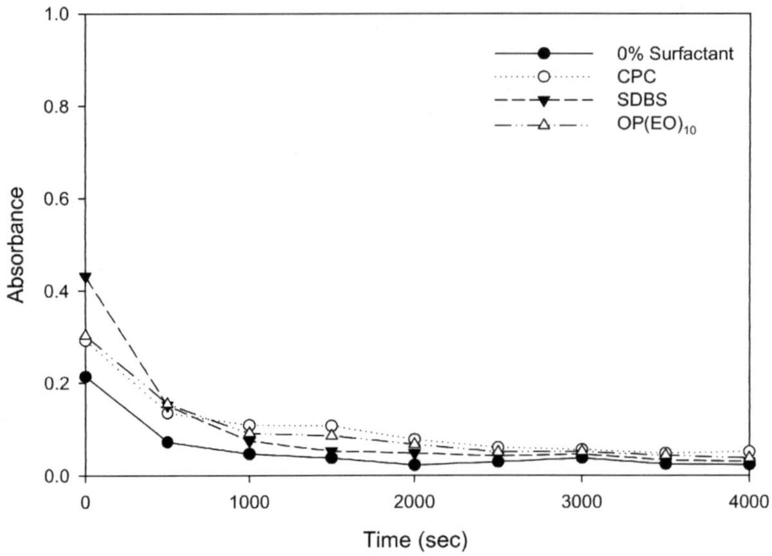

Figure 2 Colloidal stability of the dispersed MWCNTs with surfactant at 0.1 CMC and without surfactant at 25 °C.

At CMC, the anionic surfactant SDBS was favorably adsorbed by the MWCNTs because of the electrostatic attraction. In addition, the head group of SDBS is smaller than that of $OP(EO)_{10}$; its small size may increase its packing density compared to that of $OP(EO)_{10}$ causing the SDBS disperses better than $OP(EO)_{10}$. In additions, the charge stabilization of MWCNTs, which were formed from the electrostatic repulsion of SO_3^- via Coulomb interactions, may be more important than the steric repulsion of $OP(EO)_{10}$ headgroup. Thereby, the greater dispersion ability of SDBS compared to $OP(EO)_{10}$ and CPC, respectively, which can be explained in terms of charge, headgroup size and graphite-surfactant interactions.

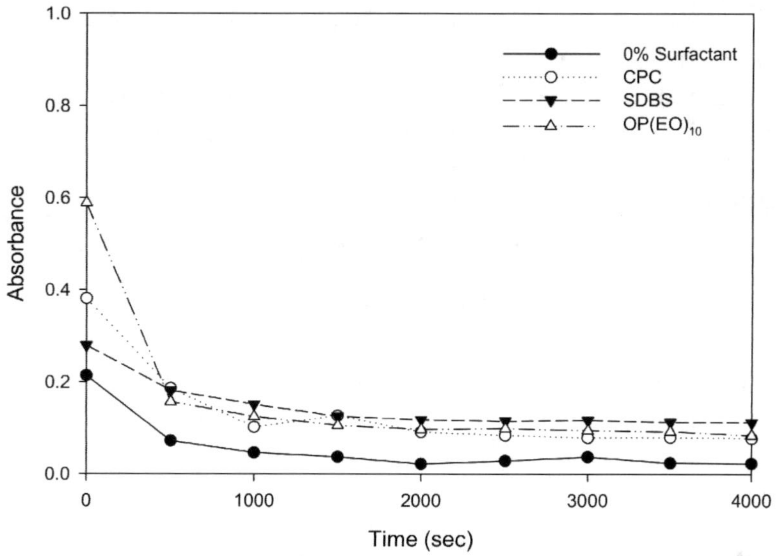

Figure 3 Colloidal stability of the dispersed MWCNTs with surfactant at CMC and without surfactant at 25 °C.

At very high surfactant concentration (10 CMC), the MWCNTs surface was presumable fully-coverage by surfactant, in addition, the micelles also formed and dispersed themselves in water. The amount of micellation affected to the dispersion stability of MWCNTs as demonstrated in Figure 4. From Figure 4, the order of dispersing ability of MWCNTs was $OP(EO)_{10}$ > SDBS > CPC. It was ascribed to the steric hindrance of the headgroup of $OP(EO)_{10}$ which can assist the MWCNTs suspension.

However, the surfaces of MWCNTs were positively charged in three surfactants. For cationic surfactant, the electrostatic repulsion between the same charges of CPC surfactant and MWCNTs surface affected to the low adsorption. Consequently, the CPC surfactant adsorption was insufficient to disperse the agglomerates of MWCNTs. This caused the low MWCNTs suspension for all CPC surfactant concentrations.

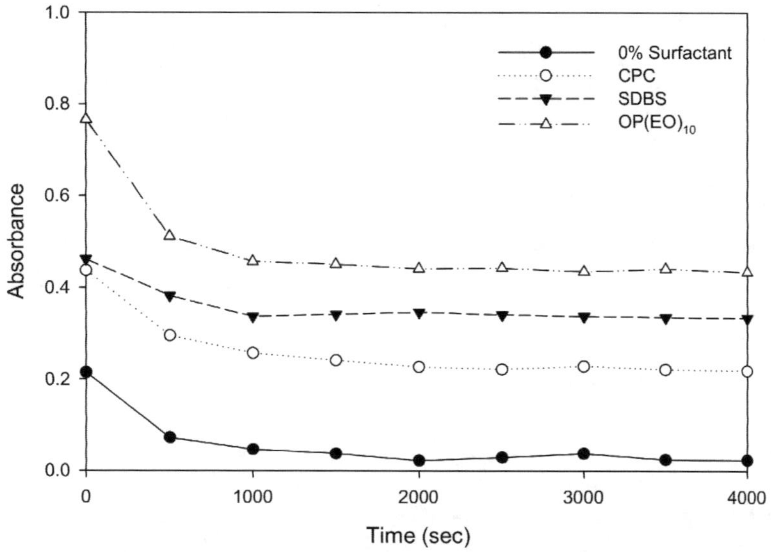

Figure 4 Colloidal stability of the dispersed MWCNTs with surfactant at 10 CMC and without surfactant at 25 °C.

Conclusions

The adsorbed amount of surfactants on MWCNTs showed an order of SDBS > $OP(EO)_{10}$ > CPC as results from the different head group, tail length and charge. The order of dispersing ability at 0.1 CMC, CMC and 10 CMC were the relative as same, SDBS > $OP(EO)_{10}$ > CPC and $OP(EO)_{10}$ > SDBS > CPC, respectively. It was arise from the charge, headgroup size and graphite-surfactant interactions.

Acknowledgements

The authors gratefully acknowledge the Petroleum and Petrochemical College, Center of Excellence for Petroleum, Petrochemicals, and Advanced Materials, Chulalongkorn University.

References

1. Bai, Y., Lin, D., Wu, F., Wang, Z. and Xing, B. (2010) Chemosphere. 79, 362-367.
2. Bai, Y., Park, I.S., Lee, A.J., Bae, T.S., Watari, F., Uo, M. and Lee, M.H. (2011) Carbon. 49, 3663-3671.
3. Bystrzejewski, M., Huczko, A., Lange, H., Gemming, T., Büchner, B., and Rümmeli, M.H. (2010) J. Colloid Interface Sci. 345, 138-142.
4. Carabineiro, S. AC., Pereira, M FR., Pereira, J. N., Caparros, C., Sencadas, V. and Lanceros-Mendex, S. (2011) Nanoscale Res. Lett. 6, 302.
5. Gupta, S.D. and Bhagwat, S.S. (2005) J. Dispersion Sci. Technol. 26, 111-120.
6. Islam, M.F., Rojas, E., Bergey, D.M., Johnson, A.T. and Yodth, A.G. (2003) Nano Lett. 3 (2), 269-273.
7. Rastogi, R., Kaushal, R., Tripathi, S.K., Sharma, A.L., Kaur, I. and Bharadwaj, L.M. (2008) J. Colloid Interface Sci. 328, 421-428.
8. Rosen, M.J. (2004). *Surfactants and Interfacial Phenomena*, 3rd Ed.; John Wiley and Sons, New York.
9. Paria, S., and Khilar, K.C. (2004) Adv. Colloid Interface Sci. 110 (3), 75-95.
10. Sabba, Y. and Thomas, E.L. (2004) Macromolecules. 37 (13), 4815-4820.
11. Soria-Sánchez, M., Maroto-Valiente, A., Guerrero-Ruiz, A. and Necskaia, D.M. (2010) J. Colloid Interface Sci. 343, 194-199.
12. Yang, K., Jing, Q., Wu, W., Zhu, L. and Xing, B. (2010) Environ. Sci. Technol. 44, 681-687.

Influence of Surfactants on Morphology Diversity of Partially Hydrolyzed Polyacrylamide Self-assembling in Aqueous Solution

Hanting Chen, Yun Fang*, Pingping Pang

School of Chemical and material engineering, Jiangnan University, Wuxi, China

Abstract Partially hydrolyzed polyacrylamide (HPAM) as a mimic of random copolymers assembled in pH response into colloidal aggregates in pure aqueous solution due to hydrogen bonding complexation. At pH 1.3, lots of giant semihexagonal sheets (450 nm × 250 nm × 50 nm) were observed by transmission electron microscope (TEM) images. The influence of cetyl trimethyl ammonium bromide (CTAB) and sodium dodecyl sulfate (SDS) on the self-assembling morphologies of HPAM was investigated, and the morphology diversity of HPAM colloidal aggregates caused by surfactants was observed. Therefore, semihexagonal sheets were transformed into rods which responsed to CTAB addition, while coagulated spherical aggregates (D=100 nm) were obtained which responsed to SDS addition. Finally the effect of surfactants on morphology transform of HPAM colloidal aggregates was discussed.

Introduction

Morphology diversity of polymer self-assembling is currently the subject of intense research interest, because of the applications in nanoreactor[1], drug delivery[2,3] and environmental protection[4] have different requirements of morphology of the aggregates. It has reported that the morphology of polymer aggregates is responsive towards stimuli such as ionic strength, temperature and pH.[5-9] Furthermore, surfactant is composed of hydrophilic group and hydrophobic group as well as amphiphilic copolymer. In the self-assembly process, the morphology of polymer aggregates can change, caused by the interactions between the hydrophilic group and the hydrophilic part of amphiphilic copolymer, and the hydrophobic group inserted into hydrophobic section of amphiphilic copolymer.[10-13]

Recently, Nambam and co-workers have investigated the interaction between PEO-PPO-PEO triblock copolymer with anionic sodium dodecyl sulfate (SDS). The addition of SDS, a "soft solid-like" microstructure was formed by aggregating self assembled triblock polymers. And strong electrostatic barrier imparted by the head group of SDS at the core-corona interface that inhibits the formation hexagonally packed layers of micelles and the packing order.[14] Herein, we report a morphology transition from sheets to rods or spherical aggregates for random copolymer partially hydrolyzed polyacrylamide (HPAM) colloidal aggregates responsed to CTAB or SDS addition. The morphology of polymer aggregates would been more diversified in the presence of surfactant.

Experiments

HCl and CTAB were purchased from Sinopharm Chemical Reagent Co. Ltd. SDS were purchased from Acros Organics Co. USA. HPAM were synthesized by ourselves. Ultra pure water was used through out to prepare HPAM, CTAB, SDS and HCl aqueous solutions. In a typical preparation of micelles, aqueous solutions of 5 mg·mL^{-1} HPAM (2 mL) and 0.1 mmol·mL^{-1} CTAB (25 mmol·mL^{-1} SDS) (2 mL) were added into the reaction vessel. A certain amount of ultra pure water [1]was added

* Corresponding author: yunfang@126.com. Supported by the National Natural Science Foundation of China (21276113).

in the solution. Under magnetic stirring, 0.1 mol·L^{-1} HCl (40 μL) were dropped per minute to adjust the solution to pH 1.3. Finally, the total amount of the solution is 10 mL. The final concentration of HPAM were 1 mg·mL^{-1}, while final concentration of CTAB and SDS were 0.02 mmol·mL^{-1} and 5 mmol·mL^{-1} respectively.

The morphology of the products were characterized using a transmission electron microscope (TEM, JEOL-JEM2100) with an accelerating voltage of 200 kV.

Results and Discussion

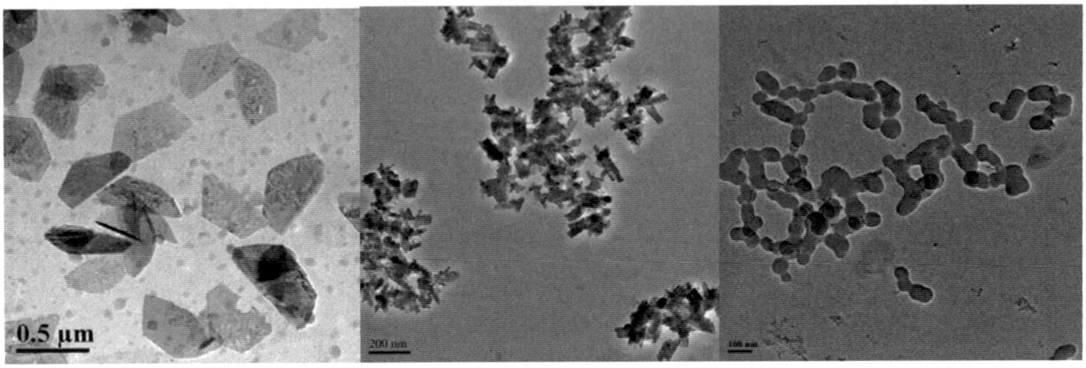

Fig.1 TEM images of HPAM aggregates without surfactants(A) and with CTAB(B) or SDS(C) at pH=1.3

Figure 1(A) showed that HPAM self-assembled into 450 nm×250 nm×50 nm of giant semihexagonal sheets at pH=1.3. When CTAB was added, rods (D=100 nm) were obtained (figure 1(B)). While SDS was added, coagulated spherical aggregates (D=100 nm) were obtained (figure 1(C)).

In the self-assembly process, the concentration of CTAB decreased and was below the critical micelle concentration. In this system, two or more CTAB aggregated driven by van der Waals force. AA segments of HPAM formed hydrophobic core due to hydrogen bonding complexation, then hydrophobic alkyl chains of CTAB inserted into the core to change the stretch of HPAM. Electrostatic barrier imparted by the head group of CTAB and protonized AM segments inhibited the formation of giant semihexagonal sheets.

As anionic surfactant SDS added, the aggregates were formed due to electrostatic interaction between the head group of SDS and protonized AM segments and hydrogen bonding complexation among AA segments. Moreover, hydrophobic alkyl chains of SDS inserted into the core changing the stretch of HAPM, so the morphology of the aggregates transformed into coagulated spherical aggregates.

Conclusions

Partially hydrolyzed polyacrylamide (HPAM) as a mimic of random copolymers assembled in pH response into lots of giant semihexagonal sheets in pure aqueous solution due to hydrogen bonding complexation. The morphology of HPAM colloidal aggregates transformed from sheets to rods or coagulated spherical aggregates responsed to CTAB or SDS addition respectively. The morphology of polymer aggregates would been more diversified in the presence of surfactant, which can be readily tailored to suit many desirable applications.

References

1. Ahn H, Park M J. Facile One-Pot Synthesis of Functional Gold Nanoparticle-Polymer Hybrids Using Ionic Block Copolymers as a Nanoreactor[J]. Macromol. Rapid Commun., 2011, 32(22): 1790-1797.
2. Lee A L Z, Venkataraman S, Sirat S B M, et al. The use of cholesterol-containing biodegradable block copolymers to exploit hydrophobic interactions for the delivery of anticancer drugs[J]. Biomaterials, 2012, 33(6):1921-1928.
3. Yang Y Q, Guo X D, Lin W J, et al. Amphiphilic copolymer brush with random pH-sensitive/hydrophobicstructure: synthesis and self-assembled micelles for sustained drug delivery[J]. Soft Matter, 2012, 8(2): 454-464.
4. Henselwood F, Liu G J. Water-Soluble Nanospheres of Poly(2-cinnamoylethyl methacrylate)-block-poly(acrylic acid)[J]. Macromolecules, 1997, 30(3): 488-493.
5. Zhang L F, Eisenberg A. Crew-Cut Aggregates from Self-Assembly of Blends of Polystyrene-b-Poly(acrylic acid) Block Copolymers and Homopolystyrene in Solution[J]. J.Polym. Sci., Part B: Polym. Phys., 1999, 37(13):1469-1484.
6. Zhang L F, Eisenberg A. Formation of Crew-cut Aggregates of Various Morphologies from Amphiphilic Block Copolymers in Solution[J]. Polym. Adv. Technol., 1998, 9(10-11): 677-699.
7. Bhargava P, Tu Y F, Zheng J X., et al. Temperature-Induced Reversible Morphological Changes of Polystyrene-block-Poly(ethylene Oxide) Micelles in Solution[J]. J. Am. Chem. Soc., 2007, 129(5): 1113-1121.
8. Khouri S, Tam K C. Complexation between r-Cyclodextrin and PEGylated-PAMAM Dendrimers at Low and High pH Values [J]. Langmuir, 2010, 26(23): 17969–17974.
9. Hua D B, Jiang J J, Kuang L J, et al. Smart Chitosan-Based Stimuli-Responsive Nanocarriers for the Controlled Delivery of Hydrophobic Pharmaceuticals [J]. Macromolecules, 2011, 44: 1298–1302.
10. Burke S E, Eisenberg A. Effect of sodium dodecly sulfate on the morphology of polystyrene-b-poly(acrylic acid) aggregates in dioxane-water mixture[J]. Langmuir, 2001, 17(26): 8341-8347.
11. Kabanov A V, Bronich T K, Yu K, et al. Spontaneous formation of vesicles from complexes of block ionomers and surfactants[J]. J. Am. Chem. Soc., 1998, 120(38): 9941-9942.
12. Sergey K, Filippov, Larisa S, et al. pH Sensitive Polymer Nanoparticles: Effect of Hydrophobicity on Self-Assembly[J]. Langmuir. 2010, 26(18): 14450-14457.
13. Pispas S. Self-assembled nanostructures in mixed anionic-neutral double hydrophilic block copolymer/cationic vesicle-forming surfactant solutions[J]. Soft Matter, 2011, 7(2): 474-482.
14. Nambam J S, Philip J. Effects of Interaction of Ionic and Non-Ionic Surfactants on Self Assembly of PEO-PPO-PEO Triblock Copolymer in Aqueous Solution[J]. J. Phys. Chem. B, 2012, 116(5): 1499-1507.

(393ag) Aerosol Manufacturing of Plasmonic Biosensors with Nanosilver Structures

Georgios A. Sotiriou, Department of Mechanical and Process Engineering, Particle Technology Laboratory, ETH Zurich, Zurich, Switzerland, **Christoph O. Blattmann**, Particle Technology Laboratory, ETH Zurich, Zurich, Switzerland and Sotiris E. Pratsinis, Particle Technology Laboratory, Institute of Process Engineering, Department of Mechanical and Process Engineering, ETH Zurich, Zurich, Switzerland

Development of biosensors for medical applications is an on-going research field of great importance. In particular localized surface plasmon resonance (LSPR) biosensors show a high potential to detect analytes at low concentrations in biological fluids [1]. The sensor's basic build-up is a usually transparent substrate onto which plasmonic nanoparticles (e.g. silver and gold) are structured. State of the art synthesis techniques are based upon lithographic methods which allow for sophisticated and highly sensitive nanoparticle constellations. Sensors will only find an application outside of the research laboratories if their manufacturing is inexpensive, scalable and reproducible. These key points reveal themselves as the major drawbacks of lithographic synthesis methods. In this project a new synthesis method, utilizing flame-spray pyrolysis [2-4], a scalable nanoparticle synthesis method, was investigated. These manufacturing issues were of primary importance. Silver, showing greater sensing capabilities and reduced costs in comparison to the conventionally used gold, were the heart of these sensors. Co-oxidizing different support materials with the silver as well as silica coating the nanoparticles allowed for the tuning of the sensitivity. Flow cell studies with adsorption of bovine serum albumin as well as the well-studied biotin-streptavidin system, showed their effectiveness and potential nevertheless their quick and facile fabrication.

References

[1] Sotiriou, G. A. & Pratsinis, S. E. Engineering nanosilver as an antibacterial, biosensor and bioimaging material. *Curr. Opin. Chem. Eng.* **1**, 3-10 (2011).

[2] Sotiriou, G. A., Hirt, A. M., Lozach, P. Y., Teleki, A., Krumeich, F. & Pratsinis, S. E. Hybrid, silica-coated, Janus-like plasmonic-magnetic nanoparticles. *Chem. Mater.* **23**, 1985-1992 (2011).

[3] Sotiriou, G. A. & Pratsinis, S. E. Antibacterial activity of nanosilver ions and particles. *Environ. Sci. Technol.* **44**, 5649-5654 (2010).

[4] Sotiriou, G. A., Sannomiya, T., Teleki, A., Krumeich, F., Vörös, J. & Pratsinis, S. E. Non-toxic dry-coated nanosilver for plasmonic biosensors. *Adv. Funct. Mater.* **20**, 4250-4257 (2010).

(393ah) Effect of Temperature, Size and Wavelength On Separation of Racemic Solutions

Nathaniel Ondeck, Carnegie Mellon University, Murrysville, PA, Nisha Shukla, Institute for Complex Engineered Systems, Carnegie Mellon Univeristy, Pittsburgh, PA and Steven Klara, Carnegie Mellon University, Pittsburgh, PA

This study involves a new method for separation of racemic mixtures with an important pharmaceutical application. The technique being explored involves creating chiral nanoparticles that will selectively adsorb one enantiomer of a racemic solution, allowing for subsequent differentiation and separation of the enantiomers. Wet chemical synthesis and the use of pressure cells were used to synthesize the nanoparticles and the enantiomer separation took place in solution. Proof of concept work has been completed proving that selective adsorption in solution is possible. Current work has shown that optimization of this process can be used by controlling the size of the nanoparticles, temperature of the solution, as well as the wavelength of the polarimeter used for analysis. Decreasing the size of the nanoparticles, increasing the temperature, and increasing the wavelength of the polarimeter, all prove to enhance the separation and analysis process. The long-term vision of the proposed work is to lay the foundation for the preparation of highly selective and stable chiral nanoparticles and for a better understanding of how novel synthesis methods for nanoparticles can be used to tailor activity and selectivity.

(393ai) Templating Nanoparticle Arrays Via Shear-Directed Assembly of Spherical and Worm-Like Block-Copolymer Micelles: A Coarse-Grained Molecular Dynamics Study

Bryan Rolfe, Chemical & Biomolecular Engineering, Cornell University, Ithaca, NY, Jaehun Chun, Pacific Northwest National Laboratory, Richland, WA and Yong L. Joo, School of Chemical & Biomolecular Engineering, Cornell University, Ithaca, NY

Spherical and worm-like block-copolymer micelles are known to attain long-range order under shear flow. Recently, these non-equilibrium structures have been exploited to template nanoparticles of diameters on order of the micelle diameter. However, the interplay between macroscopic variables and resulting mesostructure along with the shear-ordering mechanisms remains poorly understood. We present results which elucidate this connection by employing coarse-grained molecular dynamics (CGMD) – affording direct observation at a molecular scale. Critically, our model preserves the hydrodynamic interactions by use of a dissipative particle dynamics (DPD) thermostat in an explicit, coarse-grained solvent. First, the effect of nanoparticles on the formation of spherical and worm-like micelles from self-assembling block copolymers has been investigated. The micelle-nanoparticle suspensions are subjected to simple and oscillatory shear of varying strain and frequency/rate, and the rheological response is compared to experiments of similar systems published in the literature. A relationship between the rheological response and the micelle-nanoparticle structure (and transitions) is also proposed. Finally, observed structures are quantified by structure and form factors for comparison to experimental results published using small-angle neutron scattering (SANS). These scattering patterns also differentiate between zig-zag and sliding mechanisms of the ordered nanoparticle sheets.

(393aj) Streamlined RTIL Processing for Facile Nanocomposite Synthesis

James A. Throckmorton, Department of Chemical and Biological Engineering, Drexel University, Philadelphia, PA and Giuseppe R. Palmese, Chemical and Biological Engineering, Drexel University, Philadelphia, PA

Thermosetting materials and thermoset composites are widely used for structural applications, adhesives, insulation materials, electronic encapsulants, and coatings in a wide array of industries including wind energy, aerospace, sporting goods, and automotive. Recent research has shown that properties of these materials, including CTE, modulus, fracture toughness, electrical conductivity, thermal conductivity, and volatile perfusion, can be enhanced by the addition of nano-scale fillers with various chemistries and morphologies. These nanocomposites modify polymers and conventional composites by providing a moldable, flowable, polymer-like composite with advantageous properties taken from embedded materials with desirable properties over small length scales.

Achieving effective macro-scale properties from these nanofillers requires good solvent chemistry and processing for nanoparticles dispersion, effective choice of nanoparticle material and geometry for the desired property improvements, and strong interfaces between the embedded particles and the polymer matrix. Our recent work has demonstrated that well-chosen ionic liquids can simultaneously disperse nanoparticles within an epoxy matrix and initiate epoxy cure reaction. This results in materials with potentially unique interphase properties resulting from interactions among the nanoparticle surface, ionic liquid, and thermoset matrix.

This study explores the structure and properties of nanocomposites made with this unique processing method to embed graphite nanoplatelets, silica nanoparticles, and smectite clays into thermosetting resins. Experimentally and conceptually, various methods of measuring dispersion for differing particle geometries, the relationship between nano-scale dispersion and macroscopic properties, and the effect of a percolated nanofiller network are explored.

(393ak) Nanoscale Electrostatics and the Importance of Geometry

David A. Walker and Bartosz Grzybowski, Chemical and Biological Engineering, Northwestern University, Evanston, IL

The field of electrostatics is thought, by many, to be 'fully understood' - however at the nano-scale we continue to discover intriguing and counter-intuitive phenomena. Here we present the results from a series of recent fundamental studies which investigate the electrostatic interactions between non-spherically symmetric nanoparticles (NPs). We demonstrate the important role which particle geometry can play in patterning charge density (as well as other chemical properties) on NP surfaces, and how these patterns can lead to complex assemblies that would normally require the influence of external forces (such as a directing electric or magnetic field). Furthermore, we demonstrate the role which charged nanoparticles can play in templating, and mediating the assembly between highly charged, reactive NP species. This mediation lowers and interaction energy barrier between the reactive partices and 'catalyzes' their reaction into a 'nanomolecule'. The kinetics of reaction can be tuned by altering the electrostatic environments. A mixture of both experimental and theoretical evidence will be presented.

(393al) First-Principles Calculations of the Role of PVP and PEO in the Controlled Synthesis of Colloidal Ag Nanostructures

Wissam A. Al-Saidi[1], Ya Zhou[2] and Kristen Fichthorn[2], (1)Dept.of Chemical and Petroleum Engineering, University of Pittsburgh, Pittsburgh, PA, (2)Dept. of Chemical Engineering, Pennsylvania State University, University Park, PA

Inorganic colloidal nanomaterials exhibit size- and shape-dependent properties with enormous potential in a variety of applications including sensing, catalysis, and electronics. Solution-phase polyol synthesis has become a popular method for preparing noble-metal nanostructures with well-defined morphologies. The key step in the polyol process is in the presence of a polymeric capping agent that determines the relative growth rates of the different crystallographic facets of the nanoparticles. This capping mechanism depends on the nature of the structure-directing agent and its weight percentage with respect to the polyol. We use density-functional theory to elucidate the roles of polyvinylpyrrolidone (PVP) and polyethylene oxide (PEO) polymers as structure-directing agents in the shape-selective synthesis of Ag nanostructures. At the polymer segment level, we identify several different binding conformers on Ag(100) and Ag(111), and show they both have energetic preference for the {100} facet. This preference results from a surface-sensitive balance between direct binding and van der Waals attraction. At the polymer chain level, correlated segment binding leads to a strong preference for PVP bind to Ag(100) but not for PEO. Our study underscores differences between small-molecule and polymeric structure-directing agents, and explains why PVP is a better structure directing agent than PEO.

(393am) Synthesis, Self-Assembly and Biomedical Evaluation of Paclitaxel

Ran Lin, Andrew G. Cheetham, Pengcheng Zhang and Honggang Cui, Chemical and Biomolecular Engineering, Johns Hopkins University, Baltimore, MD

Paclitaxel is a taxoid currently in clinical use to treat breast, prostate, ovarian and non-small-cell lung cancers. Despite its promising therapeutic potential, paclitaxel has very low solubility in water. We recently devised a new strategy to produce supramolecular prodrugs by conjugating a hydrophilic tumor targeting peptides with cleavable linkers onto a paclitaxel moiety. The resulting paclitaxel prodrugs can self assemble into filamentous nanostructures in water and under physiological conditions. The choices of the linker chemistry provide a mechanism to tune the release rate of paclitaxel in the presence of tumor progression related biomolecules, and the peptide segments can be used to control over the size and shape of the assembled nanostructures. Our cytotoxicity results show that the paclitaxel prodrug presents a similar IC 50 against breast cancer cell lines to that of the pure paclitaxel. The low critical micellization concentration of the designed prodrug suggests these nanostructures will be very stable during the circulation.

(393an) Reactive Synthesis of Rare Earth-Based Nanostructured Materials

Gerard L. Moore, Chemical Engineering, North Carolina A&T State University, Greensboro, NC and Kenneth L. Roberts, Chemical, Biological and Bioengineering, North Carolina A&T State University, Greensboro, NC

Rare earth-based nanostructured materials were manufactured using sonolytic synthesis techniques. When specifically considering the application of bimetallic nanomaterials, additional design considerations included the control of relative metal compositions and requirements of production consistency regarding the layering of active metal on the surfaces of the nanomaterials. All nanostructured materials were characterized using x-ray diffraction (XRD), elemental analysis, BET surface area analysis, scanning and transmission electron microscopies (SEM and TEM) and superconducting quantum interference device (SQUID) magnetometric analysis. Morphology and crystallite sizes were affected by the chemical composition of the synthesis mixture and also upon the sequence of NH_3OH addition to the metal chloride solutions. The majority of interesting structural morphologies were formed in Mg/Ca, Mg/Fe, Mg/Ca mixtures and $CaCO_3$. The morphology of $CaCO_3$ precipitated microstructures indicated a dependence upon the presence of NH_3OH. When NH_3OH was included to the reaction mixture prior to sonication, $CaCO_3$ microcubics were produced. Conversely, the addition of NH_3OH after sonication was observed to produce $CaCO_3$ microdiscoids. Mg/Ca mixtures were observed to form nanorods. Mg/Fe mixtures were observed to produce flower-like peddles. Crystallite sizes were observed to range from 5.92 to 72.13 nm.

(393ao) The Immobilization of Carbonic Anhydrase (CA) with Magnetic Fe3O4 Materials Such As PVA-Fe3O4 Nanoparticles and Fe3O4-PVA Thin Films for a Biomimetic CO2 Sequestration

Joo Seob Lee, Chemical & Petroleum Engineering, University of wyoming, Laramie, WY and Patrick A. Johnson, Chemical & Petroleum Engineering, University of Wyoming, Laramie, WY

We have investigated the enzyme immobilized materials such as CA-PVA-Fe_3O_4 nanoparticles and CA-Fe_3O_4-PVA thin films, which were fabricated by an entrapment method and electrostatic interaction. The surface of PVA has hydroxyl groups to immobilize the enzyme, and positive charge of carbonic anhydrase may support the immobilization in the PVA matrix, which has a negative net charge.

First, magnetic Fe_3O_4 nanoparticles were prepared by a co-precipitation method and poly vinyl alcohol was used to coat the Fe_3O_4 surface. Fe_3O_4 and PVA-Fe_3O_4 were characterized with FT-IR, TEM, SEM and XRD to prove an immobilization environment. DSC, PPMS, DLS, and Zeta-potential measurement were used to investigate CA immobilization.

Second, CA-Fe_3O_4-PVA composite thin films were fabricated by a mixing-drying process. The fabricated PVA network on Fe_3O_4 nanoparticles provides advantages such as optical transparency, flexibility as well as easy-controlled concentrations, and magnetizations. They were characterized with XRD, DLS, TGA, and PPMS to study enzymatic materials, where fine magnetic nanoparticles and carbonic anhydrase are embedded in a series of Fe_3O_4-PVA thin films.

Finally, the enzyme activity was conducted with 4-nitrophenyl propionate (4-NPP) as a substrate, to confirm immobilized enzyme/substrate reaction. The effect of immobilized carbonic anhydrase on $CaCO_3$ precipitation was studied for using magnetic nanoparticles, and thin films, in order to develop a biomimetic CO_2 sequestration.

(393ap) Layer-by-Layer Assembly of Gel Particles for Polymer Solar Cell Applications

Jeffrey J. Richards and Danilo C. Pozzo, Chemical Engineering, University of Washington, Seattle, WA

Aqueous dispersions of colloidal gel particles containing poly(3-hexylthiophene) (P3HT) and [6,6]-phenyl-C_{61}-butyric acid methyl ester (PCBM) serve as solution-processable inks for active layers in polymer solar cells. Bulk P3HT organogels containing PCBM have been emulsified into water to produce sub-micrometer sized gel particles that are stabilized with cetyltrimethylammonium bromide (CTAB). Excess CTAB has been dialyzed from these solutions to form stable, positively charged gel particles. Glass substrates coated with tin-doped indium oxide (ITO) and with poly(3,4-ethylenedioxythiophene) poly(styrenesulfonate) (PEDOT:PSS) have been dipped into these coating solutions, and the kinetics of particle adsorption has been characterized using UV-Visible Spectroscopy. The resultant nanoparticle films have been characterized using atomic force microscopy (AFM), scanning electron microscopy (SEM), and optical microscopy. Presented here is the optimization of coating solution characteristics which maximize P3HT/PCBM gel particle surface coverage on PEDOT:PSS films. Work is ongoing to demonstrate completed solar cells using this approach. These coatings demonstrate the utility of using aqueous processing for the development of polymer solar cells and this method provides a platform for the development of coatings that incorporate lower band gap conjugated polymers.

(393aq) Ion Adsorption On Nanocrystalline TiO2 Surfaces by Using a Reactive Force Field

Sung-Yup Kim, Mechanical Engineering, Penn State University, University Park, PA, Adri van Duin, Department of Mechanical and Nuclear Engineering, Penn State University, State College, PA and James D. Kubicki, Department of Geosciences and Earth and Environmental Systems Institute, The Pennsylvania State University, University Park, PA

The reactive force field, ReaxFF, developed by van Duin and co-workers make feasible for the simulations of nanoparticles for nanoseconds with bond-making and bond-breaking reactions which lack in classical MD simulations. Previously, a new ReaxFF reactive force field has been developed to describe reactions in the Ti-O-H system. The ReaxFF force field parameters have been fitted to a quantum mechanical (QM) training set containing structures and energies related to bond dissociations, angle and dihedral distortions, and reactions between water and titanium dioxide as well as experimental crystal structure, heat of formation and bulk modulus data. Model configurations for the training set were based on DFT calculations on molecular clusters and periodic systems (both bulk crystals and surfaces). ReaxFF reproduces accurately the QM training set for structures and energetics of small clusters. Most important is that ReaxFF also describes the energetics for the rutile, brookite and anatase. The results of ReaxFF match reasonably well with those of QM for water binding energies, surface energies and H_2O dissociation energy barriers. To validate this ReaxFF description, we have compared its performance against DFT/MD simulations for 1 and 3 monolayer water interaction with a rutile (110) surface. We found agreement within 10% error between the DFT/MD and ReaxFF water dissociation levels for both coverages.

Apart from that, a new Ti-Cl force field parameter has been developed to describe reactions in Ti/O/H/Cl materials. The ReaxFF force field parameters are fitted against a quantum mechanical (QM) training set containing structures and energies related to bond dissociations, angle and dihedral distortions, and reactions between titanium and chlorine gases as well as heats of formation of $TiCl_x$ crystals. These newly developed Ti-Cl force field parameters were combined with the recently developed Ti-O-H force field. ReaxFF accurately reproduces the QM training set for structures and energetics of small clusters and $TiCl_x$ crystals. By adding Ca parameters with the previously developed Ti/O/H/Cl force field, the simulations of TiO_2 nanocrystalline surfaces in aqueous solutions with Ca and Cl ions were conducted. Before that, the force field was validated by comparing bond length between Ca and O with experimental results*. The energetic and structures of hydrated ions will be investigated. Total numbers of H_2O molecules, OH and Ca ions attached to the surfaces with respect to time will be shown. Also, the mechanism of ion adsorption caused by proton induced charge transfer and Ca ions diffusion behavior will be discussed. In addition to these, the ion adsorption characteristic behaviors of the 5 different specific faces for anatase and rutile will be presented. For another ion simulation, Na and Cl ions were added to the H_2O and the characteristic behavior will be discussed and compared with the result of Ti/O/H/Ca/Cl system.

* Markus, Y. *Chemical Reviews*, **1988**, 88,1475

(393ar) Effect of Graphene Oxide On Thermal and Morphological Behavior of Polybenzoxazine Aerogels

Almahdi A Alhwaige[1,2], Saeed Alhassan[3], Tarek Agag[2], Hatsuo Ishida[2] and Syed Qutubuddin[1,2], (1)Department of Chemical Engineering, Case Western Reserve Univesity, Cleveland, OH, (2)Department of Macromolecular Science and Engineering, Case Western Reserve Univesity, Cleveland, OH, (3)Department of Chemical Engineering, The Petroleum Institute, Dubai, United Arab Emirates

Graphene-based nanocomposites have received huge attention in many areas of science and engineering because the nanoscale dispersion of graphene enhances many properties including mechanical, thermal, and morphological behavior compared with the pure polymers or conventional composites. Polybenzoxazines are a versatile class of thermosetting polymers that possess many desirable physical properties. To the best of our knowledge, the properties of polybenzoxazine nanocomposite aerogels containing graphene oxide have not yet been investigated. Herein we report graphene-based nanocomposite aerogels of polybenzoxazine, which are prepared via freeze-drying. The preparation of graphene oxide was confirmed using Raman spectroscopy and Fourier transform infrared (FTIR). Benzoxazine precursors were synthesized using 4-hydroxybenzaldehyde as phenolic component, paraformaldehyde, and two different functional polyetheramines. The chemical structures of the benzoxazine precursors were confirmed by nuclear magnetic resonance spectroscopy (^{1}H-NMR) and (FT-IR).The morphological, surface, and thermal properties of the aerogels obtained were examined and compared with the neat polybenzoxazine aerogels. The structure of the aerogels determined using X-ray diffraction (XRD), scanning electron microscopy (SEM), and transmission electron microscopy (TEM) will be reported. The thermal behavior of the hybrid aerogels was characterized by thermogravimetric analysis (TGA). The effect of graphene oxide on the ring-opening polymerization of benzoxazine was also evaluated using differential scanning calorimetry (DSC). Potential applications of these novel highly porous and low density aerogels will be discussed.

(393as) Preparation of Au Nanoparticles Decorated Graphene Sheet

Jaewook Lee, Nano fusion technology, Pusan National University, Mirayng, South Korea, Jeonghyo Kim, Pusan National University, Mirayng, South Korea and Jaebeom Lee, Nano Fusion Technology, Pusan National University, Mirayng, South Korea

Recently, graphenes functionalized with metal nanoparticles (NPs) have attracted great interest because of their potential properties. In particular, many researchers are interested in Au-NP-hybridized graphene sheets that show electrical properties, Raman scattering signal, catalytic behavior, and surface plasmon resonance (SPR). In this study, we successfully synthesized Au-NP-decorated graphene sheets (AuN-GrS) by using phytochemicals as reducing agents at room temperature. The synthesis reaction occurred in two steps: the first step was the intercalation of Au ions in graphene flake multilayers and the second step was the reduction of Au ions, resulting in the conversion of Au NPs in the graphene layers, accompanied by a simultaneous exfoliation process. The physiochemical properties of AuN-GrS were characterized by TEM, SEM, UV-Vis spectroscopy, FT-IR spectroscopy, XRD, and a semiconductor parameter analyzer. It was observed that Au NPs were well dispersed on the graphene layers. In addition, the electrical conductivity of AuN-GrS was enhanced. Therefore, this novel synthesis route for AuN-GrS by using phytochemical reducing agents may be effectively exploited for non-thermal-assisted reactions and one-pot processes.

(393at) First Principles Models for Developing p-Type Transparent Conductors

Maria Stoica and Cynthia S. Lo, Energy, Environmental and Chemical Engineering, Washington University in St. Louis, St. Louis, MO

Nanostructured inorganic materials have in the past ten years seen a rise in interest due to the enhanced performance exhibited at low dimensions. In our work, we are interested in studying the effect of dimensionality on optoelectronic properties of ZnO and $CuAlO_2$-based materials. In the field of transparent conductors, it is of interest to find high performance p-type materials that can be used in combination with n-type indium tin oxides (ITOs) or doped zinc oxides (ZOs) to expand the array of applications for these materials. Specifically, we investigate the "dark" electrical conductivity and optical absorption of these materials by first setting a baseline utilizing Molecular Dynamics (MD) and Density Functional Theory (DFT) ab initio computational methods in combination with post-processing of the momentum/optical matrix elements to predict properties of experimentally tested materials and determining the error in absolute value and trends of our results. We calculate properties by integrating over the entire first Brillouin zone as well as performing a single calculation at the Γ point. We then predict properties for the novel materials using the previously tested methods. We are interested in transition metal doping of ZnO with oxygen vacancies as well as single replacement and charge-compensated replacement doping and $CuAlO_2$ compounds, both of which have shown potential as p-type materials. We perform computations for both the bulk and 2-D nanowires of each group of materials, and show the enhancement of properties as a result of nanostructuring.

(393au) Enhancing Photoluminescence Intensity of Quantum Dots in Polymer Matrix by Using Hollow Silica Particles

Hyun Chang Kim[1], Taegyeong Kang[2], Cheolsang Yoon[3], Hyun-Guk Hong[4], Young-Joo Kim[4] and Kangtaek Lee[5], (1)Department of Chemical & Biomolecular Engineering, Yonsei University, Seoul, South Korea, (2)Department of Chemical & Biomolecular Engineering, Yonsei University, Soeul, South Korea, (3)Chemical and Biomolecular Engineering, Yonsei University, Seoul, South Korea, (4)Mechanical Engineering, Yonsei University, Seoul, South Korea, (5)Department of Chemical Engineering, Yonsei University, Seoul, South Korea

Enhancing photoluminescence intensity of quantum dots in polymer matrix by using hollow silica particles

Hyun Chang Kim[1], Taegyeong Kang[1], Cheolsang Yoon[1], Hyun-Guk Hong[2], Young-Joo Kim[2], and Kangtaek Lee[1]

[1]Department of Chemical and Biomolecular Engineering, Yonsei University

[2]Department of Mechanical Engineering, Yonsei University

ABSTRACT

Achieving high photoluminescence intensity of quantum dots in polymer matrix is of interest in display field. In this study, we improve dispersion of quantum dots in polymer matrix, and subsequently enhance the photoluminescence intensity by encapsulating a homogeneous quantum dot layer at a radial equidistance on silica nanoparticle. Previous studies on encapsulation of quantum dots on silica nanoparticle were conducted in aqueous phase, which exhibited substantially lower photoluminescence intensity and stability than in organic phase. A new method of encapsulating quantum dots from organic phase onto silica particles is introduced, and to further increase the photoluminescence intensity, hollow silica nanoparticles are used. Hollow silica particles have higher transparency than silica particles, and should increase the photoluminescence intensity of the encapsulating quantum dots more significantly due to their unique near field optics and light trapping properties.

Keywords: *Materials, Nanotechnology, Hollow Silica, Polymer Film, and Quantum dot*

(393av) Effects of PLGA Nano Patterns of Various Dimensions On Osteoblast and Osteosarcoma Cell Responses

Yongchen Wang[1], Lijuan Zhang[1], Linlin Sun[2] and Thomas J. Webster[2], (1)Department of Chemistry, Brown University, Providence, RI, (2)School of Engineering, Brown University, Providence, RI

Effects of PLGA Nano Patterns of Various Dimensions on Osteoblast and Osteosarcoma Cell Responses

Yongchen Wang[1], Lijuan Zhang[1], Linlin Sun[2], and Thomas J. Webster[2]

[1]Department of Chemistry, Brown University, Rhode Island 02912, USA

[2]School of Engineering, Brown University, Rhode Island 02912, USA

Abstract: The interactions between substrates and cells, and especially the use of nanotopographies on the substrates to manipulate cell functions have been widely studied. Among these studies, research on the effects of nanotopographies on cancer cell behaviors can help develop new strategies for cancer therapy. In this work, nano patterns were fabricated on a biocompatible and biodegradable co-polymer, poly(lactic-co-glycolic acid) (PLGA), with a template method [1, 2]. The sizes of nano patterns included 27nm, 190nm, 300nm, 400nm, and 520nm, and a flat surface feature was also made as a control. The surface features, root-mean-square (RMS) roughness and wettability of PLGA nano patterns were characterized. Both human healthy osteoblasts (OB) and human cancerous osteosarcoma cells (OS) were cultured on these PLGA surface features and the effects of PLGA nano patterns on their cell adhesion was evaluated by 4 hour cell adhesion assays. In the cell adhesion assays, 27nm PLGA nano patterns enhanced both osteoblast and osteosarcoma cell adhesion the most compared with other surface features and there were relatively strong correlations between either osteoblast density or osteosarcoma cell density and the RMS roughness of PLGA nano patterns. More interestingly, 27nm PLGA nano patterns exhibited a significant increase in osteoblast density compared with osteosarcoma cell density cultured on the same surface features and therefore could be further developed for a novel bone cancer therapy instead of the traditional cancer therapy methods.

Experiments: The nano patterns were initially created by the self-assembly of polystyrene (PS) beads of 27nm, 190nm, 300nm, 400nm, and 520nm. These patterns were first transferred from PS beads to polydimethylsiloxane (PDMS) molds, which polymerized in close contact with the PS beads, and then were transferred from PDMS molds to PLGA films, which formed in close contact with the PDMS molds after solvent volatilization. After synthesis, the surface features of PGLA nano patterns were studied by AFM. RMS roughness values were calculated from flattened z-sensor scan AFM images and the wettability were characterized with a contact angle goniometer to measure the water contact angles of the PLGA nano patterns.

Human healthy osteoblasts (CRL-11372, ATCC) and human cancerous osteosarcoma cells (CRL-1427, ATCC) were cultured in their respective complete culture medium under a standard environment. The cell adhesion of both osteoblasts and osteosarcoma cells were evaluated by culturing these two types of cells on the PLGA nano patterns for 4 hours and investigating the cell density by means of MTT assays. Cell numbers were counted with a hemocytometer. All cell studies were accomplished in triplicate and the data were analyzed with a one-tailed student t-test.

Results and Discussion: From height scan AFM images (Scanning size: 5μm°Á5μm), nano patterns of 27nm, 190nm, 300nm, 400nm and 520nm and a flat PLGA surface were fabricated (Fig. 1). The RMS roughness and wettability of different PLGA surface features were characterized (Table. 1). 27nm PLGA nano patterns exhibited the largest RMS roughness and its surface was the most hydrophobic. Flat PLGA surfaces showed the smallest RMS roughness and its surface was the most hydrophilic.

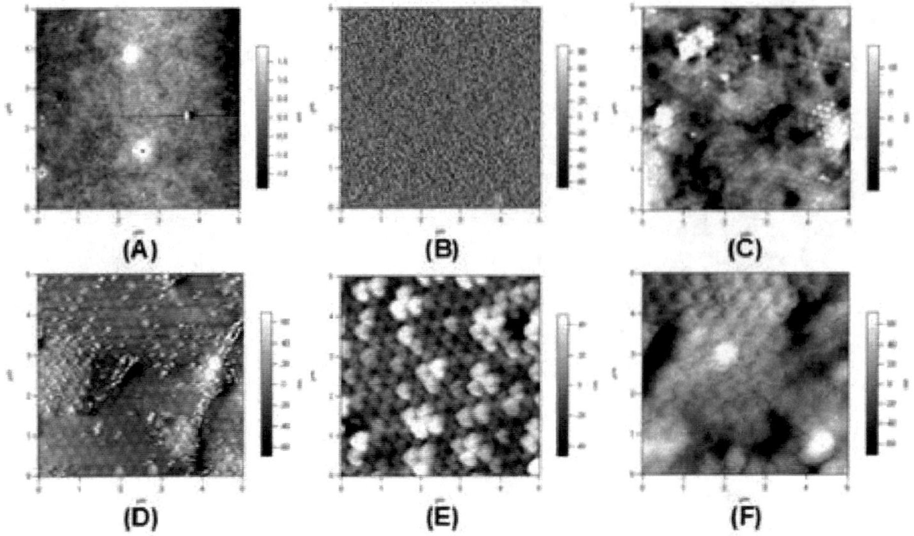

Figure. 1. AFM images of PLGA films with (A) flat surfaces; (B) 27nm; (C) 190nm; (D) 300nm; (E) 400nm; and (F) 520nm nano patterns.

Table. 1. RMS roughness and water contact angles of the flat PLGA film and PLGA films with 27nm, 190nm, 300nm, 400nm, and 520nm nano patterns.

Subgroup	Flat	27nm	190nm	300nm	400nm	520nm
RMS	0.710nm	40.123nm	4.535nm	25.553nm	18.950nm	24.973nm
Average Deviation	0.539nm	35.068nm	3.492nm	15.379nm	15.607nm	18.436nm
Average Contact Angle	84.28°	97.18°	89.56°	95.54°	91.01°	95.47°
Standard Deviation	1.97°	1.13°	2.21°	2.09°	2.29°	1.99°

For osteoblasts alone or osteosarcoma cells alone, the 27nm nano patterns increased cell adhesion the most (Fig. 2). For osteoblasts and osteosarcoma cells cultured on the same surface feature, only 27nm nano patterns showed a significant osteoblast cell density increase compared with osteosarcoma cell density. From the correlation plots between cell density of osteoblasts or osteosarcoma cells, and the physical properties including RMS roughness and wettability (Fig. 3; Table. 2), there were relatively strong correlations between either osteoblast density or osteosarcoma cell density and the RMS roughness of PLGA nano patterns, and relatively weak correlations between either osteoblast density or osteosarcoma cell density and the wettability of PLGA nano patterns. These results could be explained in respect to several aspects. First, larger RMS roughness means larger surface area and more sites for the adsorption of proteins regulating cell adhesion. Second, the RMS roughness of 27nm PLGA nano patterns were the closest to the roughness of the real bone matrix [3]. Third, 27nm is the closest to the dimensions of the extracellular matrix proteins participating in cell adhesion [4, 5]. However, currently, the mechanism of the effects of PLGA nano patterns on the cellular responses of osteoblasts and osteosarcoma cells is not known. Further study is needed to fully understand this.

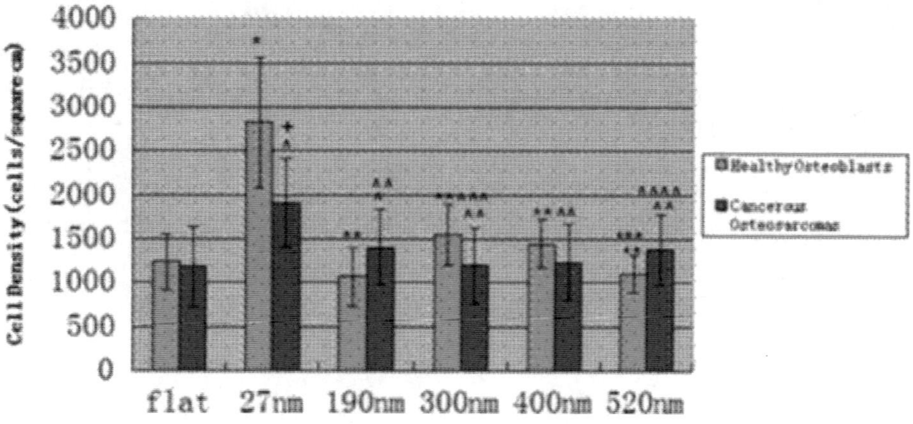

Figure. 2. 4 hr cell adhesion results of OB and OS (N=3; seeding density: 3500 cells/cm^2). The cell densities are average values +/- S.E.M. For OB alone, *, ** and *** represents p<0.05 compared with the flat, 27nm and 300nm surface features, respectively; for OS alone, ^, ^^, ^^^ and ^^^^ represents p<0.05 compared with the flat, 27nm, 190nm and 300nm surface features, respectively; and + represents p<0.05 for OS cell density compared with OB density on the same surface features.

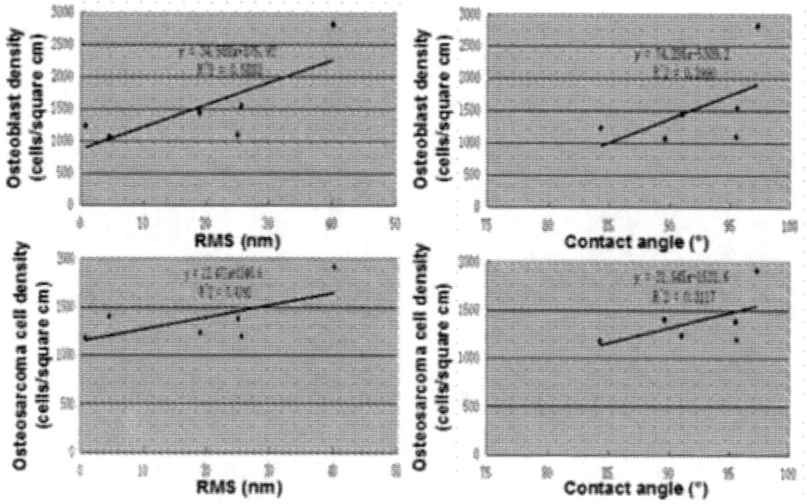

Figure. 3. Correlation plots between osteoblast density/RMS values obtained from AFM, osteoblast density/ contact angle, osteosarcoma cell density/RMS values obtained from AFM and osteoblast cell density/contact angle.

Table. 2. R^2 values and R values of the correlation plots between osteoblast density/RMS values obtained from AFM, osteoblast density/contact angle, osteosarcoma cell density/RMS values obtained from AFM, and osteosarcoma cell density/contact angle.

Subgroup	OB density and RMS	OB density and contact angle	OS cell density and RMS	OS cell density and contact angle
R^2 values	0.5882	0.2990	0.4392	0.3117
R values	0.7669	0.5468	0.6627	0.5583

Conclusions: PLGA nano patterns of various dimensions could be fabricated with a template method. For cell studies, 27nm PLGA nano patterns not only increased the cell density most for osteoblasts alone or osteosarcoma cells alone, but it also significantly increased osteoblast density compared with osteosarcoma cell density cultured on the same surface features. It was also found that there were relatively strong correlations between either osteoblast or osteosarcoma cell densities, and RMS roughness of PLGA nano patterns. Thus, 27nm PLGA nano patterns should be further studied as a potentially novel bone cancer therapy with few adverse effects.

Acknowledgements: We would like to thank Hermman Foundation for their financial support, and Vera Fonseca and Professor Eric Darling for their technical support with AFM use.

References:

[1] Zhang L. et al. Inter J Nanomed. 5, 269, 2010.

[2] Zhang L. et al. J Biomed Mater Res A. 100A, 94, 2012.

[3] Liu, H. et al. J Biomed Mater Res A. 78A, 798, 2006.

[4] Elias, K. L. et al. Biomaterials. 23, 3279, 2002.

[5] Koteliansky, V. E. et al. Eur J Biochem. 119, 619, 1981.

PHASE EQUILIBRIUM PREDICTION OF DIFFERENT PHOSPHOLIPIDS IN SUPERCRITICAL CO_2 WITH EHANOL AS A CO-SOLVENT

L. R. C. FERRAZ[1], R. G. NEDIA[1], I. ESPÍRITO SANTO[1], R. FIALHO[1], G. M. COSTA[1], E. CABRAL-ALBUQUERQUE[1]

[1] Programa de Pós graduação em Engenharia Industrial (PEI), Federal University of Bahia, Brazil.

Abstract

Over the last decades, diverse types of drug delivery systems have been under intense investigation. One of them is liposomes, a colloidal association of amphiphilic lipids that organize themselves spontaneously in bilayer vesicles as result of unfavorable interactions between phospholipids and water. As these lipid vesicles present lipophilic and hydrophilic portions, they can entrap substances with different lipophilicities in the phospholipid bilayer, in the aqueous compartment or either at the bilayer interface, which can modify physicochemical properties and can enhance biological activity of entrapped compounds. A wide variety of conventional techniques exists to produce liposomes. However, conventional techniques may not be suitable for the preparation of formulations because involve additional steps such as sonication or removal of the organic. So, special interest has been given to the application of supercritical fluid to obtain liposomes due to the possibility of exploiting the peculiar properties of supercritical fluids, and in particular of supercritical carbon dioxide (sc-CO_2), the most used supercritical fluid. Therefore, working with sc-CO_2 it is possible to carry out the process at near-ambient temperatures, avoiding the degradation of thermolabile substances. Additionally, sc-CO_2 also provides an inert medium and eliminates or reduces the use of toxic or contaminant organic solvents in the process. Knowledge of description of the solid-liquid equilibrium involving phospholipids, supercritical CO_2 as solvent and ethanol as co-solvent is essential for the design of these processes. In this study, the solubility modeling of different phospholipids (phosphatidylcholine, phosphatidylglycerol and phosphatidylethanolamine) in supercritical CO_2 and ethanol as co-solvent was analyzed to determine the best conditions to initialize the process of liposome production. The thermodynamic model chosen to study the equilibrium between the phospholipids and the supercritical solvent was the model that uses the Peng-Robinson equation of state with the LCVM mixing rule and the UNIFAC method (PR-LCVM). In the literature there is no information on critical properties, molar volume of the solid and sublimation pressure of this solution required in the modeling. Using an alternative technique, values of those properties were obtained through group contribution method and showed good agreement with experimental data using one parameter fit model. In the present study, predicted results showed that this model enables the prediction of the solubility of different phospholipids in supercritical carbon dioxide and ethanol as co-solvent. The results were very consistent as compared with low solubility data reported in the literature. In addition, it is also possible to predict solubilities of phospholipids in several temperature and pressure conditions expanding experimental data.

Introduction

Recently, the use of supercritical technology in producing nanoparticles with pharmaceutical and medical interests, such as liposomes, has been the subject of several studies. Liposomes are colloidal associations of amphiphilic lipids capable of transporting substances with different polarities that can modify their physicochemical characteristics and increase the biological activity of bioactive compounds [1, 2, 3]. The most common phospholipid used in liposome preparations is phosphatidylcholine (PC), a neutral charged lipid that can be found in cell membranes. However other phospholipids can also be used to produce liposomes, as phosphatidylethanolamine (PE) and phosphatidylglycerol (PG) [4].

All conventional methods for producing these vesicles involve the use of large amounts of organic solvents, which are usually toxic to humans and the environment. Moreover, these techniques are multi-stage, which is a limitation to the process scale up. So, the use of supercritical fluids has been studied as an alternative route for producing liposomes proved to be promising, environmentally friendly and versatile [5, 6, 7, 8]. Liposomes can be produced by several techniques using supercritical. One of them was developed by Otake *et al.* [9], which consists in a process that uses supercritical CO_2 (sc-CO_2) and ethanol. However, to ensure the formation of liposomes, it is necessary to perform further studies on the equilibrium established between the phospholipids and the mixture sc-CO_2 - ethanol.

In this work, a prediction of solubility of different phospholipids in the mixture sc-CO_2 – ethanol was made. Peng-Robinson equation of state (PR) [10] was selected and the model of corporate activity coefficient was obtained through UNIFAC with group contribution. The mixing rules incorporated in the equation were a linear combination of Vidal and Michelsen mixing rules (LCVM) [11]. Due to the lack of thermodynamic parameters of PC and PE in the literature, data as sublimation pressure and critical volume of the solute were estimated initially using group contribution technique, which its effective was tested with other substances with relatively complex chemical structure, such as carnosic acid.

Experimental

In this study, different methods were applied in order to obtain thermodynamic data of PC, PG and PE, since these parameters are not available in the literature. In order to determine PG solubility, experiments were performed with this compound in a mixture of sc-CO_2 and ethanol. PG was purchased from Lipoid (Ludwigshafen, Germany), CO_2 from White Martins (Rio de Janeiro, Brazil) and ethanol 99,5% from Merck (Darmsdadt, Germany).

Bubble point measurements of sc-CO_2 + PG ethanolic solutions (1000 and 5000 ppm) were conducted in an apparatus equipped with a variable volume cell that was detailed in previous studies [12, 13]. The used systems composition (PG + ethanol/CO_2 ratio) were 50% and 70% for the 1000 ppm solution, and 50% and 65% for the 5000 ppm solution.

Few extraction data of PC and PE [14] were used as reference for the solubility studies of these compounds. The extraction amount of the solutes was obtained from the variation of the amount of phospholipid that was extracted versus time for a given condition of temperature, pressure, solvent-cosolvent ratio and its flow rate. Sc-CO_2 – ethanol flow, extraction time and density of CO2 and ethanol mixture under the specified conditions of pressure, temperature and composition [15] were utilized for determining the mass of the supercritical fluid at these conditions. Using molecular weights of the phospholipids, it was possible to obtain the molar fractions of PC and PE in the supercritical fluid at each condition.

Solubility model

Thermodynamic modeling studies

To model the solubility of a solid solute in a supercritical fluid (CO2 + ethanol), it is often assumed that the supercritical fluid does not dissolve in the solid phase, and the phase rich in the supercritical fluid behaves live a dense gas. So, the solubility of PC in the supercritical phase, in terms of its molar fraction (ys), can be expressed by:

$$y_s = \frac{P_s^{sat}}{P} \cdot \frac{\varphi_s^{sat} \exp[v_s^{solid}(P - P_s^{sat})/RT]}{\varphi_s^{SCP}} \qquad (1)$$

where the P_s^{sat} is the sublimation pressure of the solid, *P* is the system pressure and P_s^{sat}*/P* ration is the ideal solubility. In the Eq. (1), it is assumed that the molar volume of the solid is constant. As the saturation pressure is too small, the fugacity coefficient of the solute at this pressure is very close to the unity ($\varphi_s^{sat} \approx$ 1). Thus, key factors in thermodynamic modeling refer to variables P_s^{sat} and φ_s^{SCF}, corresponding to the saturation pressure and the fugacity coefficient of the solute in the supercritical phase, respectively.

Peng-Robinson equation of state was used to calculate the solubility (y_s)

$$P = \frac{RT}{v-b} - \frac{a}{v(v+b)+b(v-b)}$$

(2)

where *T* is the temperature, *P* is the pressure, *v* is the molar volume, *R* is the gas constant, *a* and *b* are specific parameters of the mixture. The parameters a_i and b_i of the pure components are describes in terms of critical temperature Tc_i, critical pressure Pc_i and acentric factor w_i.

In this work, LCVM mixing rules was employed incorporating the UNIFAC method. To calculate the parameter *b*, LCVM mixing rules use the linear relationship expressed by:

$$b = \sum_i x_i b_i \qquad (3)$$

Regarding the parameter *a,* initially the parameter α, which is related to the parameters *a* and *b*, is calculated by:

$$\alpha = \frac{a}{bRT} = \left(\frac{\lambda}{A_V} + \frac{1-\lambda}{A_M}\right)\frac{G^E}{RT} + \frac{1-\lambda}{A_M}\sum_i x_i \ln\frac{b}{b_i} + \sum_i x_i \alpha_i \quad (4)$$

Employing the mixing rule LCVM, φ_s^{SCF} can be calculated as:

$$\ln \varphi_s^{SCF} = \frac{b_i}{b}\left(\frac{Pv}{RT} - 1\right) - \ln\frac{P(v-b)}{RT} - \frac{\overline{\alpha}_i}{2\sqrt{2}}\ln\left[\frac{v(1+\sqrt{2})b}{v(1-\sqrt{2})b}\right] \quad (5)$$

where

$$\overline{\alpha}_i = \left(\frac{\partial n\alpha}{\partial n_i}\right)_{T,P,n_j \neq n_i} = \left(\frac{\lambda}{A_V} + \frac{1-\lambda}{A_M}\right)\ln\gamma_i + \frac{1-\lambda}{A_M}\left(\ln\frac{b}{b_i} + \frac{b_i}{b} - 1\right) + \alpha_i \quad (6)$$

For Peng-Robinson equation of state, A_V = -0.623 e A_M = -0.52. γ_i is the activity coefficient calculated by the group contribution of UNIFAC method and λ is the LCVM mixing rule parameter. In this current work, the value of λ is adjusted from experimental data.

Physical properties of the phospholipids

In order to use PR-LCVM-UNIFAC method, it is necessary calculate the properties of the pure compounds, such as critical properties, acentric factor, molar volume of the solid and vapor pressure of the solute. There are no experimental data of these properties for PC and PE, which have their chemical structures depicted in Figure 1, due to the difficulty of experimental determination. Therefore, a good alternative to resolve this is to estimate these properties by group contribution methods.

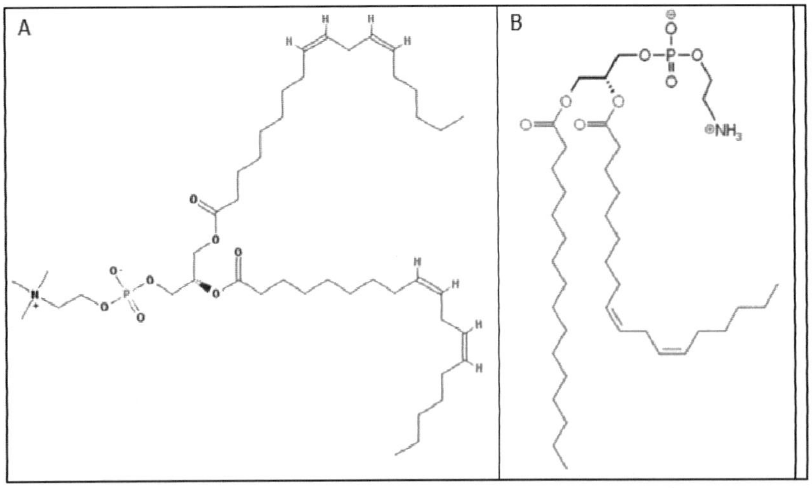

Figure 1. Chemical structure of (a) PC ($C_{44}H_{80}NPO_8$) and (b) PE ($C_{39}H_{74}NPO_8$)

A different systematic with good results for a solute-solvent system was developed by Melo *et al.* [16]. This systematic allows adapting predictive values to values closer to real ones, from properties obtained by contribution group, employing few experimental solubility data. For this systematic, the parameter λ of the LCVM mixture rule is the only predictive value that has to be obtained from comparison between the experimental and the calculated solubility of the solute in the fluid phase.

The critic volume (V_c) is calculated by Vetere method, the critical pressure (P_c) and the relation between the normal boiling temperature and the critical temperature ($\theta = T_b/T_c$) are obtained by Lydersen method [17] using the following equations:

$$V_c = 33.04 + \left[\sum_i \Delta V_i M_i \right] \tag{7}$$

where ΔV_i corresponds to each constituent group of the molecule and M_i is the molecular weight of each group.

$$P_c = M\left(0.34 + \sum \Delta P_i\right)^{-2} \tag{8}$$

$$\theta = \frac{T_b}{T_c} = 0.567 + \sum_i \Delta T_i - \left(\sum_i \Delta T_i \right)^2 \tag{9}$$

Applying the procedure proposed by Subra *et al.* [18], there is an initial estimate for T_c using the following equations:

$$\beta = \frac{(1-\theta)^{2/7} - 0.048 \ln V_c + (1-\theta)^{2/7} \ln P_c + 1.225}{(1-\theta)^{2/7}} \tag{10}$$

$$T_b = 0.01286 \theta \exp(\beta) \tag{11}$$

$$T_c = \frac{T_b}{\theta} \tag{12}$$

As shown by Vieira de Melo *et al.* [16], a good estimate of the molar volume of the solute (v_s^{solid}) is to consider it as being 1.5 times the value of the van der Waals volume (V_w) obtained by group contribution method [19].

Therefore, the sublimation pressure P_s^{sat} is estimated by:

$$\ln P^{sat} = A - \frac{B}{T} \tag{13}$$

$$B = \frac{\Delta H^{sub}}{R} \tag{14}$$

$$A = \frac{B}{T_f} \tag{15}$$

where ΔH^{sub} is the sublimation enthalpy obtained by the group contribution method [19] and T_f is the melting temperature of the solute, estimated by Joback method [17].

Results and discussion

In order to verify the efficiency of the proposed methodology, extended on this work to solvent and cosolvent, it was selected in the literature a compound of relatively complex chemical structure that already had several experimental solubility data for a mixture of sc-CO_2 and ethanol. So, the carnosic acid, which has its chemical structure depicted in Figure 2, was chosen for this verification [20].

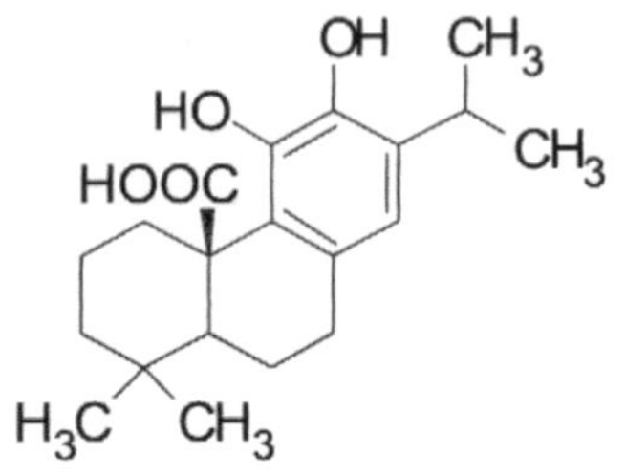

Figure 2. Chemical structure of carnosic acid ($C_{20}H_{28}O_4$).

Applying the methodology proposed by Cháfer *et al.* [20], the estimate of initial and final values of carnosic acid physical properties calculated in relation to the experimental values are shown in Table 1. The initial and final values for each parameter λ temperature are shown in Table 2.

Table 1. Estimative of initial and final values of carnosic acid physical properties.

Value	T_C (K)	P_C (MPa)	ω	A	B
Initial	933.39	1.91	0.99548	25.84	24161.65
Final	1255.33	4.84	1.19687	5.05	15104.51

Table 2. Estimative of initial and final values of the carnosic acid parameter λ.

Value	Temperature (K)		
	313.15	323.15	333.15
Inicial	-1.0	-1.0	-1.0
Final	-1.70	-1.12	4.92

Comparing the experimental solubility values [20] with the estimated ones from the physical properties, the efficiency of the proposed method could be demonstrated with low mean deviation, as shown in table 3.

Table 3. Mean deviation between experimental and the estimated values of carnosic acid in sc-CO2 and ethanol, where N is the number of experimental data.

Temperature (K)	N	Pressure (MPa)	EtOH/sc-CO_2 ratio (%)	Average deviation (%)
313.15	6	27.5 – 40.0	6	4.62
323.15	4	27.5 – 40.0	6	11.74
333.15	4	27.5 – 40.0	6	9.33

As demonstrated the efficiency of the methodology in the estimation of these parameters, this process was applied to study the solubility of PC and PE in the mixture sc-CO_2 – ethanol. However, there is no explicit solubility data of these phospholipids in terms of molar fraction.

For using the UNIFAC method in the LCVM mixture rules for the phospholipids, it was necessary to adapt one of the values. As it can be seen in Figure 1, the phospholipids have a PO_4^- group in their structure. Neverthless, there are no available data of group contribution for molecules that have phosphorus in their structure. So, it was considered that the nitrogen was the central element of this group, since these chemical elements are part of the same periodic group and have some similar chemical properties. The values of each group for the three phospholipids are shown in Table 4.

Table 4. UNIFAC groups for each phospholipid, where NMG: number of main groups, NSG: number of secondary groups, NGM: number of groups in the molecule, R*: group volume, Q: superficial area of the group.

Compound	Group	NMG	NSG	NGM	R^*	Q
PC	CH_3	1	1	4	0.9011	0.848
	CH_2	1	2	24	0.6744	0.540
	CH	1	3	1	0.4469	0.228
	CH=CH	2	6	4	1.1167	0.867
	CH_2COO	11	23	2	1.6764	1.420
	CH_2O	13	26	2	0.9183	0.780
	CH_3N	16	35	1	1.1865	0.940
	NO_2	26	57	1	1.1075	1.020
PE	CH_3	1	1	2	0.9011	0.848
	CH_2	1	2	25	0.6744	0.540
	CH	1	3	1	0.4469	0.228
	CH=CH	2	6	2	1.1167	0.867
	CH_2COO	11	23	2	1.6764	1.420
	CH_2O	13	26	2	0.9183	0.780
	CH_2NH_2	14	30	1	1.3692	1.236
	NO_2	26	57	1	1.1075	1.020

Analyzing these values and applying the methodology used with the carnosic acid to the phospholipids, it was possible to estimate the initial and final values of the physical properties of the parameter λ of the compounds for LCVM method. Some of these values for PC and PE are shown in Table 5. In order to facilitate the comprehension, from now on, just these two phospholipids will be analyzed.

Table 5. Estimative of initial and final values of physical properties of the parameter λ for LCVM method.

Compound	Value	T_c (K)	P_c (MPa)	ω	A	B	Λ
PC	Initial	593.38	0.60	-2.29856	65.58	53922.97	-0.1
	Final	3061.60	2.15	0.48278	88.37	56704.66	0.18
PE	Initial	2331.67	0.67	-0.4165	57.98	49346.52	4.9
	Final	2346.03	1.90	0.0250	24.50	28904.86	4.86

It also can be observed that the solubility behavior of the phospholipids with temperature and pressure variations are in agreement with what was observed experimentally, as demonstrated in Table 6. As it can be noticed, when the pressure is increased, it also increased the solubility of the solute in the supercritical phase, since under these circumstances the passage of solute for this phase is more favored. Moreover, when the temperature was increased, the solubility of the solute in the supercritical mixture was reduced. The enhancing of the temperature leads to a decrease in the density of supercritical mixture and, as a consequence, there is a loss of ability to solubilize some solutes, since there is less contact between the particles of solute, solvent and cosolvent.

Table 6. Comparison between the experimental and the estimated solubility values of PC and PE.

Compound	Temperature (K)	Pressure (MPa)	Experimental value (10^6)	Estimated values (10^6)
PC	333.15	17.2	4.329	2.885
	333.15	20.7	7.108	7.108
	343.15	20.7	6.989	6.989
	353.15	17.2	0.685	0.685
	353.15	20.7	5.885	5.130
	Mean deviation (%)			9.24
PE	333.15	17.2	36.475	36.476
	333.15	20.7	45.153	48.723
	343.15	20.7	50.292	42.311
	353.15	17.2	11.512	11.512
	Mean deviation (%)			5.94

Analyzing the results demonstrated in Table 6, it can be seen that there is a similarity between the experimental and the estimated solubilities, which is evidenced by satisfactory values of average percentage deviations.

From the physical properties values obtained with this methodology, it was possible to estimate the solubility of PC and PE in different conditions, as depicted in Figure 3. Despite the small amount of experimental data, they were obtained at different temperatures. Because the parameter λ of LCVM mixture rule is strongly dependent of the temperature, it was necessary an additional estimation of this parameter to get the description presented in Figure 3a (λ = 0.30, 0.32, 0.36 for 3313.15, 343.15, 353.15 K, respectively) and Figure 5b (λ = 4.82, 4.86, 4.90 for 333.15, 343.15, 353.15 K, respectively). It is important to know that all the other proprieties were presented as final values and are listed in Table 5.

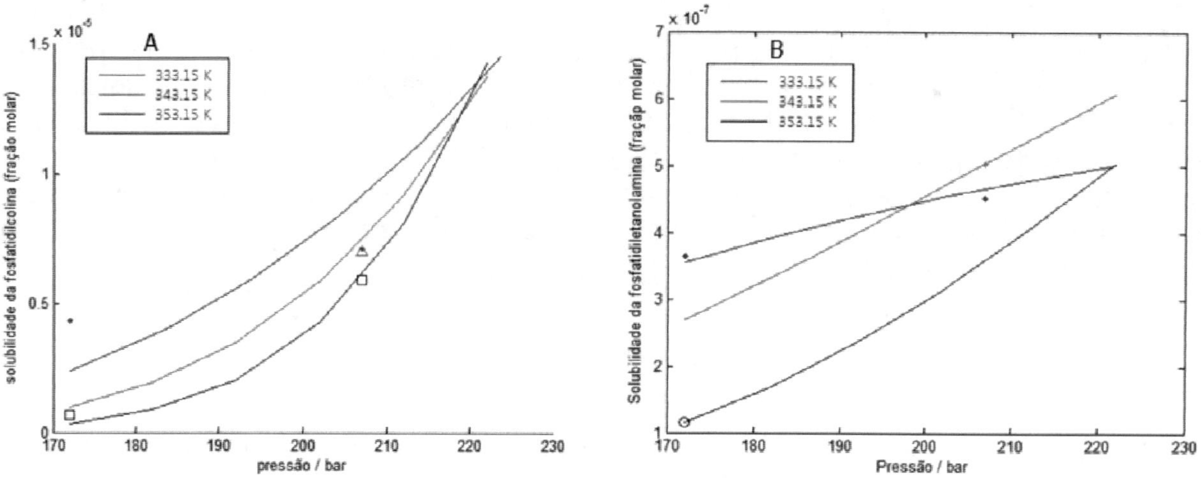

Figure 3. Solubility of (a) PC and (b) PE in a supercritical mixture (10% ethanol, in weight) at different temperatures: (♦) 333.15, (Δ) 343.15, (□) 353.15 K

It was also evaluated the solubility of PC for a supercritical mixture with ethanol concentration of 12.50% (in weight). All previously estimated physical properties were maintained, as well as the values of parameter λ for each studied temperature. The results for this ethanol concentration for two different temperatures are shown in Figure 4, as well as the only one known single experimental value. It can be seen that increasing the weight percentage of ethanol in the supercritical mixture from 10% to 12.5% produces an increase in the solubility of PC, as expected.

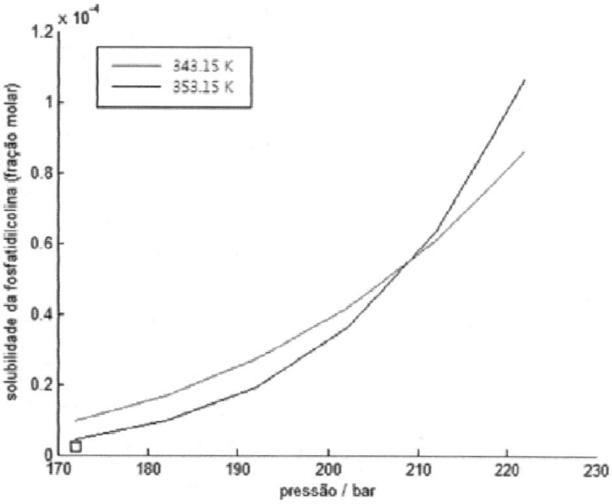

Figure 4. Solubility of PC in a supercritical mixture (12.50% ethanol, in weight) at different temperatures: (□) 353.15 K.

Conclusions

In this study, it was evaluated the modeling of PC and PE in supercritical CO_2 and ethanol, as a cosolvent. The thermodynamic model used (PR-LCVM-UNIFAC) requires the knowledge of the physical properties of the pure compounds and the estimation of only one parameter. A system, previously developed for solvent and solute systems, was extended also considering a cosolvent. This methodology has been successfully tested initially for the system carnosic acid, supercritical CO_2 and ethanol, as the literature provides several experimental data. Physical properties of the solutes, initially obtained by contribution of group, were adequate to describe the experimental data using only one tuning parameter. Since there is almost no information in the literature for the solubility of phospholipid in a supercritical mixture, this methodology has been successfully applied. So, an predictive evaluation of the solubility of PC and PE in different temperatures and pressures were made, allowing the expansion of the experimental data. However, other parameters will be evaluated in order to understand the behavior of these and other phospholipids in supercritical mixtures.

References

1. Gortzi, O.; Lalas, S.; Chinou, I.; Tsaknis, J. (2007), 'Evaluation of the antimicrobial and antioxidant activities of *Origanum dictamnus* extracts before and after encapsulation in liposomes", Molecules 12, pp. 932-945.
2. Joshi, M. D. and Muller, R. H. (2009), "Lipid nanoparticles for parenteral delivery of activities", European Journal of Pharmaceutics and Biopharmaceutics 71, pp. 161-172.
3. Sharma, A. and Sharma, U. (1997), "Liposomes in drug delivery: progress and limitations", International Journal of Pharmaceutics 154, pp. 123-140.
4. New, R. R. C. (1990), "Liposomes: a practical approach", ed. New York, NY, Oxford University Press.
5. Cardoso, M. A. T. (2008), "Supercritical micronization of biological therapeutics", Lisboa, Superior Technical Institute.
6. Daar, J. and Poliakoff, M. (1999), "New directions in inorganic metal-organic coordination chemistry in supercritical fluids", Chemical Reviews 99, pp. 495-541.
7. Reverchon, E. and Adami, R. (2006), "Nanomaterials and supercritical fluids", Journal of Supercritical Fluids, 37, pp. 1-22.
8. Yeo, S.-D. and Kiran, E. (2005), "Formation of polymer particles with supercritical fluids: a review", Journal of Supercritical Fluids 34, pp. 287-308.
9. Otake, K.; Imura, T.; Sakai, H.; Abe, M. (2001), "Development of a new preparation method of liposomes using supercritical carbon dioxide", Langmuir 17, pp. 3898-3901.
10. Peng, D. Y. and Robinson, D. B. (1976), "A new two constant equation of state", Industrial and Engineering Chemistry: Fundamentals 15, pp. 59-64.
11. Bourkouvalas, C.; Spiliots, N.; Coutsikos, N.; Tassios, D. (1994), "Prediction of vapor-liquid equilibrium with the LCVM model: a linear combination of the Vidal and Michelsen mixing rules coupled with the original UNIFAC and the t-mPR equation of state", Fluid Phase Equilibria 92, pp. 75-106.
12. Benazzi, T. L. (2005), "Estudo do comportamento de fases de sistemas contendo glicerol e óleo de oliva em propano na presença de surfactantes", Erechim, Brazil.
13. Ndiaye, P. M.; Franceschi, E.; Oliveira, D.; Dariva, C.; Tavares, F. W.; Oliveira, J. V. (2006), "Phase behavior of soybean oil, castor oil and their fatty acid ethyl esters in carbono dioxide at high pressures", Journal of Supercritical Fluids 37, pp. 29-37.
14. Teberikler, L. Caoseoglu, S.; German, A. (2001), "Selective extraction of phosphatidylcholine from lecithin by supercritical carbon dioxide – ethanol mixture", Journal of the American Oil Chemists' Society 78 (2), pp. 115-118.
15. Kiran, E. and Pohler, H. (1997), "Volumetric properties of carbon dioxide + ethanol at high pressures", Journal of Chemical Engineering 42, pp. 384-388.

16. Melo, S. A. B. V; Costa, G. M. N.; Melo, R. L. F. V.; Alves, T. L. M. (2005), "Solubility of L-Dopa in supercritical carbono dioxide: prediction using a cubiq equation of state", Journal of Supercritical Fluids 34, pp. 231-236.
17. Reid, R. C.; Prausnitz, J. M.; Poling, B. E. (1987), "The properties of gases and liquids", New York, ed. McGraw-Hill.
18. Subra, P.; Castellani, S.; Ksibi, H.; Garrabos, Y. (1997), "Contribution to determination of the solubility of beta-carotene in supercritical carbone dioxide and nitrous oxide: experimental data and modeling", Fluid Phase Equilibria 131, pp. 269-286.
19. Bondi, A. (1968), "Physical properties of molecular crystals liquids and glasses", New York, ed. John Wiley & Soans Inc.
20. Cháfer, A.; Fornari, T.; Berna, A.; Ibañez, E.; Reglero, G. (2005), "Solubility of solid carnosic acid in supercritical CO_2 with ethanol as a co-solvent", Journal of Supercritical Fluids 34, pp. 323-329.

(393ax) Liquid Ethanol Simulated On Crystalline Alpha Alumina

Anh T. Phan, School of Chemical, Biological and Materials Engineering, The University of Oklahoma, Norman, OK, David R. Cole, Chemical Sciences Division, Oak Ridge National Laboratory, Oak Ridge, TN and Alberto Striolo, School of Chemical Biological and Materials Engineering, The University of Oklahoma, Norman, OK

Equilibrium molecular dynamics simulations were employed to investigate the structural properties of interfacial ethanol molecules on two alumina surfaces – the (0001) and the (102) terminations of a-Al
the CLAYFF force field. The results were quantified in terms of density profiles in the direction perpendicular to the substrate as well as along planes parallel to the substrate, in-plane radial distribution functions, residence times in contact with the substrate, orientational distribution functions for the methyl and the OH groups of ethanol, and reorientation autocorrelation functions. The density profiles show that ethanol molecules in the first monolayer have a well-organized arrangement reminiscent of the structure of the substrates. The orientation distribution in the second layer is rather broad and similar to that observed in the bulk. On the contrary, significant orientational order is observed within the first adsorbed layer, in which case the orientational distribution obtained for the methyl group of ethanol on a-Al
Lett. **2**, 1831 (2011)]. Complementing experimental observations, our simulations show long residence times and slowly decaying reorientation autocorrelation functions for ethanol molecules in the first adsorbed layer, suggesting that these molecules are strongly coordinated with both a-Al_2O_3 (0001) and a-Al_2O_3 (102) surfaces.

(393ay) Degradation of Organic Contaminants in Eco Solutions by Using Pulsed Corona Discharges and Electrostatic-Sensitive Nanoparticles: Experiments and Modeling

Chinyere P. Mbachu, Chemical Engineering, Tennessee Technological University, Cookeville, TN and Dr. Pedro E. Arce, Chemical Engineering, TENNESSEE TECHNOLOGICAL UNIVERSITY, Cookeville, TN

Non-thermal high oxidation methods, a novel high oxidation technique based on localized electrical discharges have been used successfully used to degrade high molecular weight organic compounds (dye solutions, kappa casein, humic acids,and fulvic acids) and also pharmaceutical compounds (carbameapazine).Organic contaminants are tested experimentally to determine the ability of pulsed corona power system in degrading these contaminants and leading to a safe drinking product. The role of electrostatic-sensitive nanoparticles (TiO2, activated carbon particles, and glass spheres) will be investigated in this research from both experimental and modeling aspects for the elimination of selected organic contaminants. The key focus of the research is on the impact of degradation efficiency of the corona discharge reactor's different physical parameters, for example, frequency, pH, concentration, electrode shapes, and gap distances under a variety of conditions of the nanoparticles: concentrations, surface characteristics, sizes, etc. Modeling efforts will focus on the formulation and solution of predictive models for "global" reactor behaviors.

(393az) Role of Engineered Nanoparticles in Ultrafiltration of Drinking Water

Milad R.Esfahani, Chemical Engineering, Tennessee Technological University, Cookeville, TN, Holly A. Stretz, Department of Chemical Engineering, Tennessee Technological University, Cookeville, TN and Martha J. M. Wells, EnviroChem Services, Cookeville, TN

The research objective of the overall project is to conduct a fundamental study of adsorption behavior in complex mixtures of nanoparticles, humic acids (HA) and polysulfone-flocs (PSF). The results will inform the nano-scale design of modern nanocomposite ultrafiltration (UF) water membrane separations. Water treatment can be an energy-intensive process requiring costly regeneration of the filtration media when fouled. The model foulant used in this study was humic acid. The experimental design was to stir flocked polysulfone in combination with humic acid and nanoparticles (gold, 50 nm diameter) in water to understand competitive adsorption phenomena. The mixture is stirred 24 hours, then filtered. The filtrate is analyzed by fluorescence spectroscopy (EEMS) for HA and by UV-visible spectroscopy for nano-gold. The relative amounts remaining in the filtrate will be used to determine adsorption coefficients with fit to a modified Freundlich isotherm. Results will be compared to previous studies performed using granular activated carbon instead of the polysulfone.

(393b) Laser Induced Photothermal Heating and Bubble Generation of Gold Nanoparticles

Ioannis H. Karampelas[1] and Edward P. Furlani[1,2], (1)Dept. of Chemical and Biological Engineering, University at Buffalo (SUNY), Buffalo, NY, (2)Dept. of Electrical Engineering, University at Buffalo (SUNY), Buffalo, NY

Laser Induced Photothermal Heating and Bubble Generation of Gold Nanoparticles

I. H. Karampelas[1] and E. P. Furlani[1,2]

[1]Dept. of Chemical and Biological Engineering, [2]Dept. of Electrical Engineering
University at Buffalo, SUNY, NY 14260 USA

In this study, we present an analysis of photothermal effects associated with nanosecond-pulsed, laser-illuminated gold nanoparticles immersed in aqueous solutions. We use combined computational electromagnetic and fluid analysis to model the fundamental aspects of this photothermal process. More specifically, we study energy conversion within the nanoparticles at plasmon resonance, heat transfer to the fluid, homogenous bubble nucleation and the dynamic behavior of the bubble and surrounding fluid. In addition, we demonstrate the theory via application to various nanoparticle geometries such as spheres, rods and tori. We show that process parameters such as laser intensity, pulse duration, wavelength, polarization and particle geometry can be tuned to control the size and behavior of nucleated bubbles. Moreover, we present results on multi-particle systems and demonstrate that cooperative heating allows for the generation of larger bubbles using significantly less energy as compared to single-particle systems. We discuss details of the modeling approach and specific applications including recent advances in photothermal tissue therapy at the cellular level.

(393ba) Electrospun Metal Oxide Nanofibers for Selective High Temperature CO Detection

Yixin Liu and Yu Lei, Chemical, Materials & Biomolecular Engineering, University of Connecticut, Storrs, CT

A high temperature CO sensor was developed based on electrospun metal oxide nanofibers. The morphology of the nanofibers was investigated by scanning electron microscopy (SEM) and transmission electron microscopy and the cystal structure and composition was examined by X-ray diffaction (XRD). The high temperature stability was investigated by SEM and XRD. The impedance spectroscopy of the as-prepared nanofibers based sensor was investigated in N_2 and different gas mixtures (O_2, CO, CO_2, NO and SO_2 balanced by N_2 in varied concentrations) at 800 °C to study the interaction between the sensing materials and different gases. Sensitive, selective and stable detection of CO in high temperature environment was achieved, suggesting that the as-prepared nanofiber is a promising material for the application of high temperature CO sensors. The interference responses from other gases (O_2, CO_2, NO and SO_2) are examined. The analytical characteristics such as limit of detection, reproducibility, and accuracy will also be discussed.

(393bb) Application of Novel Spectroscopy Methods in Electrochemical Studies

Aria Kahyarian, Chemical and Biomolecular Engineering Department, Ohio University, Athens, OH, Damilola A. Daramola, chemical and biomolecolar engineering, ohio university, Athens, OH and Gerardine G. Botte, Chemical and Biomolecular Engineering Department, Center for Electrochemical Engineering Research, Ohio University, Athens, OH

Application of novel spectroscopy methods in electrochemical studies

Aria Kahyarian, Damilola A. Daramola, and Gerardine G. Botte

Center for Electrochemical Engineering Research

Department of Chemical and Biomolecular Engineering

Ohio University

Athens, OH 45701

In the past few decades, there has been a growing interest in minimizing the dimensions of engineered devices. Electrochemical systems are not exempted from this general trend. Specifically in electrochemical engineering, theoretical framework, principles and equations for design, understanding and diagnostics of electrochemical devices had been developed at the continuum scale. Although these principles have been accepted and used, their validation has been cumbersome due to difficulties associated with measuring properties.

This study is focused on the characterization of electrochemical reactions at the molecular level. In other words, is there any way to validate or modify known electrochemical principals at the macro scale and to extend that to the nano scale systems? This understanding is necessary for the appropriate design of electrochemical devices like super capacitors, micro fluidic and nano fluidic cells, and also, electrochemical reaction kinetics and surface behavior of electrodes.

Recently developed cavity ring down spectroscopy (CRDS) is considered as a very precise and powerful method due to its outstanding measurement limit in the order of ppm and ppb. CRDS was first introduced in 1988 [1] and it has become a well-established method in gas phase concentration measurement after just a few years.[2,3] Unique characteristics of this method over conventional spectroscopy methods include its independence on light source intensity fluctuations, which allows the use of lasers, and the long exposure lengths between the light and the substrate, which is due to the use of optical cavity.

Although CRDS has shown outstanding detection limits in gas phase, its accuracy is limited in liquid phase in liquid phase due to light absorption being orders of magnitude higher in the liquid phase. This fact limits the exposure length and consequently the detection limits. Despite this limitation, this method is still able to measure concentrations in the order of micro and nano moles with a repetition rate of up to MHz and spatial resolution of a few hundred nanometers.[4,5,6,7]

The main goal of this project is the fundamental study of electrochemical principles at the nano scale. Additionally, new methods based on CRDS will be introduced as an alternative way to study electrochemical reaction kinetics during this study. As an example, capillary-based microelectrochemical cells are now well established and showed the importance of the ability to probe electrochemical activity of surfaces on the micro scale. However, this method exhibits complications like the limiting effect of resolution of the potentiostat on the size of tip diameter. These complications are due to natural behavior of electrochemical systems in such dimensions such as high absolute impedance caused by micrometer area of working electrode.[8] Using new methods based on CRDS in order to direct measurement of concentration profile in such a device can be an alternative solution to avoid such limitations.

References

[1] Anthony O'Keefe and David A. G. Deacon," Cavity ring-down optical spectrometer for absorption measurements using pulsed laser sources", Rev. Sci. Instrum. Vol. 59, p 2544-2551, 1988

[2] Giel Berden, Rudy Peeters, and Gerard Meijer, "Cavity ring-down spectroscopy: Experimental schemes and applications", Reviews in Physical Chemistry, Vol. 19, No. 4, p 565-607, 2000

[3] J. J. Scherer, J. B. Paul, A. O'Keefe, and R. J. Saykally, "Cavity Ringdown Laser Absorption Spectroscopy: History, Development, and Application to Pulsed Molecular Beams", Chem. Rev. Vol. 97, p 25-51, 1997

[4] Kate L. Snyder and Richard N. Zare, "Cavity Ring-Down Spectroscopy as a Detector for

Liquid Chromatography", Anal. Chem., Vol. 75, p 3086-3091, 2003

[5] Shucheng Xu, Guohe Sha, and Jinchun Xie," Cavity ring-down spectroscopy in the liquid phase", Rev. Sci. Instrum.Vol. 73, p 255-258, 2002

[6] Helen Waechter, Jessica Litman, Adrienne H. Cheung, Jack A. Barnes and Hans-Peter Loock," Chemical Sensing Using Fiber Cavity Ring-Down Spectroscopy", Sensors,Vol. 10, 1716-1742, 2010

[7] Alexander J. Hallock, Elena S. F. Berman, and Richard N. Zare," Direct Monitoring of Absorption in Solution by Cavity Ring-Down Spectroscopy", Anal. Chem., Vol. 74, p 1741-1743, 2002

[8] N. Birbilis, B.N. Padgett, R.G. Buchheit," Limitations in microelectrochemical capillary cell testing and transformation of electrochemical transients for acquisition of microcell impedance data", Electrochimica Acta, Vol. 50, p 3536–3544 , 2005

(393bc) Electronic Platform to Assess Toxicity of Carbon Nanotubes and Associated Cellular Behavior in Real Time

Reem Eldawud, Chemical Engineering, West Virginia University, morgantown, WV

Carbon nanotubes (CNTs) exhibit many unique properties that allow their use in a wide spectrum of biomedical applications, such as delivery of therapeutically molecules, genes and drugs. However, advances in the biomedical applications of CNTs are being hindered by many uncertainties regarding their cellular uptake mechanisms and fate inside the body. Reports have shown that CNTs toxicity can be attributed to their physical and chemical properties. However, there are no compelling reports that provide fundamental understanding of the toxicological and pharmacological profiles of cellular systems exposed to CNTs. In this research, an electronic platform is used to assess the toxicity of CNTs when incubated with human lung epithelial cells *in vitro*. This non-invasive technique relies on an electrical cell impedance sensing system (ECIS) used as a proxy to assess changes in cellular morphology and cell-cell interactions upon exposure to different concentrations and functionalities of CNTs. Our real time measurements are further correlated with the mechanisms associated with CNTs cytotoxic and apoptotic events by different cellular and optical assays. CNTs were functionalized in the lab to have different chemical and physical properties (i.e. hydrophilicity, length, diameter etc.). Atomic Force Microscopy and Scanning Electron Microscopy along with Raman spectrometry were used to analyze the wide spectrum of functionalized CNTs in an attempt to correlate length, surface chemistry, size distribution, and degree of dispersion with CNTs toxicity as observed using ECIS. Our results provide novel means to investigate cellular toxicity associated with CNTs exposure in real time using a high throughput electronic platform; moreover, our platform and data analysis can be extended to investigate other types of nanomaterials in real time.

(393bd) Magnetic Carbon Nanocomposites for Environmental Remediation

Sowjanya B. Rapole[1], Hongbo Gu[1], Jiahua Zhu[1], Suying Wei[2] and Zhanhu Guo[1], (1)Integrated Composites Laboratory (ICL), Dan F Smith Department of Chemical Engineering, Lamar University, Beaumont, TX, (2)Chemistry and Biochemistry, Lamar University, Beaumont, TX

In-situ fabricated Fe@Fe_2O_3/graphene magnetic nanocomposites (MGNCs) and Fe@Fe_2O_3/polypropylene magnetic nanocomposites (MPPNCs) are tested for the Cr (VI) removal from polluted water and their removal performances and efficiencies are compared. Both nanocomposites have demonstrated extremely fast Cr (VI) removal performances from polluted water, with the MGNCs exhibiting relatively higher removal efficiency with an almost complete removal of Cr (VI) within 5 min as compared to that of 10 min by MPPNCs. The saturation magnetization of the MGNCs (96.3 emu/g) is also found to be greater than that of the MPPNCs (32.5 emu/g). The pH effect on the Cr(VI) removal efficiencies of the NCs is also investigated and the adsorbents have been found to exhibit much higher adsorption capacities in acidic solutions than in alkali solutions. Kinetic investigations reveal the pseudo-second order adsorption of Cr (VI) in both the cases. The significantly reduced treatment periods and easy recycling of the adsorbents from the liquid suspensions in a more energetically and economically sustainable way owing to their large saturation magnetizations makes these nanocomposites promising for the efficient removal of hazardous Cr (VI) from polluted water.

(393be) Magnetic Polymer Nanocomposites for Environmental Remediation

Hongbo Gu[1], Sowjanya B. Rapole[1], Yudong Huang[2], Suying Wei[3] and Zhanhu Guo[4], (1)Integrated Composites Laboratory (ICL), Dan F Smith Department of Chemical Engineering, Lamar University, Beaumont, TX, (2)School of Chemical Engineering and Technology, State Key Laboratory of Urban Water Resource and Environment, Harbin Institute of Technology, Harbin, China, (3)Chemistry and Biochemistry, Lamar University, Beaumont, TX, (4)Dan F. Smith Department of Chemical Engineering, Lamar University, Beaumont, TX

Magnetic Polymer Nanocomposites for Environmental Remediation

Hongbo Gu, Sowjanya B. Rapole, Yudong Huang, Suying Wei and Zhanhu Guo

In this talk, the toxic Cr(VI) removal from polluted water by the magnetic conductive polyaniline (PANI) polymer nanocomposites (PNCs) synthesized using a facile surface initiated polymerization (SIP) method was investigated. The effects of initial Cr(VI) concentration, PNC concentration and pH value on the Cr(VI) removal have been studied. The synthesized PNCs showed unique capability to remove Cr(VI) in solutions with a wide pH range. The saturation magnetization of these PNCs was observed to have no obvious decrease after treatment with Cr(VI) solutions and they can be easily recycled by a permanent magnet from the solution. The Cr(VI) removal kinetics in the different pH solutions was determined. The Cr(VI) removal mechanism was investigated by Fourier transform infrared spectroscopy (FT-IR), X-ray photoelectron spectroscopy (XPS), energy-filter transmission electron microscopy (EFTEM) and temperature dependent resistivity as well.

(393bf) Magnetic Polyolefin-Based Nanocomposites

Qingliang He[1], Suying Wei[2] and Zhanhu Guo[1], (1)Dan F. Smith Department of Chemical Engineering, Lamar University, Beaumont, TX, (2)Chemistry and Biochemistry, Lamar University, Beaumont, TX

Magnetic polyolefin-based nanocomposites were fabricated through a facile low-temperature (~ 140 °C) thermal decomposition of organo-metallic precursor, i.e. $Fe(CO)_5$ or $Co_2(CO)_8$ in solution condition. The whole fabrication includes dissolution of polyolefin-based hosting matrix in refluxing organic solvent followed by the injection of metallic precursor to perform the in-situ thermal decomposition step. The particle sizes, morphology and dispersion quality of these in-situ synthesized magnetic nanoparticles were investigated by transmission electron microscopy (TEM). Room temperature mössbauer spectrum analysis was used to determine the exact species of these magnetic nanoparticles. Room temperature magnetic property investigation was utilized to further reveal the magnetic behaviors for these nanocomposites by specify the saturation magnetization and coercivity. Thermal gravimetric analysis (TGA) was used to determine the thermal stability of these as-prepared nanocomposites and these particle loadings. The formation mechanisms of these magnetic particles were proposed by the evidence of TEM observations and detailed evolutions will be presented as well.

(393bg) Multifunctional Epoxy Nanocomposites

Xi Zhang[1], Johnnie Stewart IV[1], Sowjanya B. Rapole[1], Suying Wei[2] and Zhanhu Guo[3], (1)Integrated Composites Laboratory (ICL), Dan F Smith Department of Chemical Engineering, Lamar University, Beaumont, TX, (2)Chemistry and Biochemistry, Lamar University, Beaumont, TX, (3)Dan F. Smith Department of Chemical Engineering, Lamar University, Beaumont, TX

Three different kinds of nanofillers, polyaniline (PANI), barium titanate ($BaTiO_3$) and $BaTiO_3$ nanoparticles coated with PANI layer are applied to enhance the dielectric permittivity of epoxy resin. The conductive $BaTiO_3$-PANI polymer nanocomposites (PNCs) were synthesized by a surface initiated polymerization (SIP) method, the amino group of PANI can react with epoxy monomer and improve the dispersion of the $BaTiO_3$ nanoparticle in epoxy resin. For all the series of samples, mechanical property, rheological property, thermal stability, electrical conductivity and dielectric property are systematically studied. It is observed that all three nanofillers can help to increase the real permittivity of epoxy resin, however, at same loading, the increment resulted from the $BaTiO_3$-PANI polymer nanocomposites is much higher than the increment caused by the other two and the mechanism of the difference are analyzed theoretically. In addition, reduced viscosity is found in the Rheology test based on liquid samples, at lower loading liquid epoxy samples with both PANI and $BaTiO_3$ nanoparticles has lower viscosity than pure epoxy, however, at higher loading, the viscosity increases with the loading increase. PANI as the soft nanofillers can also increase the conductivity of matrix, for epoxy with PANI nanocomposites, both nanoshpere and nanofiber are applied to explore the morphology effect, at same loading, epoxy with PANI nanofibers has lower volume resistance than epoxy with nanospheres, which is resulting from the contact resistance and a 3-d variable range hopping (VRH) electron transport mechanism.

(393bh) Synthesis, Dispersion and Toxicological Studies of ZnO Nanoparticles for Agricultural Applications

Ankit Goyal[1], **Nitin Roy**[1], Ramamurthy Nagarajan[2], Swayamprava Dalai[3] and Amitava Mukherjee[3], (1)Department of Chemical Engineering, Indian Institute of Technology,Madras, Chennai, India, (2)Department of Chemical Engineering, Indian Institute of Technology, Madras, Chennai, India, (3)Vellore Institute of Technology, Vellore, India

Zinc is an important micro nutrient both for plants and humans. In recent years zinc deficiency in humans has received increasing attention both by nutritional and medical scientists. A very good answer to this problem can be replenishment of zinc content in agricultural soil. Application of zinc containing fertilizers represents a quick and useful solution to bio-fortification of cereal grains with Zn that can be easily practiced. Among all available zinc sources, zinc oxide(ZnO) is the cheapest and contains highest percentage of zinc. This has tempted us to use ZnO as source of zinc. Though zinc oxide seems to be the best option but it has some drawbacks like inability to form stable suspension with fertilizers. Micron sized and crystalline Zinc oxide particles shows resistance to absorption by plants. Zinc oxide in nano size range has the potential to overcome these issues. Due to its impressive properties like small size, high surface to volume ratio, ability to form stable dispersions nano zinc oxide can be panacea to these problems. The whole strategy behind the work was to synthesize nano-particles , test dispersion of commercially available nano zinc oxide and to study toxicology due to reduced size. Top down approaches were adopted for synthesis process. Size reduction was achieved both by ball milling and ultrasound. Traditional methods like ball milling are time consuming, inefficient and come with penalty of contamination. Fragmentation was also performed using ultrasound. Sono-fragmentation is an ecofriendly and facile technique for size reduction as compared to other bottom up approaches which are complex and require use of various chemicals under highly controlled environment. Acoustic waves can tear off the liquid which can create cavitation bubbles. Implosion of these bubbles can accelerate particles at high speeds, can generate shock waves or can form micro jets on the surface of particles with speeds as high as 100m/s. These mechanisms can break the particles into smaller pieces or colliding particles can melt locally at point of contact and rapid cooling can fuse them. The same can also be used efficiently for fast and uniform dispersion of the nano zinc oxide particles. Dispersion stabilization is provided by two major mechanisms: electrostatic and steric stabilization. But since zinc oxide have high surface energy and the its aqueous dispersion is highly sensitive to pH, it is unsuited for electrostatic stabilization. In our studies we varied various parameters to find optimum conditions for fragmentation. Samples were characterized by SEM, laser particle counter and XRD. Results showed size reduction and agglomeration in some cases. In some experiments XRD studies have also shown increased reactivity of ultrasound treated ZnO. To avoid agglomeration, studies were also performed using surfactants. SEM images showed change in morphology of particles from highly crystalline to spherical. Thus increased reactivity of particles due to alteration of physical properties, like shape, will facilitate formation of chelates with organic matter in soil, which can be easily absorbed by plants. Aqueous dispersion of commercial nano ZnO was also prepared and analyzed quantitatively using different surfactants and at different pH levels of the solution. The stable dispersions of zinc oxide prepared can be easily stored and used for uniform impregnation of zinc in the fertilizers. Toxicological studies of ZnO nano-particles on soil bacteria showed reduced viability even at very low concentrations like 1 μg/ml which was later confirmed to be due to Zn^{2+} ion toxicity. Interestingly, the same ionic concentration was found to be growth promoting in plants, causing faster seed germination and root elongation whereas ZnO nanoparticles showed toxic response. Hence, formulating stable nano-particle dispersion with controlled release of Zn^{2+} ions from the nano-particles will be helpful in agricultural aspects.

(393bj) Aggregation of Humic Substances: Effects of Filtration

Milad R.Esfahani, Chemical Engineering, Tennessee Technological University, Cookeville, TN, Vasanta Pallem, Department of Chemical Engineering and Center for the Management, Utilization and Protection of Water Resources, Tennessee Technological University, Cookeville, TN, Holly A. Stretz, Department of Chemical Engineering, Tennessee Technological University, Cookeville, TN and Martha J. M. Wells, EnviroChem Services, Cookeville, TN

Humic acids are complex polyelectrolytes which exhibit various conformations and aggregation states depending on the solution chemistry. In the current research, dynamic light scattering and fluorescence spectroscopy were used to investigate the size distributions and fluorescence quenching behavior for the dissolved organic fractions of commercial humic acid (Sigma Aldrich grade). The size distribution results suggested that the HA molecules re-aggregate after filtration through 0.45 mm nylon fiber filters, and form significantly large micron sized aggregates even at low concentrations around 8 ppm dissolved organic carbon (DOC). The size distributions for the total organic carbon (TOC) were compared with the filtered samples as a function of carbon concentration. These results suggest that humic substances are dynamic in nature and their size is constantly changing due to interactions with other HA molecules. The fluorescence of DOC at different carbon concentrations was examined to detect potential fluorescence quenching behavior due to re-aggregation of HA molecules after filtration.

(393c) Enzyme Immobilization and Stabilization Into Polyaniline Nanofibers for Biofuel Cells

Inseon Lee, Hyeongseok Kim and Jungbae Kim, Chemical and Biological Engineering, Korea University, Seoul, South Korea

A model enzyme, glucose oxidase (GOx), was immobilized and stabilized into polyaniline nanofibers (PANFs) via a three-step process, consisting of enzyme adsorption, precipitation, and crosslinking (EAPC). EAPC of GOx showed 9.6 and 4.2 times higher enzyme activities per unit weight of PANFs than those of enzyme adsorption (EA) and enzyme adsorption/crosslinking (EAC), respectively. Under rigorous shaking at room temperature for 56 days, the relative activities of EA, EAC and EAPC, defined as the percentage of residual activity to the initial activity, were 22%, 19% and 91%, respectively. The potential application of EAPC-GOx in enzyme-based biofuel cells was checked by preparing the enzyme anode via Nafion-entrapment of EAPC-GOx/PANFs onto the carbon paper. The two critical problems of biofuel cells, low power densities and short lifetime, could be solved in part by using highly-loaded and stable EAPC-GOx. The maximum power density could be maintained without any loss even after incubation for 2 months or thermal treatment at 60°C for 4 hours, suggesting the good performance stability of the EAPC anode. It is anticipated that highly-stable EAPCs/PANFs will be employed in various other enzyme applications such as bioconversion and bioremediation as well as electrochemical biocatalysis.

(393d) Enzyme Stabilization in Magnetically-Separable Mesoporous Silica Via Ship-in-a-Bottle Approach for Uses in Nonaqueous Solvent

Byoungsoo Lee[1], Ee Taek Hwang[2], Man Bock Gu[2] and Jungbae Kim[1], (1)Chemical and Biological Engineering, Korea University, Seoul, South Korea, (2)College of Life Sciences and Biotechnology, Korea University, Seoul, South Korea

Magnetically-separable mesocellular silica (Mag-MSU-F), with larger mesocellular pores of 38 nm connected by smaller window mesopores of 18 nm and deposited by magnetic nanoparticles, were used for the immobilization and stabilization of subtilisin Carlsberg (SC) from Bacillus licheniformis in the form of nanoscale enzyme reactors (NERs). NERs of SC (NER-SC) were prepared via a simple two-step process: enzyme adsorption and crosslinking. The crosslinked enzymes in large mesocellular pores cannot leach out through smaller window pores, creating a ship-in-a-bottle approach as one of the stabilization mechanisms. The other stabilization mechanism would be the multi-point covalent linkages that can effectively prevent the denaturation of enzymes. As a result, NER-SC maintained 66% of initial activity even after incubation under shaking (200 rpm) for 15 days while free and adsorbed SC showed less than 1% and 2.8% of initial activities, respectively. Stable NER-SC was successfully used for the transesterification of N-acetyl-L-phenylalanine ethyl ester (APEE) with n-propanol in isooctane. Magnetic separation of Mag-MSU-F facilitated the repeated uses of stable NER-SC. This is the first demonstration for the uses of stable and magnetically-separable NERs in organic solvents, which will create a great potential for various synthetic reactions including the biodiesel production.

(393e) Immobilization and Stabilization of Lipase in Polyaniline Nanofibers for Esterification of Racemic Ibuprofen

Sung-Gil Hong[1], Hansol Kim[1] and Jungbae Kim[2], (1)Chemical and biological engineering, Korea University, Seoul, South Korea, (2)Chemical and Biological Engineering, Korea University, Seoul, South Korea

Lipase (LP) is one of useful enzymes in various fields such as food, oil and fat, detergent, cosmetics and pharmaceutical industries. We immobilized and stabilized lipase into polyaniline nanofibers (PANFs) via a three-step process of enzyme adsorption, precipitation and crosslinking (EAPC). EAPC-LP/PANFs resulted in high enzyme loading/activity and stability. EAPC showed 5.1 and 5.9 times higher activity than enzyme adsorption (EA) and enzyme adsorption/crosslinking (EAC), respectively. After incubation under shaking at room temperature for 84 days, EA, EAC and EAPC retained 11%, 24% and 74% of initial activities, respectively. The EAPC approach successfully increased both stability and loading/activity of enzymes, which can be employed in various fields of enzyme applications. As a potential application, EAPC-LP/PANFs were employed for the resolution of racemic ibuprofen via esterification with n-propanol and water in isooctane including AOT. For 102 hours under 100mM AOT, EAPC showed 42% conversion of racemic ibuprofen, whereas EA and EAC showed 1.2% and 1.4% of conversion, respectively.

(393f) Nanoscale Enzyme Reactors in Mesoporous Carbons for Sensitive and Reliable Biosensors

Jae Hyun Kim, Chemical & Biological Engineering, Korea University, Seoul, South Korea and Jungbae Kim, Chemical and Biological Engineering, Korea University, Seoul, South Korea

It has been demonstrated that the approach of nanoscale enzyme reactors (NERs), consisting of enzyme adsorption and crosslinking, can improve both the enzyme stability and loading in mesoporous materials with a bottle-neck structure. These advantageous features of NERs result from the ship-in-a-bottle mechanism, in which crosslinked enzymes in larger-sized mesocellular pores cannot be leached out through smaller-sized bottle-neck mesopores. MSU-F-C, a mesoprous carbon, has a bottleneck pore structure with mesocellular pores of 26nm connected with window mesopores of 17 nm. When compared to simple adsorption with no crosslinking, the NER approach proved to be effective in improving the stability and sensitivity of electrochemical biosensors. The performance of biosensors is governed by the combination of high enzyme loading and follow-up electron generation. The improved performance with NERs of glucose oxidase (GOx) and organophosphorus hydrolase (OPH) in conductive mesoporous carbons will be presented in comparison with free and adsorbed enzyme. The excellent stability of the NER creates new potential of practical, environmental, and electrochemical applications of biosensors.

(393g) Interfacial Synergy Between Energy and Nanomaterials

Randy L. Vander Wal, Energy and Mineral Engineering & The EMS Energy Institute, Pennsylvania State University, University Park, PA

Introduction

Energy and Nanomaterials are intimately related. Most energy processes occur at the interface. Nanomaterials, when used as interfacial modifiers have the potential to significantly alter the energy landscape. Although carbon nanomaterials have been explored for a suite of energy applications, to-date the approach has been based on replacing bulk materials, not as interfacial modifiers. The work presented here will highlight this latter approach by showing results for a suite of energy applications from the author's own work.

Experimental

Experimental details will be presented at the presentation with regards to the synthesis and utilization of the nanomaterials. Our applications include energy conversion, generation, storage, efficiency, conservation and control.

Results and Discussion

Nanoscale materials are redefining the relation between material composition, size and properties. Chemical properties (e.g. reactivity) and physical properties (e.g. surface area) become a strong function of size at the nanoscale. Applications illustrated include the following selected examples from the author's work: catalysis, composite materials, energy storage, sensors, thermal management, and tribology. The key concept is that the nanomaterials serve as interfacial modifiers. Since most energy related processes are dominated by interfacial reactions, nanomaterials have the potential to dramatically affect energy conversion rates and magnitudes. Examples follow.

Catalysis is central to particulate and NO_x after-treatment systems. A prime example of reducing materials to nanoscale and realizing new properties is catalysis by nanoscale gold. Au nanoparticles supported on oxides such as CeO_2, TiO_2 and Fe_2O_3 offer ambient temperature oxidation of CO, volatile organic compounds (VOCs) and potentially exhaust hydrocarbons. Examples will be presented.

Lightweight materials will reduce weight significantly yielding substantial benefits in fuel efficiency with reduced emissions. Composites with substantial gains in Young's modulus, tensile strength, and EM shielding may be realized in polymeric composites using carbon nanotubes as an interfacial modifier rather than bulk filler [1]. As will be described, the interfacial modification of a metal foil using carbon nanotubes directly grown upon the foil. The purpose of the foils is to serve as a gas impermeable barrier layer within the polymer composite.

Advances in energy storage include batteries, ultra-capacitors. These can support lighting, appliances, a starter, cooling fans, transmission and hydraulic systems, fuel and air handling systems and ultimately enable hybrid systems. Towards these goals, substantial gains in Li ion battery cathode and anode materials have been realized using carbon nanofibers, coating processes and including elements such as tin and silicon. Modifications of the CNT surfaces increases the Li ion capacity beyond the theoretical limit of normal graphite. As an illustration, CNTs were directly fabricated for this end-application use upon ultra-fine SS mesh. Significantly no harvesting, purification or processing (using binders) was required. The CNTs could be used directly as synthesized. TEM and SEM images will be presented of the CNTs upon the SS mesh and discharge (de-intercalation) curves [2].

With regards to overall system integration and control, sensors will play a prominent role [3]. With ultrahigh surface exposure relative to bulk material, nanoscale materials are exceedingly sensitive to gas adsorption. Exploitation of nanoscale properties will lead to new NO_x sensors and in-cylinder oxygen sensors. Examples include catalyst coated metal oxide semiconductors capable of ambient temperature operation, as will be illustrated in the talk.

Thermal management will benefit from nanofluids. Nanofluids can increase thermal conductivity and reduce radiator and heat exchanger size. Carbon-based nanofluids using nano-onions and carbon nanotubes have increased water conductivity by ~ 20% [4]. Examples as shown by HRTEM images of such nanocarbon additives will be compared during the presentation.

Lubrication is critical to many engine components and powertrain systems. Nanolubricants can bridge the gap between fluid and solid materials [5]. As additives with liquids or greases, synergistic properties may be realized, particularly in boundary-phase lubrication. Improved coating formulations and properties can reduce or eliminate fretting and pitting. Results with nanocarbons show superior performance relative to graphite, diamond like carbon (DLC) and even Teflon, as will be detailed.

Conclusions

Nanoscale materials are redefining the relation between material composition, size and properties. Their applications to energy processes include selected examples from the author's work. Selected technologies include catalysis, composite materials, energy storage, sensors, thermal management, and tribology. Interfacial modification serves to affect changes in energy conversion rates, and transfer magnitudes. This appears to be a rather universal concept and suggests R&D directions for nano-engineering of interfaces.

Acknowledgements

Funding through The Penn State Institutes for Energy and the Environment (PSIEE) and the Pennsylvania Keystone Innovation Starter Kit (KISK) is gratefully acknowledged.

References

(1) Vander Wal, R. L., and Hall, L. J., Advanced Engineering Materials **2004,** 6, 48-52.

(2) Luo, Y., Vander Wal, R. L., and Scherson, D. A., Electrochemical and Solid State Letters **2003,** 6, A56-A58.Vander Wal, R. L., Mozes, S. D., and Pushkarev, Nanotechnology ***2009***, 20, 105702-10.

(3) Hunter, G. W., Vander Wal, R. L., Liu, C. C., Xu, J. C., and Berger, G. M., Sensor and Actuators B ***2009***, 138, 113-119.

(4) Vander Wal, R. L., Mozes, S. D., and Pushkarev, Nanotechnology ***2009***, 20, 105702-10.

(5) Street, K. W., Miyoshi, K., and Vander Wal, R. L., "Application of Carbon Based Nano-particles to Aeronautics and Space Lubrication", in *Superlubricity*, Chapter 19, pp 311-340. Edited by Drs. Jean-Michel Martin and Ali Erdemir, Elsevier, Amsterdam (2007). (also NASA/TM-2007-214473).

SYNTHESIS AND CHARACTERIZATION OF TEMPERATURE-RESPONSIVE VO_2/PNIPA NANOCOMPOSITE HYDROGELS

Mi Wang, [1,2] Jingxing Feng, [1] Xuhong Guo, [1,*] Yanfeng Gao [2]

[1] *State Key Laboratory of Chemical Engineering, East China University of Science and Technology, Shanghai, China.*

[2] *Shanghai Institute of Ceramics (SIC), Chinese Academy of Sciences (CAS), Shanghai, China*

[*] *Corresponding authors. Tel.: +86-21-64253491; E-mail: guoxuhong@ecust.edu.cn*

Introduction

Thermochromic materials are perfectly suited for application in smart windows, which show obvious advantages, such as no additional energy consumption, high switching range between off- and on-state and low effort for installation. In particular, it is amazing that the solar energy itself is used as a promoter against sun heat. As a typical thermochromic material, vanadium dioxide (VO_2) has been widely studied for energy-efficient windows because of its high color contrast, flexible switching speed and low response time. VO_2 exhibits a phase transformation from a semiconductor to a metal at Tc (68 °C for bulk VO_2), accompanied by the optical properties in the near-infrared region from a low-temperature transparent state to a more reflective state at high temperatures. However, the high cost of material synthesis limits their practical applications, so organic alternatives have also attracted much attention. Poly(N-isopropyl acryl amide) (PNIPA) is the first reported non-ionic hydrogel exhibiting a temperature dependent phase transition at 32 °C (the lower critical solution temperature, LCST). PNIPA hydrogels show a temperature dependent phase separation, which can change from a hydrophilic phase to a hydrophobic phase, accompanied by a transparent state at low temperature to a turbid state at high temperature. To avoid the strong volume reduction during phase transition, much research has been concentrated on producing water-soluble microgels. Recently, organic/inorganic composites often possess unique properties to organics and inorganics, so it is of great importance to composite VO_2 and PNIPA hydrogels.

Experimental

Thermosensitive hydrogels with the lower critical solution temperature (LCST), were prepared by free radical polymerization based on PNIPA, using N, N'-methylenebisacrylamide (BIS) as crosslinker and potassium persulfate (KPS) was used as thermal initiator. By adjusting the ratio of the crosslinker to the monomer, PNIPA microgels with LCST at about 32 °C were prepared (**Fig.1**). When the amount of BIS was 2 wt% of NIPA, the product showed a visible transmittance value more than 50% at 25 °C and less than 10% at 80 °C (**Fig.2**).

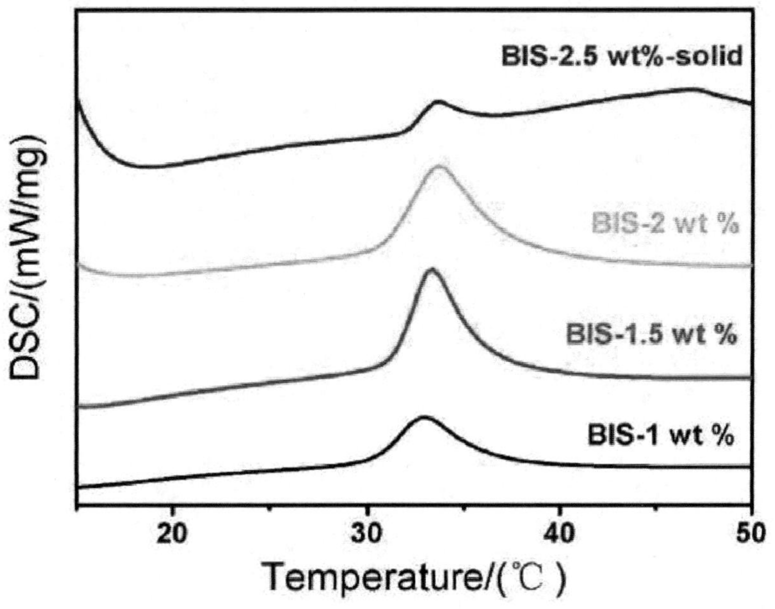

Figure 1. DSC curves of the synthesized hydrogels.

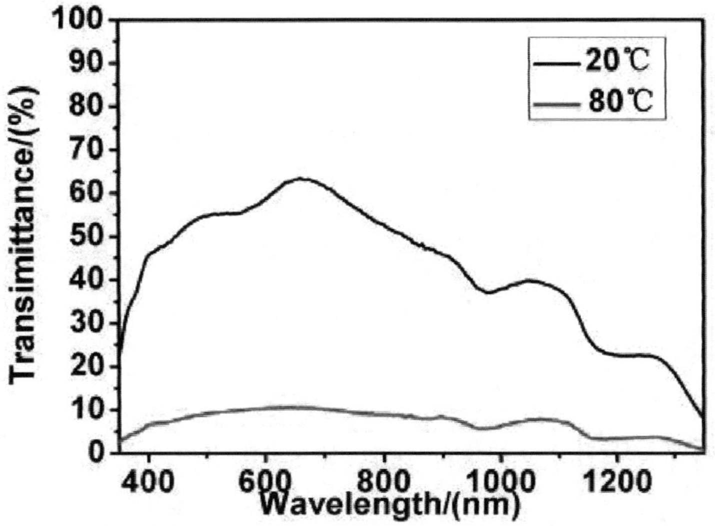

Figure 2. Transmittance (T%) for the microgels with BIS 2 wt% of NIPA.

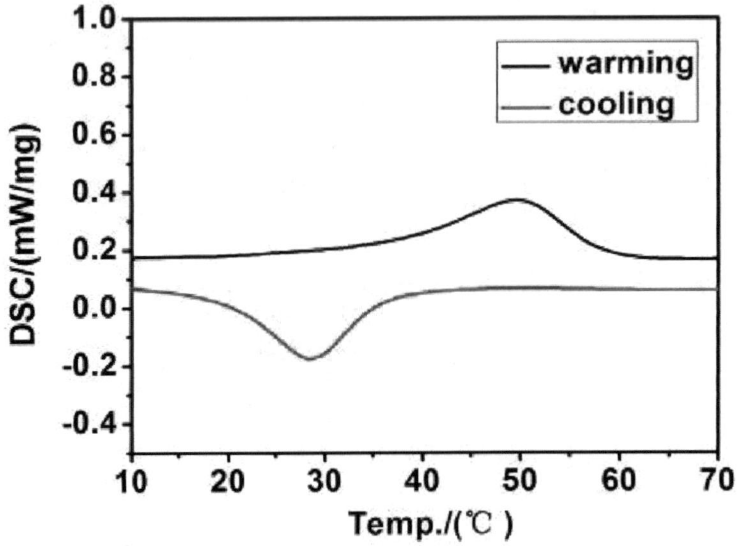

Figure 3. DSC curves of the synthesized VO_2 nanoparticles.

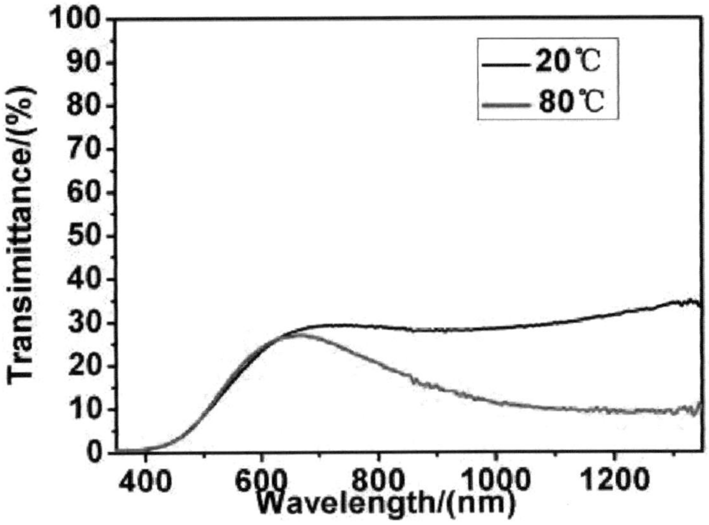

Figure 4. Transmittance (T%) of the synthesized VO_2 nanoparticles.

VO_2 nanoparticles (~ 50 nm) were synthesized by the hydrothermal method using V_2O_5 as initial materials. A small amount of WO_3 was added to adjusted the phase transition temperature of VO_2. The prepared VO_2 nanoparticles showed the phase transition temperature at about 50 °C (**Fig.3**) and exhibited a light regulation capability of about 14% in the wavelength range from 350 nm to 1350 nm **(Fig.4)**.

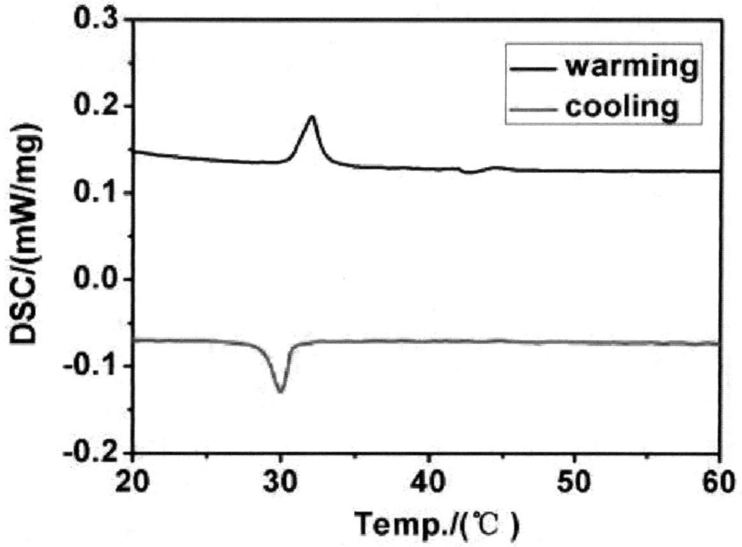

Figure5. DSC curves of the VO_2/PNIPA composite hydrogel with VO_2 1 wt% of NIPA.

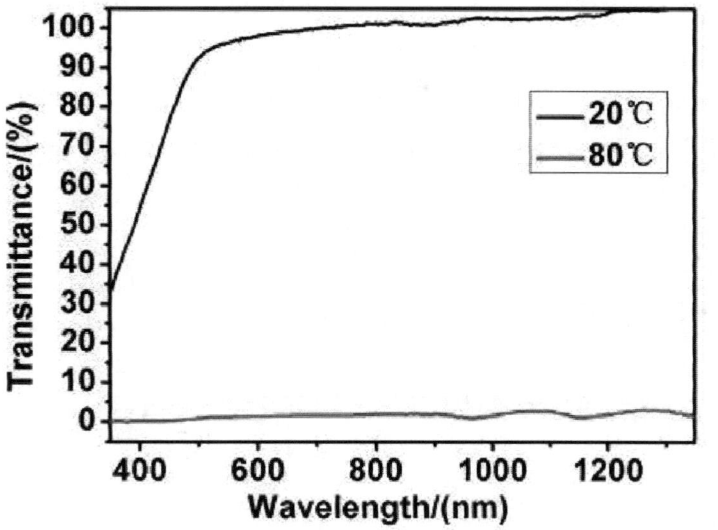

Figure 6. Transmittance (T%) of the VO_2/PNIPA composite hydrogel with VO_2 1 wt% of NIPA.

VO_2 with the weight 0.5%, 1% and 2% of NIPA were respectively added and dispersed in water before polymerization. Results show that VO_2 does not influence the LCST of the PNIPA hydrogel

(Fig.5). As increasing VO_2 content, the crosslinking density decreases and the viscosity firstly falls down and then rises up. It is supposed that VO_2 plays a role as crosslinker in the reaction due to its negatively charged surface. When the VO_2 weight is 2% of NIPA, the product possesses visible transmittance less than 50% at low temperature and VO_2 nanoparticles precipitates at high temperature. The other two samples both show excellent stability and satisfactory light modulation efficiency, about 88% (350 ~ 1350 nm), but the sample containing VO_2 with the weight 1% of NIPA flows more freely **(Fig.6)**.

References

1. Stuart MAC, Huck WTS, Genzer J, et al. Emerging applications of stimuli-responsive polymer materials. *Nature Materials.* 2010;9(2):101-113.
2. Zhang Z, Gao Y, Luo H, et al. Solution-based fabrication of vanadium dioxide on F: SnO_2 substrates with largely enhanced thermochromism and low-emissivity for energy-saving applications. *Energy Environ. Sci.* 2011;4(10):4290-4297.
3. Wang Z. Phase and shape controlled VO_2 nanostructures by antimony doping. *Energy Environ. Sci.* 2012.
4. Gao Y, Luo H, Zhang Z, et al. Nanoceramic VO_2 Thermochromic Smart Glass: A Review on Progress in Solution Processing. *Nano Energy.* 2011.
5. Dai L, Cao C, Gao Y, Luo H. Synthesis and phase transition behavior of undoped VO_2 with a strong nano-size effect. *Sol Energ Mat Sol C.* 2011;95(2):712-715.
6. Jin J, Xue JP, Cui SS, Cao J, Zhang DY, Gu YQ. Synthesis and Characterization of a Novel Folate-Conjugated Themoresponsive Micelles. *Advanced Materials Research.* 2012;529:555-559.
7. Elsaeed SM, Farag RK, Maysour NS. Synthesis and characterization of pH - sensitive crosslinked (NIPA - co - AAC) nanohydrogels copolymer. *J Appl Polym Sci.* 2012.
8. Wei HL, Yao K, Yang Z, et al. Preparation of thermosensitive hydrogels by means of tandem physical and chemical crosslinking. *Macromol Res.* 2011;19(3):294-299.
9. Seeboth A, Ruhmann R, Mühling O. Thermotropic and Thermochromic Polymer Based Materials for Adaptive Solar Control. *Materials.* 2010;3(12):5143-5168.

(393i) Effects of the Nanoscale in Interfaces Solid-Gas-Liquid

Laura M. Calle and Watson L. Vargas, Department of Chemical Engineering, Universidad de los Andes, Bogotá, Colombia

It is well established that reducing the sizes of particles leads to a dramatic increase in the portion of surface/interface atoms. Since the properties of a solid are essentially controlled by related surface/interface energies such effects affect many physicochemical properties that can be observed at the macroscale. Although such effects are believed to be present every time that nanoparticles are added to a given system, particularly at interfaces where gas-solid-liquid coexist; there are few experimental experiences for the expected phenomena that we are ought to "see" at the macroscale.

This work aims to analyze the influence of the nanoscale properties (e.g particle size) in thermodynamics and physicochemical interactions of substances, in a mixture that contains nanoparticles, through the study of the dynamics of a falling drop and the observation of changes during the drop impact by introducing nanoparticles of known size. The evolution of the drops has been recorded by a high-speed camera and image analysis techniques are used for obtaining quantitative data from the details of drop impact, initial drop deformation and further spreading (a phenomena which is driven mainly by surface tension) and where nanoscale effects will manifest more evidently.

Recent theoretical studies predict that solid–vapor interface energy, liquid–vapor interface energy, solid–liquid interface energy, and solid–solid interface energy of nanoparticles and thin films fall as the diameter or thickness decreases to several nanometers while the solid–vapor interface energy ratio is size-independent and equals that of the bulk [1]. Some results and discussion will be provided regarding the validity of these theoretical predictions.

Keywords: Nanoscale effects, Nanoparticles, surface energy.

References:

[1] Q. Jiang and H.M. Lu Size dependent interface energy and its applications. *Surface Science Reports* 63 (2008) 427–464

Microstructure of magnetic nanoparticles in spherical polyelectrolyte brushes observed by Small Angle X-Ray scattering

*Weihua Wang, Xuanji Yu, Fangfang Chu, Yan Zhu, Haoya Han, Kaimin Chen, Xuhong Guo, Li Li**

Introduction

The research of spherical polyelectrolyte brushes (SPB) concerning about synthesis, characterization and biotechnological applications has been one of the most attractive fields in material chemistry and physics in the recent decades[1-5]. SPBs are dispersed in water and consist of solid polystyrene (PS) core and linear chains of polyacrylic acid (PAA), which densely grafted onto the PS core. Due to the particular properties, SPBs can be used as nanoreactors for the synthesis or adsorption of tiny nanoparticles, such as proteins, enzymes, metal nanoparticles, etc. Magnetic Fe_3O_4 nanoparticles (MNPs) are the most widely studied magnetic particles generated in SPBs which has been reported for several times[6-10]. Containing MNPs made it possible for SPBs to be recyclable easily by appending and removing outer magnetic field. The MNPs used in this investigation are successfully synthesized in situ using SPB as a nanoreactor as shown in Figure 1.

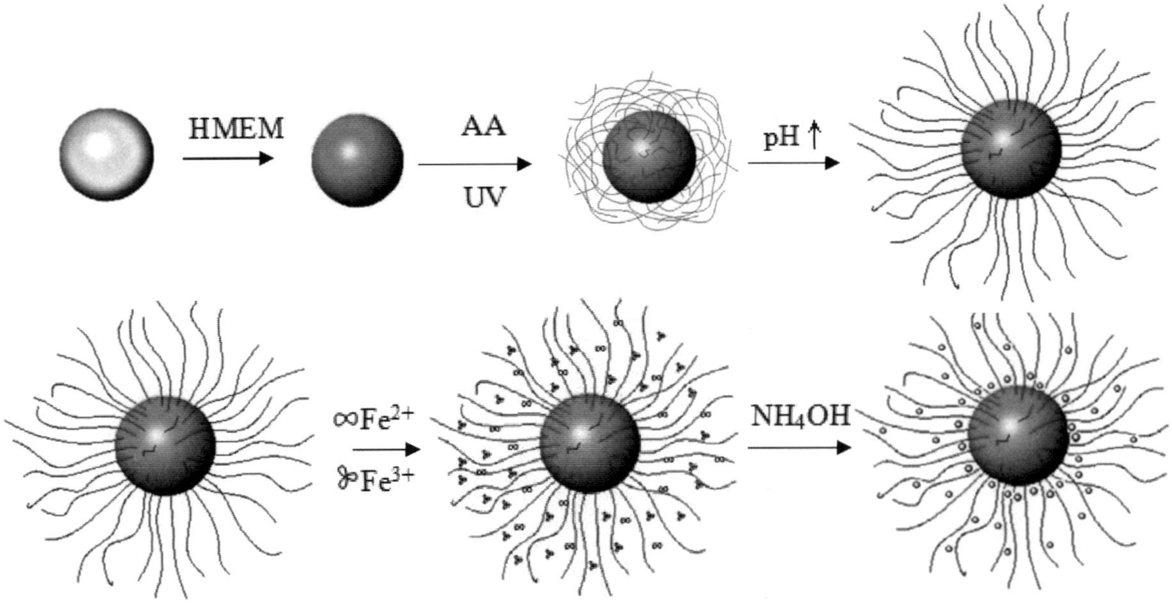

Figure 1. Synthesis of magnetic spherical polyelectrolyte brushes.

Experimental

SPBs were prepared by photo-emulsion polymerization (Figure 1)[11, 12]. In the first step, the surface of polystyrene (PS) was covered by the photoinitiator 2-[p-(2-hydroxy-2-methylpropiophenone)]- ethyleneglycol-methacrylate (HMEM). In the second step, poly(acrylic acid) (PAA) chains were grafted from the surface of PS core under UV radiation. Then Fe^{2+} and Fe^{3+} solution (with the mole ratio 2:3) were added into SPB emulsion simultaneously with the protection of N_2 atmosphere followed by addition of aqueous ammonia (1 wt%). Finally MNPs were prepared within SPB.

Small angle x-ray scattering (SAXS) is a suitable method to observe the microstructure and interaction of nanoparticles, especially for SPBs.[13-15] In this work, SAXS was employed to investigate the structure and interactions among SPBs with MNPs immobilized. Generated MNPs can be adsorbed strongly into the PAA chains due to the intense interaction of chemical coordination - between carboxyl groups in PAA and MNPs. The SAXS curves legibly reveal that the PAA chains are virtually uncharged at low pH. With increasing pH the chain length increases gradually until getting full extended at high pH (>8). The influence of salt concentration is mainly reflected in the interaction among SPBs. High salt concentration can shield the interaction between PAA chains completely. The influences of these factors make obvious changes of the radius of gyration. Since the obvious electron density contrast ($\Delta\rho$) between PAA, PS, MNP(Fe_3O_4) and the medium (shown in table 1), SAXS is a especially suitable for microstructure investigation of SPBs and the generated MNPs onto SPBs[16-17].

Table 1. Electron density contrast of studied chemicals

Material	ρ^e	$\Delta\rho^e$ (ρ^e-ρ^e_{H2O})
Water(H_2O)	333.3	0
Poly(styrene)(C_8H_8)	326.9	6.4
Poly(acrylic) acid($C_3H_4O_2$)	429.3	96
MNP(Fe_3O_4)	1476.3	1143

In addition, transmission electron microscope (TEM) was used in order to intuitively detect the morphology of SPB and MNP. Dynamic Light Scattering (DLS) was applied to test hydrodynamic radius which would be compared with the radius of gyration by SAXS as shown in Figure 2[18].

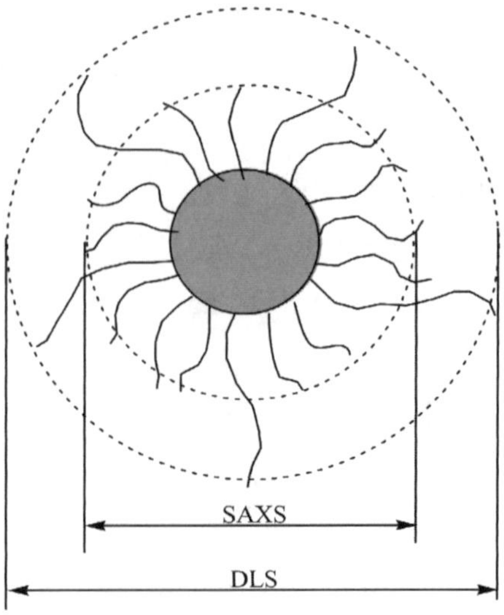

Figure 2. Schematic representation of SPB size determination by SAXS and DLS

Results and discussion

The morphology of SPB and MSPB can be observed by TEM shown in Figure 3. Scattering curves of MSPB and SPB can be obtained by SAXS due to the enough electron density contrast between MNP, SPB and the medium at the same polymer concentration and pH (Figure 4). The model of electron density excess distribution of PAA shell was established in the inset of Figure 4. We changed the parameters of distribution until the fitting results match the experimental data. It is observed in Figure 6 that there is an obvious shift of the maximum position of MSPB scattering curves toward the smaller q region compared with the scattering curve of SPB. The radius of gyration of MSPB decreased.

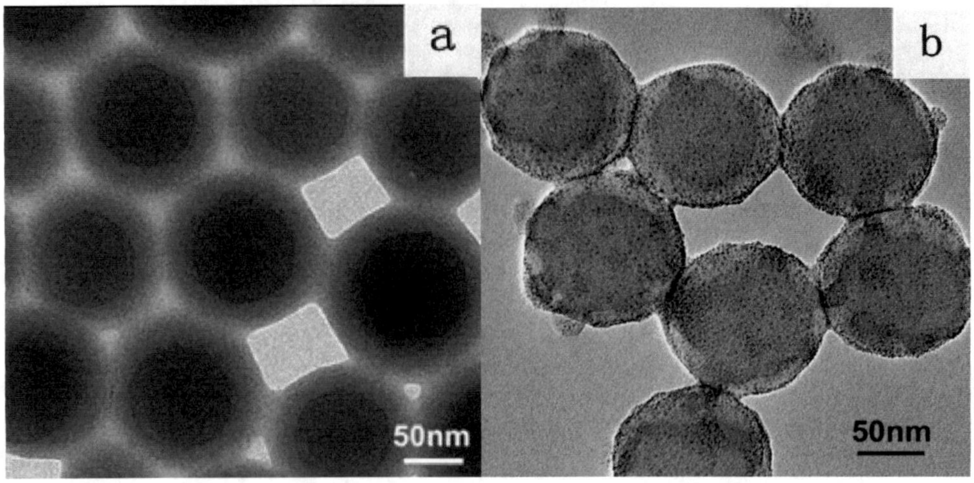

Figure 3. TEM photographs of SPB and MSPB at pH=11. (a) SPB; (b) MSPB.

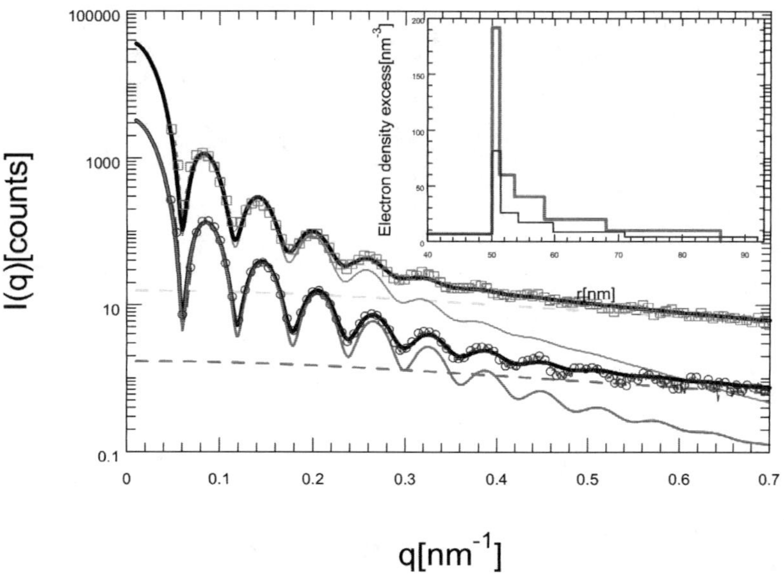

Figure 4. Scattering intensity curves of MSPB and SPB at pH=11. The circles and squares present the measured scattering intensities of SPB and MSPB, respectively. The profiles of electron density contrast Δρ of SPB and MSPB were shown in the inset with light and heavy line, respectively. Both intensity curves are normalized to the mass fraction of SPB.

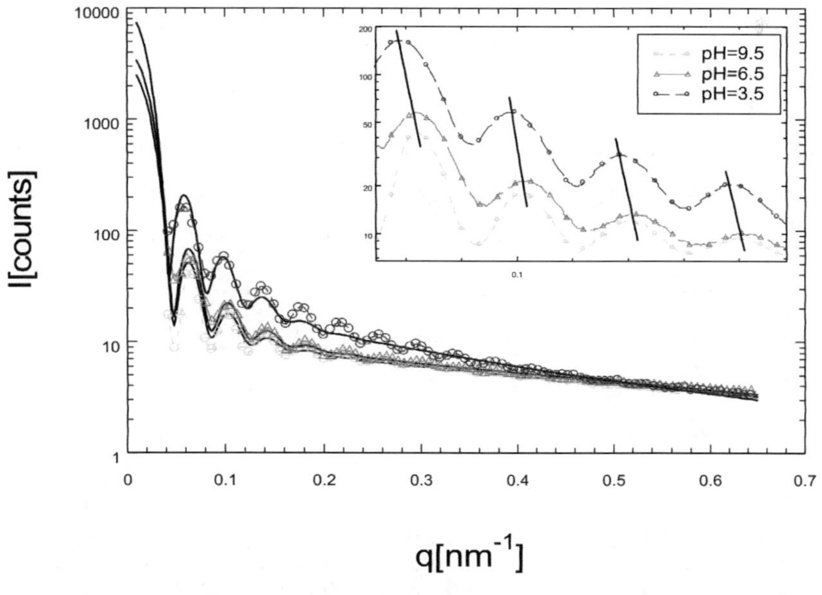

Figure 5. Scattering curves of MSPB at different pH (circle, pH=3.5; inverse triangle, pH=6.5; fitting square, pH=9.5) are shown in the picture. The inset shows locally zoomed scattering intensity curves in small q region. The radius of the PS core of all samples is 50 nm.

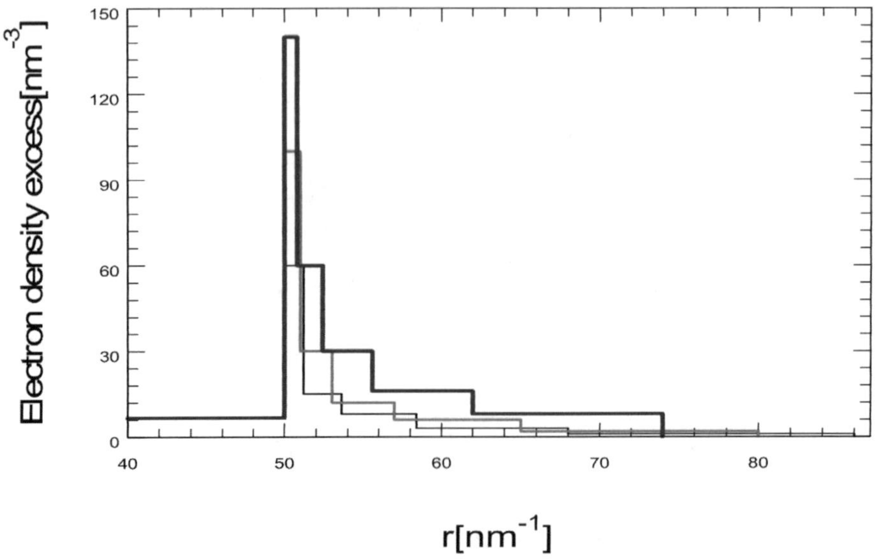

Figure 6. The profiles of electron density contrast distribution of MSPB, which were used to fit the experimental data shown in Figure 5, respectively. The thick line is the high electron density excess distribution of PAA chains at pH=3.5 with apparent radius of gyration 54 nm. The thin line and the dashed line are the distribution of $\Delta\rho$ at moderate pH=6.5 and high pH=9.5 with the apparent radius of gyration of 53 nm, 52 nm respectively.

Figure 5 displays the scattering curves of MSPB at different pH from 3.5 to 9.5. Good fitting results are obtained according to the model in which the electron density excess of PAA chains decreases gradually in Figure 6. With the increase of pH the PAA chains increase gradually and get fully stretched state finally, which can be proved by the change of apparent size of established model from the thick line (L=24nm) to the tiniest line (L=36nm). From the fitting results of Figure 5, we can also obtain the specific changing parameters. The distribution degree of PAA statistical length changes from 0.04 to 0.05 towards the direction of more inhomogeneous particle size. What's more, from the fitting parameters we can conclude that the effect of increasing electron density excess is more significant than the reduction on particle size. This offers a favorable proof that MNPs have entered into the brushes layer and distributed mostly on the interface of PS core and PAA shell.

References

1. Physical Chemistry of Biological Interfaces; Baskin, A., Norde, W., Eds.; Marcel Dekker: New York, 1999.
2. Seigel, R. R.; Harder, P.; Dahint, R.; Grunze, M.; Josse, F.; Mrksich, M.; Whitesides, G. M. Anal. Chem. 1997, 69,3321.
3. Czeslik, C.; Winter, R. Phys. Chem. Chem. Phys. 2001, 3, 235.
4. Proteins at Interfaces II: Fundamentals and Applications; Horbett, T.A., Brash, J. L., Eds.; ACS Symposium Series 602; American Chemical Society: Washington, DC, 1995.
5. Gray, J. J. Curr. Opin. Struct. Biol. 2004,14,110.
6. Teja, A. S.; Koh, P. Prog. Cryst. Growth Charact. 2009, 55, 22-45.
7. Tartaj, P.; Morales, M. P.; Gonzalez-Carreqo, T.; Veintemillas-Verdaguer, S.; Serna, C. J. J. Magn. Magn. Mater. 2005, 290-291, 28-34.
8. Laurent, S.; Forge, D.; Port, M.; Roch, A.; Robic, C.; Elst, L. V.; Muller, R. N. Chem. Rev. 2008, 108, 2064-2110.
9. Tartaj, P.; Morales, M. D. P.; Veintemillas-Verdaguer, S.; Gonzalez-Carreno, T.; Serna, C. J. J. Phys. D Appl. Phys. 2003, 36, R182- R197.
10. Landfester, K. Adv. Mater. 2001, 13, 765-768.
11. X.Guo, M.Ballauff, Langmuir 2000, 16, 8719-8726.
12. X. Guo, A. Weiss, and M. Ballauff, Macromolecules 32, 6043 (1999).
13. K. Henzler, S. Rosenfeldt, A. Wittemann, L. Harnau, S. Finet, T. Narayanan and M. Ballauff, Phs. Rev. Lett., 2008, 100, 158301.
14. S. Rosenfeldt, A. Wittemann, M. Ballauff, E. Breininger, J. Bolze and N. Dingenouts, Phys. Rev. E: Stat., Nonlinear, Soft Matter Phys., 2004, 70, 061403.
15. K. Henzler, A. Wittemann, E. Breininger, M. Ballauff and S. Rosenfeldt, Biomacromolecules, 2007, 8, 3674.
16. M. Ballauff, Prog. Colloid Polym. Sci. 1998. 110, 76 and references therein.
17. T. Kawaguchi, T. Hamanaka, J. Appl. Crystallogr. 1992, 25. 778.
18. M. Ballauff. Macromol. Nanoscopic Polymer Particles with a Well-Defined Surface: Synthesis, Characterization and Properties.Chem. Phys. 2003, 204, 220–234.

(393k) Synthesis of Highly Concentrated Gold Nanoparticles: A Perspective for Industrial Batch Production Processes Based On Green Chemistry

Laura M. Hernandez, Angel A. Galvis, Juan R. Reyes and Watson L. Vargas, Department of Chemical Engineering, Universidad de los Andes, Bogotá, Colombia

Controlled synthesis of nanostructured materials with functional properties has been the center of attention during the last decade because of their unique size and shape-dependent properties (*e.g.* optical, electrical, magnetic, chemical, etc.). Therefore, several applications on diverse disciplines such as medical diagnosis based on imaging and cancer treatment research have recently been directed towards the use of nanomedicine.

Results have been successful in spite the fact that industrial scalability of these processes nowadays is still in development, for instance the colloidal gold production with high reproducibility and remarkable control of shape and size at concentrations in the range of parts per million has been established; nonetheless, reproducible and facile synthesis processes in order to obtain more concentrated colloidal dispersions of gold are yet to be developed.

The present study evaluates three different techniques for production of highly concentrate colloidal gold nanoparticles (hundreds of parts per million). The first synthesis is a *green chemistry approach* based on the use of essential oils extracted from plants such as *pelargonium graveolens geranium* and explores their use as both capping and reductive agents. The second synthesis uses sodium citrate as the reductive agent, polyvinylpyrrolidone (PVP) is used as capping agent, in addition, the process is carried out at room temperature (293.15 K) and the effect of ultrasonic dispersion is also studied. These two syntheses are contrasted with the well-known Turkevich method. It´s known that this last traditional synthetic pathway may generate colloidal instabilities when certain concentrations are required (> 40 ppm) due mainly to the electrical properties of the solid-liquid interface and its relation with the zeta potential. Finally, for these three procedures hydrogen potential (pH) and Temperature were followed rigorously in order to develop kinetic models. Stability was monitored over time using UV-VIS-NIR spectroscopy.

The nanoparticles obtained from these techniques have been proven to be stable over relatively long periods of time (months), making these products very useful for biomedical applications and low cost of the reductive and capping agents used.

Nanoparticles were characterized by Transmission Electron Microscopy (TEM), Scanning Electron Microscopy (SEM), Dynamic Light Scattering, Zeta potential, Fourier transform infrared spectroscopy (FT-IR), X-ray diffraction (XRD) and UV-VIS-NIR spectroscopy. Biocompatibility of nanoparticles and their capping agents was tested *in-vitro* using an MTT protocol in Chinese hamster ovarian cells (CHO), (*i.e.* by a cytotoxicity standard assay).

(393l) Diagnostic Na^+ Sensor for the Real-Time Screening and Diagnosis of Cystic Fibrosis

Evan K. Wujcik, Daniel Trowbridge and Chelsea N. Monty, Department of Chemical and Biomolecular Engineering, The University of Akron, Akron, OH

We present the conception of an electrochemical sodium ion sensor for the real-time screening and early diagnosis of cystic fibrosis (CF), a genetic disease that attacks the lungs, pancreas, liver, and intestines. CF is characterized by the abnormal transport of sodium and chloride ions across the epithelium and therefore is most commonly diagnosed by sweat testing. In this work, a flexible electronic sensor has been developed to quantify the amount of sodium ions in sweat in real-time alleviating the wait time, large sample size, possible contamination, and equipment needed currently associated with sweat testing. This sensor is composed of Nylon 6, a high strength polymer possessing a high elastic modulus that is chemical and abrasion resistant and is used in such materials as surgical sutures and toothbrush bristles. This non-conductive polymer has been functionalized with conductive carbon nanotubes (CNTs) or graphene to allow for charge carrier transport to the electrodes. For transduction, it has been functionalized with a cyclo-oligomeric calix[4]arene, which has been shown to form a supramolecular complex with sodium ions. Upon complex formation with the sodium analyte, the current is impeded and the sodium ion detection is prevalent at levels appropriate for accurate diagnosis of CF. This device could also be applied to common diabetic neurological complications, including autonomic /peripheral neuropathy. This poster will discuss the optimization of carbon- and calix[4]arene- functionalization, nylon conditions, as well as the selectivity and sensitivity of the sensor.

(393m) Self-Assembly of Nanorods in Photosensitive Binary Mixture

Ya Liu, Olga Kuksenok, Egor Maresov and Anna Balazs, Department of Chemical & Petroleum Engineering, University of Pittsburgh, Pittsburgh, PA

Computer simulations reveal how photo-induced chemical reactions in polymeric mixtures can be exploited to create long-range order in materials whose features range from the sub-micron to the nanoscale. The process is initiated by shining a spatially uniform light on a photosensitive AB binary blend, which thereby undergoes both a reversible chemical reaction and phase separation. When a well-collimated, higher intensity light is rastered over the sample, the system forms defect-free, spatially periodic lamella structures. We now build on this approach by introducing nanorods that have a preferential affinity for one phases in a binary mixture. The rods form percolating network in the uniform light source. By introducing a secondary light with higher intensity and rastering over the sample with the proper velocity, we can create ordered arrays of rods within periodically ordered materials in essentially one processing step. Our method has the advantage to control lamella width and ordering of rods dynamically and can be applied to design complex hierarchically ordered structures.

(393n) Simulation of Heat Transfer in Double-Walled Carbon Nanotubes in Silica

Khoa N. Bui[1], Huong Nguyen[2], Alberto Striolo[2] and Dimitrios V. Papavassiliou[2], (1)School of Chemical, Biological and Materials Engineering, The University of Oklahoma, Norman, OK, (2)School of Chemical Biological and Materials Engineering, The University of Oklahoma, Norman, OK

Double-walled carbon nanotubes (DWCNTs) have been found to be more thermally stable and mechanically stronger than single-walled carbon nanotubes (SWCNTs) for various applications. [1-3] Our previous work has demonstrated thermal rectification behavior of DWCNTs for heat transfer in the radial direction in vacuum. In other words, the resistance to heat transfer from the inner nanotube to the outer has been found to be smaller than the resistance to heat transfer from the outer to the inner. The expansion/contraction of the two walls of the DWCNT during the heating and/or cooling and the relative rotation of the inner wall versus the outer wall play an important role to heat transfer.[4] The unanswered question is whether these interesting phenomena are also observed when DWCNTs are imbedded in another medium. In this work, thermal transfer between the walls of DWCNTs surrounded by amorphous silica is studied by means of molecular dynamics simulations to address this question. More importantly, the enhancement of the thermal conductivity of DWCNT-silica nanocomposites by utilizing the thermal rectification behavior will be discussed here.

References

[1] R. Saito, R. Matsuo, T. Kimura, G. Dresselhaus, and M. S. Dresselhaus, Chemical Physics Letters 348 (2001) 187.

[2] Y. A. Kim, H. Muramatsu, T. Hayashi, M. Endo, M. Terrones, and M. S. Dresselhaus, Chemical Physics Letters 398 (2004) 87.

[3] K. Koziol, J. Vilatela, A. Moisala, M. Motta, P. Cunniff, M. Sennett, and A. Windle, Science 318 (2007) 1892.

[4] K. Bui, H. Nguyen, C. Cousin, A. Striolo, and D. V. Papavassiliou, The Journal of Physical Chemistry C 116 (2012) 4449.

(393p) Molecular Dynamics Simulations of Nanoparticle Self-Assembly At Ionic Liquid-Based Interfaces

Denzil S. Frost and Lenore L. Dai, School for Engineering of Matter, Transport and Energy, Arizona State University, Tempe, AZ

Molecular Dynamics Simulations of Nanoparticle Self-Assembly At Ionic Liquid-Based Interfaces

Denzil S. Frost and Lenore L. Dai

Liquid-liquid interfaces are able to provide 2D templates for the self-assembly of small particles. By the virtue of inter-particle interactions alone, intricate patterns and structures spontaneously form and may be fused together to maintain their shape in the absence of the interfacial template. So far, particle self-assembly has been studied primarily at oil/water interfaces. Ionic liquids (ILs) are a new class of liquids with an unfathomed wealth of useful properties that may revolutionize this field. Particle behavior at IL-based liquid-liquid interfaces, however, has not yet been investigated. It is the objective of this study to pioneer a fundamental understanding of this phenomenon in comparison to oil/water interfaces and provide reference examples of particle self-assembly using this new class of liquids.

We have studied the self-assembly of hydrophobic nanoparticles at ionic liquid (IL)-water and IL-oil (hexane) interfaces using molecular dynamics (MD) simulations. For the 1-butyl-3-methylimidazolium hexafluorophosphate ([BMIM][PF_6])/water system, the nanoparticles rapidly approached the IL-water interface and equilibrated more into the IL phase although they were initially in the water phase. In contrast, when the nanoparticles were dispersed in the hexane phase, they slowly approached the IL-hexane interface and remained primarily in the hexane phase. Consequently, the IL-hexane interface was rather undisturbed by the nanoparticles whereas the IL-water interface changed significantly in width and morphology to accommodate the presence of the nanoparticles. Potential of mean force (PMF) calculations supported the equilibrium positions of the nanoparticles.

The effect of particle charge on self-assembly was also investigated. In the IL/water system, nanoparticles equilibrated at the interface, somewhat favoring the IL, but this preference for the IL diminished with increased nanoparticle charge. In the IL/hexane system, all charged nanoparticles interacted with the IL to some degree, whereas the uncharged nanoparticles remained primarily in the hexane phase. PMF calculations provided insight into the particle-interface interaction. In particular, these calculations may suggest that macroscopic theories underestimated the range and magnitude of the particle-interface interaction due to low interfacial tension, thus large capillary waves, of the ionic liquid based interfaces (See Figure 1).

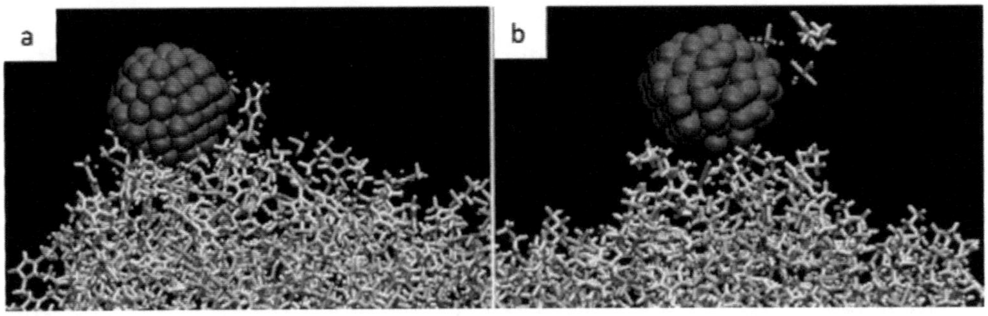

Figure 1 Interfacial deformations in the (a) IL/water and (b) IL/hexane systems with +4 charged particles. Solvent molecules were removed so as to create a clear view of the particles and the deforming interface.

Interesting ordering and charge distributions were observed at the IL-liquid interfaces. At the IL-hexane interface, the [BMIM] cations preferentially oriented themselves so that they immersed more into the hexane phase and packed efficiently to reduce steric hindrance. The ordering likely contributes to a heightened IL density and slightly positive charge at the IL-hexane interface. In contrast, the cations at the IL-water interface were oriented isotropically unless in the presence of nanoparticles, where the cations aligned across the nanoparticle surfaces.

(393q) Noble Metal-Based 1-D Nanomaterials in the Application of Direct Alcohol Fuel Cells

Liang Su, Yixin Liu, Dan Manuzzi, Wenzhao Jia and Yu Lei, Chemical, Materials & Biomolecular Engineering, University of Connecticut, Storrs, CT

Direct alcohol fuel cells (DAFCs) are envisaged to be a new generation of portable power sources which convert chemical energy into electrical energy with, theoretically, both economical and environmental benefit. Noble metal-based composite nanomaterials have shown great applicability as the cathode and anode catalysts in application of DAFCs. Furthermore, 1-D bimetallic nanomaterials have shown improved performance towards oxygen reduction reaction and alcohol oxidation reaction attributed to the ameliorated durability from the 1-D nanostructure and the enhanced catalytic activity from the incorporation of the second metal.

In this work, Pt-based nanomaterials with 1-D hollow nanostructure prepared from solution-based Cu nanowires or solid state electrospun Cu nanofibers have been successfully fabricated through a versatile galvanic replacement reaction process. The reaction condition was systematically investigated to optimize the chemical and morphological properties of the as-prepared nanomaterials according to SEM, TEM, XRD, and XPS. Moreover, the electrocatalytic activities of the Pt-based and Pd-based nanowires towards oxygen reduction reaction and ethanol oxidation reaction, respectively, have also been demonstrated and compared with the state-of-the-art Pt/C catalysts.

(393r) Multiscale Thermal Transport Model for Nano Scale Graphene Systems

Sesha Hari Vemuri[1], Young In Jhon[2], Pil Seung Chung[1], Lorenz T. Biegler[1] and Myung S. Jhon[1], (1)Chemical Engineering, Carnegie Mellon University, Pittsburgh, PA, (2)School of Advanced Material Science and Engineering, Sungkyunkwan University, Suwon, South Korea

Graphene [1], a monolayer of hexagonally sp^2-bonded carbon atoms, has been attracting great interest because of its elegant properties, such as superior thermal conductivity and excellent mechanical strength [2,3]. These properties are extremely useful in the nano-electronics industry, where the relevant dimensions are steadily decreasing into the nano-scale region with stringent operating conditions including high thermal gradients. Thus, a suitable strategy for thermal management is urgently required to introduce a paradigm shift in this industry. However, the thermal phenomena in graphene are not well understood once it is bonded with a substrate, as the thermal conductivity falls to about one-fifth of its suspended state conductivity, which will have major effects on its viability as a next generation material [4].

In this study, we attempt to develop a multiscale model spanning from the molecular to the mesoscale to understand the process of the thermal propagation in graphene sheets at the nano-scale to obtain successful design criteria for optimized graphene based devices. We constructed a preliminary molecular model to explore the thermo-mechanical properties of suspended graphene sheets, illustrating that defects such as grain boundaries influence heat transfer compared to pristine graphene structures [5]. We also developed a mesoscale thermal model utilizing a descritization of Boltzmann transport equation called the lattice Boltzmann method (LBM) for primary heat carriers (phonons and electrons) in the nano-scale thermal modeling for various test materials including silicon nano-wires [6]. This work will focus on extending and hybridizing the molecular/mesoscale models to multi-layered graphene films of practical interest including realistic grain boundary orientations and distributions. The molecular model will act as a bridge between nano and micro scale heat transfer by transferring accurate material information to implement the rules for LBM. This study will provide a deeper understanding of the integrated thermal performance of hierarchically arranged materials, which greatly depends not only on the properties of their individual constituents but also on their morphology and interfacial characteristics. The results will also allow the use of composite structures to precisely control thermal conductivity depending on the system requirements.

[1] A. K. Geim, K. S. Novoselov, Nature Materials, **6**, 183 (2007).

[2] R. Grantab, V. B. Shenoy, and R. S. Ruoff, Science, 12, 946 (2010).

[3] S. Ghosh, W. Bao, I. Calizo, D. Teweldebrhan, F. Miao, C. N. Lau, and A. A. Balandin, Nano Letters, 8, 902 (2008)

[4] Y.I. Jhon, S.E. Zhu, J.H. Ahn, and M.S. Jhon, Carbon (accepted)

[5] J.H. Seol, I. Jo, A.L. Moore, L. Lindsay, Z. H. Aitken, M.T. Pettes, X. Li, Z. Yao, R. Huang, D. Broido, N. Mingo, R.S. Ruoff, L. Shi, Science, 328, 213 (2010)

[6] S.S. Ghai, W.T. Kim, R.A. Escobar, C.H. Amon, and M.S. Jhon, J. Appl. Phys., **97**, 10P703 (2005)

(393s) Detection of Cancer Specific EML4-ALK Fusion Gene Based On FRET Between Quantum Dots and Gold Nanoparticles

Taegyeong Kang, Department of Chemical & Biomolecular Engineering, Yonsei University, Soeul, South Korea, Hyun Chang Kim, Department of Chemical & Biomolecular Engineering, Yonsei University, Seoul, South Korea, Kyong-Ah Yoon, National Cancer Center, Goyang, South Korea and Kangtaek Lee, Department of Chemical and Biomolecular Engineering, Yonsei University, Seoul, South Korea

Detection of Cancer Specific EML4-ALK Fusion Gene Based on FRET between Quantum Dots and Gold Nanoparticles

One critical area of activity is the development of robust and accurate diagnostics for the routine identification of ALK translocations in lung adenocarcinoma. However, the diagnosis of oncogenic EML4-ALK fusion gene is challenging due to a large number of different EML4-ALK variant. In this study, we have developed a detection method for variants3 of EML4-ALK fusion gene based on Föster resonance energy transfer (FRET) between quantum dots and gold nanoparticles. To fabricate this sensor, CdSe/ZnSe/ZnS quantum dots and gold nanoparticles were functionalized using four different types of probe oligonucleotides were designed to be a perfect match to EML4-ALK fusion gene. Addition of complementary target DNAs into the suspensions of quantum dots-DNA conjugates and AuNPs-DNA conjugates, fluorescence quenching of quantum dots occurred due to energy transfer from quantum dots to gold nanoparticles. This method could be successfully used to simultaneously detect presence of two types of EML4-ALK variant3. This system provides a simple, fast, inexpensive, and highly sensitive way to detect fusion genes.

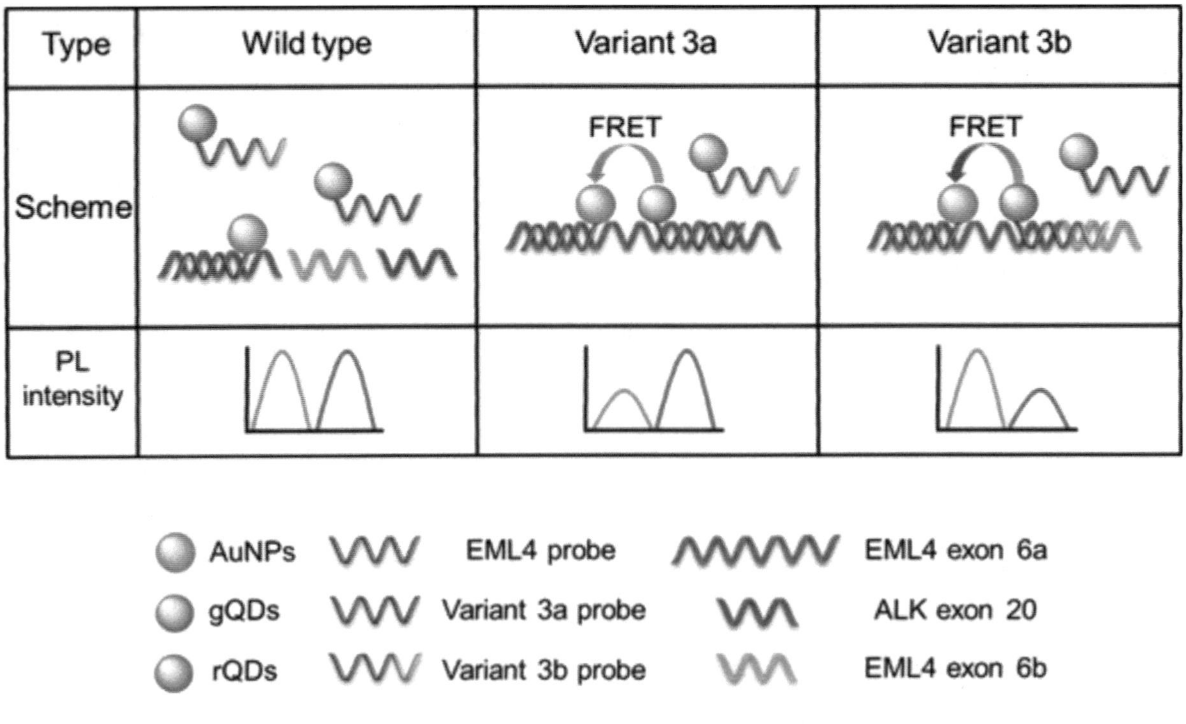

Figure 1. Schematics of detection method

Experimental Study On the Methane Adsorption by Activated Carbons and Metal Organic Frameworks

*Narumol Kumpoomee[a], Pramoch Rangsunvigit[a,b], Santi Kulprathipanja[c] and Boonyarach Kitiyanan[a,b]**

[a)]*The Petroleum and Petrochemical College, Chulalongkorn University, Thailand.*
[b)]*Center of Excellence on Petrochemical and Materials Technology*
[c)]*UOP, A Honeywell company, USA.*

ABSTRACT

Compressed natural gas (CNG) is currently being utilized as one of the fuels for vehicles in Thailand. However, the CNG still has much lower energy density than other fuels such as gasoline or diesel. To improve the storage capacity of CNG, the use of an adsorbent that can adsorb natural gas is suggested. The purpose of this study is to determine the amount of methane adsorbed on Metal Organic Frameworks (MOFs); Basolite C300 and Basolite Z1200; and activated carbons derived from coconut shell and eucalyptus. The volumetric adsorption apparatus was used to determine the amount of methane adsorption. Methane was pressurized up to 900 psia, and the temperature was set at 303, 308, and 313 K. The adsorbents were also characterized for physical properties. The results show that the gravimetric amount of methane adsorption is highest when using Basolite C300, followed by Baolite Z1200, Eucalyptus Activated Carbon, and Coconut Shell Activated Carbon. The experimental data were best fitted by Sips isotherm model.

INTRODUCTION

The consumption of natural gas as a vehicular fuel has continuously increased due to cheaper cost than other fossil-based fuels, and the compatibility to be used in the existing internal combustion engine. Furthermore, from the environmental point of view, the natural gas can be completely combusted, emitting low pollution–dust, particulate, and smoke. Typically, during transportation, natural gas can be stored by three methods which are LNG (Liquefied Natural Gas), CNG (Compressed Natural Gas) and ANG (Adsorbed Natural Gas). CNG is currently being implemented as the mean to store natural on vehicle in Thailand. The natural gas contains methane as the major component and cannot be stored with the energy density as high as gasoline [11 MJ/L for CNG at 24.8 MPa compared with 32 MJ/L for gasoline). Therefore, the CNG vehicle is required to re-fuel more often. To increase the amount of natural gas storage, the use of adsorbent is suggested. Due to their high surface area, the porous materials, such as activated carbon and Metal Organic Frameworks (MOFs) are promising adsorbents for this application. Many studies have reported the methane adsorption on activated carbon derived from renewable materials such as coconut shell (Azevedo *et al.*, 2007; Prauchener *et al.*,2008; Wei *et al.*, 2008) and eucalyptus (Tancredi *et al.*, 1996). It has been suggested that the amount of methane uptake strongly depends on the physical properties of activated carbon such as BET surface area, micropore volume, packing density and pore size distribution (Salehi et al., 2007; Lozano et al., 2002). The purpose of this study is to investigate methane adsorption on MOFs and activated carbons. Basolite C300 and Basolite Z1200 are the commercial MOFs used in this investigation. Activated carbons are those derived from indigenous sources including coconut shell, and eucalyptus. The experiment was done by volumetric method at various temperature (303, 308, 313 K) and the pressurise up to 900 psia.

EXPERIMENTAL

A. Sample Preparation

All adsorbents- Coconut Shell Activated Carbon, Eucalyptus Activated Carbon, Basolite C300, and Basolite Z1200 were dried in an over at 383 K for 1 day. After drying, the sample were tested for amount of methane adsorption by volumetric apparatus.

B. Methane Adsorption Measurement

The measurement of pure methane gas adsorption can be carried out by volumetric method (Salehi *et al.*, 2007). The schematic diagram of the volumetric apparatus is shown in Fig. 1. A high pressure and constant volume sample holder was made from stainless. Inside the reactor, a K-type thermocouple was placed to measure the temperature. The pressure regulator was also installed to control the gas flow rate into the whole system. The pressure transducer was used to measure pressure of the system for measuring range of 0-3,000 psig and 0.13% global error. The known amount of adsorbents filled into sample holder. The pressure of methane was recorded before and after methane expansion.

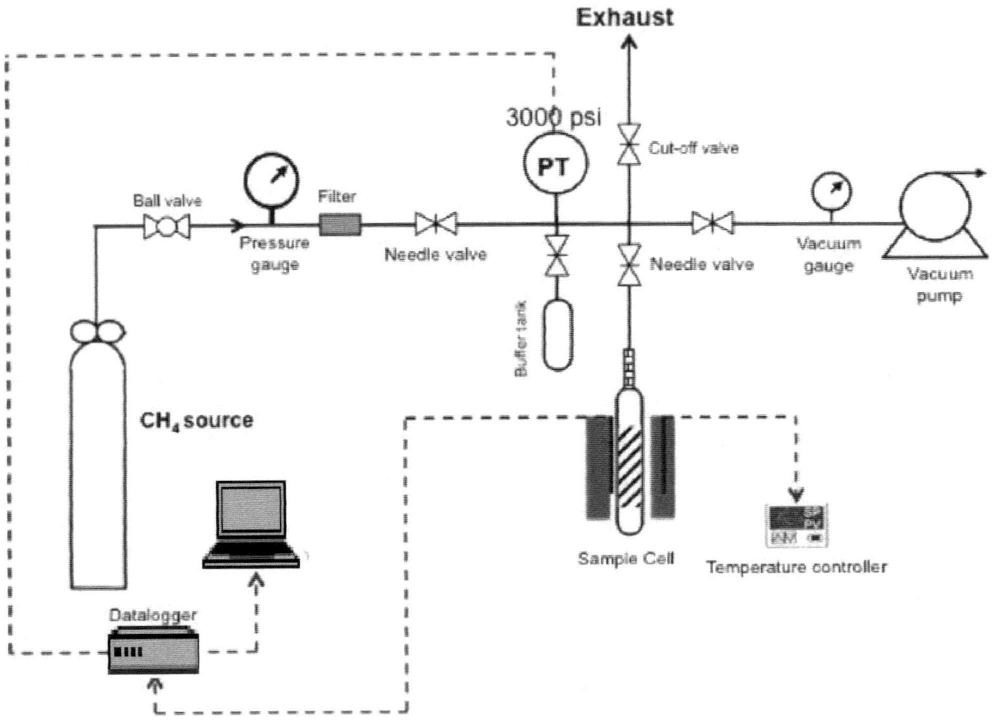

Figure. 1 Schematic diagram of the volumetric apparatus

RESULTS AND DISCUSSION

Adsorbent Characterizations

The physical properties of adsorbents were characterized by using a nitrogen adsorption instruments at 77 K. Table 1 presents the physical characteristic of adsorbents used, contains S_{BET} (BET surface area), micropore volume (V_{DR}), total pore volume, and average pore diameter. The result shows that Basolite C300 has highest BET surface area, micropore volume and total pore volume, following by Basolite Z1200, Eucalyptus Activated Carbon, and Coconut Shell Activated Carbon. The MOFs- have higher BET surface area, micropore volume and total pore volume than activated carbons.

Table 1 Physical properties of adsorbents

Adsorbents	S_{BET} m^2/g	V_{DR} cc/g	**Total pore volume** cc/g	**DA pore diameter** Å
Basolite C300	2434	1.27	1.33	9.6
Basolite Z1200	1449	0.78	0.88	16.8
Eucalyptus Activated Carbon	861	0.47	0.61	14.8
Coconut Shell Activated Carbon	713.5	0.39	0.42	11.2

Methane Adsorption Isotherm

For methane adsorption experiments, the volumetric method can be used to obtain the amounts of adsorbed methane. The methane gas was pressurized up to 900 psia at various 303, 308, and 313 K. Figure 1 shows methane adsorption isotherms of the activated carbons- (a) coconut shell (b) eucalyptus. Figure 2 shows methane adsorption isotherms of the MOFs- (a) Basolite C300 (b) Basolite Z1200. The adsorption isotherms were the plot of mole of methane adsorbed per gram of adsorbent against equilibrium pressure of methane. The result shows that the increase in pressure causes the amount of adsorbed methane to increase. However, the increasing tendency of methane adsorption isotherms does not significant at high pressure. Salehi *et al.* (2007) reported that, the situation approached when it reaches saturation of adsorbent bed. It is also shows that the amount of adsorbed methane decreases with an increase in temperature as observed. It was confirmed by Delavar *et. al.* (2010). Basolite C300 has the highest gravimetric amount of methane adsorbed following by Basolite Z1200, Eucalyptus Activated Carbon, and Coconut Shell Activated Carbon. It also concluded that metal organic frameworks have higher gravimetric amount of methane adsorbed than activated carbons. This is because they have higher BET surface area and micropore volume.

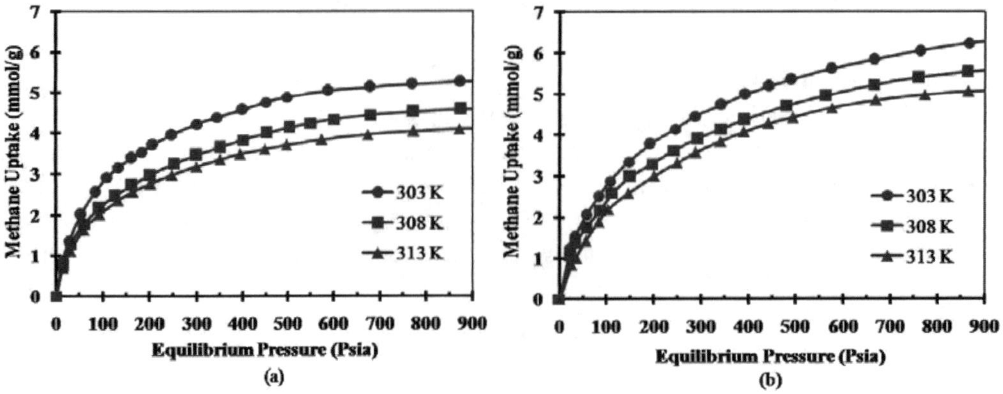

Figure 2 The experimental adsorption isotherm of methane by activated carbons (a) coconut shell (b) eucalyptus

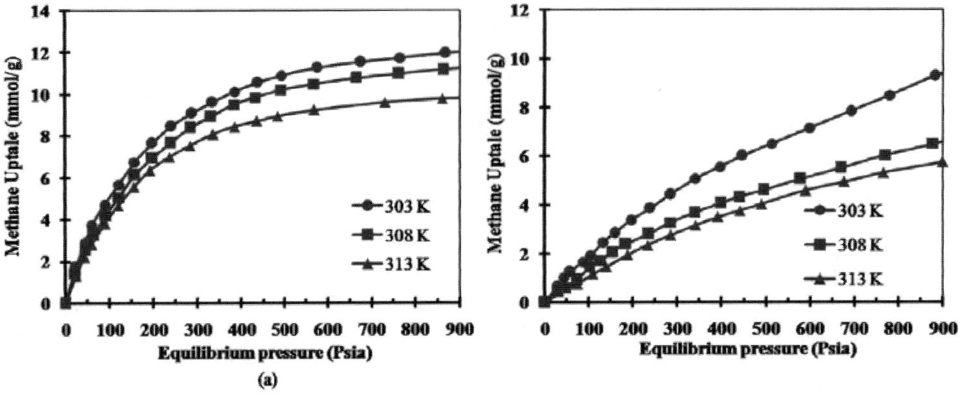

Figure 3 The experimental adsorption isotherm of methane by metal organic frameworks (a) Basolite C300 (b) Basolite Z1200

Methane Adsorption Isotherm Models

The adsorption isotherm model is important to estimate how much methane is adsorbed at various temperatures and to understand adsorption behavior of different adsorbents of interest. In this experiment, 5 adsorption isotherm models; including Langmuir, Freundlich, Sips, Toth, and Unilan isotherm models; were studied. Sips is the best model to explain the methane adsorption isotherm of all adsorbents in this experiment. Figures 3 show the methane adsorption on Coconut Shell Activated Carbon and Eucalyptus Activated Carbon from the experiment and Sips isotherm model. Figures 3 show the methane adsorption on Basolite Z1200 and Basolite C300 from the experiment and Sips isotherm model. The symbols represent the experimental data while the lines represent Sips isotherm model which can fit all experimental data very well. The experimental data show the nonlinearity, which is the characteristic of both activated carbons (Balathanigaimani *et al.*, 2003) and MOFs.

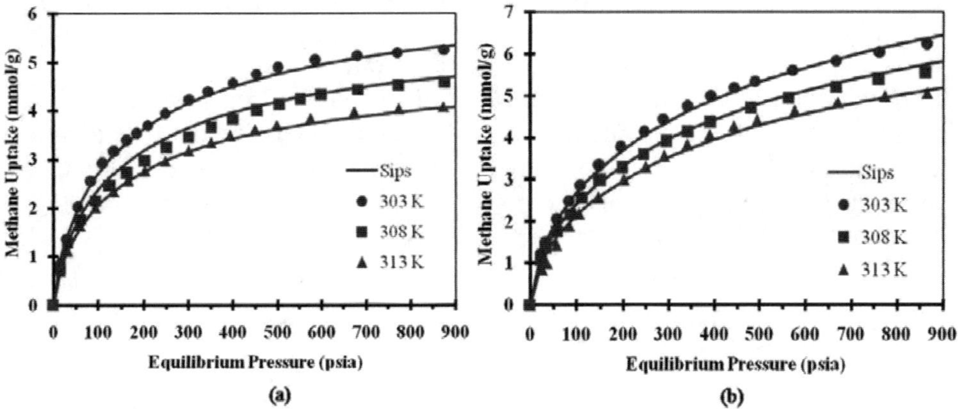

Figure 4 The experimental and Sips adsorption isotherm of methane by activated carbons (a) coconut shell (b) eucalyptus

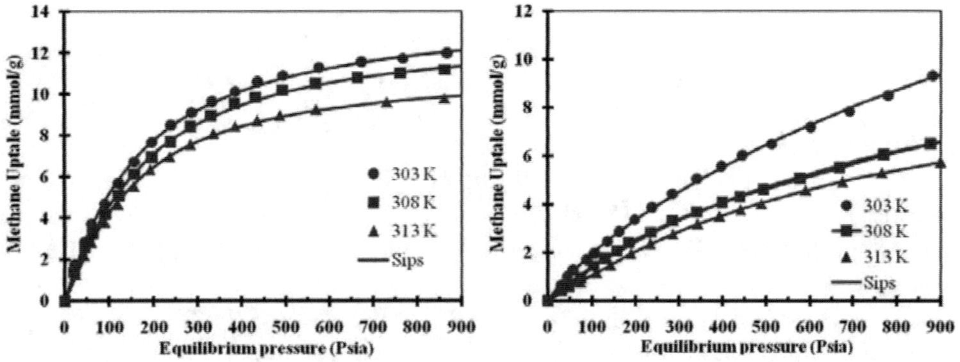

Figure 5 The experimental and Sips adsorption isotherm of methane by metal organic frameworks (a) Basolite C300 (b) Basolite Z1200

Table 2 The adsorption capacity of methane on adsorbents at 303 K and 900 psia in. storage vessel 70 L

Adsorbents	Methane Uptake at 303 K and 900 psia (kg)	Percentage (%)
Empty	2.78	-
Basolite C300	3.10	+ 11.32
Basolite Z1200	3.06	+ 9.98
Eucalyptus Powder Activated Carbon	4.08	+ 46.79
Coconut Shell Powder Activated Carbon	3.08	+ 10.44

CONCLUSION

Experimental results show that Metal Organic Frameworks (MOFs); Basolite C300 and Basolite Z1200; give much more gravimetric amount of methane adsorption than those of activated carbons; Eucalyptus Activated Carbon and Coconut Shell Activated Carbon. The physical properties of adsorbents, including BET surface area, micropore volume, total pore volume, and apparent density (packing density), are key success factors of the methane adsorption capacity. The addition of the adsorbent could increase the methane storage capacity in the CNG vessel. The experimental data were best fitted by Sips isotherm model.

ACKNOWLEDGEMENT

We would like to acknowledge the supports from National Metal and Materials Technology Center (MTEC), The Petroleum and Petrochemical College, Chulalongkorn University, and National Center of Excellence for Petroleum, Petrochemical, and Advance Materials.

REFERENCES

1. Azevedo, D.C.S., A,raújo, J.C.S., Bastos-Neto, M., Torres, A.E.B., Jaguaribe, E.F., and Cavalcante, C.L. (2007). Microporous and Mesoporous Materials 100, pp. 361-364.
2. Balathanigaimani, M.S., Kang, H-C., Shim, W-G., Kim, C., Lee, J-W., and Moon, H. (2006). Korean Chemical Engineering 23(4), pp. 663-668.
3. Chen, Y., Zhu, Y., Wang, Z., Li. Y., Wang, L., Ding, L., Gao, X., Ma, Y., and Guo, Y. (2011). Advances in Colloid and Interface Science 163, pp. 39-52.
4. Delavar, M., Ghoreyshi, A.A., Jahanshahi, M., and Irannejad, M. (2010). Engineering and Technology 62, 47-50.
5. Dreisbach, F., Staudt, R., and Keller, J.U. (1999). Adsorption 5, pp. 215-227.
6. Lozano-Castelló, D., Cazorla-Amorós D., and Linares-Solano, A. (2002) Energy & Fuels 16, pp. 1321-1328.
7. Salehi, E., Taghikhani, V., Ghotbi, C., Lay, E.N., and Shojaei, A. (2007). Journal of Natural Gas Chemistry 16, pp. 415–422.

(393u) Organic Nanoparticle Deposition in Organosilane Nano-Arrays (Poster)

Sunxi Wang, Pedram Jahanian, Juxhin Xhahysa and Guangzhao Mao, Chemical Engineering and Materials Science, Wayne State University, Detroit, MI

The ability to make nano-arrays enables high-throughput screening and nanofluidics. This paper describes a method to generate octadecyltrichlorosilane (OTS) nano-arrays on oxidized silicon wafers for nanoconfining small organic molecules. Particle lithography was carried out on 300 and 900 nm polystyrene particle array templates by chemical vapor deposition of OTS, which reacted with the substrate in the water menisci to form a ring pattern. We discovered reaction conditions that favor bulk polymerization resulting in ~14 nm thick rings for the 900 nm particles and ~9 nm thick rings for the 300 nm particles, both of which exceed the OTS monolayer thickness of 2.6 nm. It is estimated that a 1×1 cm^2 wafer (900 nm particles) contains more than 100 million ordered "nano-flasks" with volume $\sim3\times10^{-10}$ nL. The nano-array pattern is stable against common organic solvents thus allowing repeated use in solution crystallization screening experiments. Several hydrophobic compounds including docosane and aspirin were deposited on the nano-array pattern for the study of nanoconfinement effect on molecular crystallization. The conditions for preferential deposition of the compounds insider the "nano-flasks" will be discussed. The resultant organic nanoparticle arrays on the nanopattern were characterized by AFM and field-emission SEM.

(393v) Tunable Steric Stabilization Effects On Iron Oxide Nanoparticle Dispersability in Gas Expanded Liquid Systems

Pranav S. Vengsarkar[1], Jennifer N. Duggan[2] and Christopher B. Roberts[1], (1)Department of Chemical Engineering, Auburn University, Auburn, AL, (2)Chemical Engineering, Auburn University, Auburn University, AL

Magnetic nanoparticles have properties which are extremely important in applications like catalysis, biomedicine, environmental remediation and data storage. The most widely researched magnetic nanoparticles are iron-oxide and cobalt nanoparticles, which are useful because of their relatively high magnetic susceptibility and easy synthesis procedures. Since the magnetic properties of materials are heavily size dependent, the sizes of the nanoparticles are of critical importance to the application for which they are being used. While synthesis procedures to create monodisperse nanoparticles are available in the literature, most of these methods require expensive reagents, high temperatures and are useful only in lab scale preparations. These simple synthesis procedures usually generate particles having a large size distribution and hence nanoparticles obtained from these procedures require some post-processing before use in demanding applications. While magnetic separation might initially seem straightforward, it is not recommended at these small length scales due to particles of comparable size having very similar magnetic susceptibility values which make it difficult to get efficient separation without using extremely accurate and expensive instrumentation. Our lab has developed a technique to size-selectively fractionate nanoparticles from organic phase dispersions which utilize the pressure tunable properties of CO_2 Gas-eXpanded Liquids (GXLs). This size-selective fractionation technique is based on the controlled reduction of the solvent strength through increases in the concentration of CO_2(a known nonsolvent for aliphatic stabilizing ligands) via pressurization. The advantages of this process are the facts that it is reversible and it allows facile recycle of solvent and antisolvent. These subtle changes in solvent strength affect the balance between the osmotic repulsive forces and the Van der Waals forces of attraction between differently sized nanoparticles necessary to maintain a stable dispersion, thereby allowing fractionation of the nanoparticles into multiple narrowly sized populations.

The aim of this particular study is to investigate how the steric nature of a solvent or ligand affects the ligand-solvent interaction and the size dependent precipitation of iron oxide nanoparticles. Iron oxide nanoparticles were synthesized by the co-precipitation method using Fe^{2+} and Fe^{3+} salts and ammonium hydroxide. These particles were then coated with fatty acids which acted as the lipophilic ligands and were dispersed in various solvents of our choice. Characterizations of these nanoparticles involve the use of transmission electron microscopy (TEM) and dynamic light scattering (DLS) to analyze their size distribution and hence judge the efficacy of the size-selective fractionation. The pressure of incipient precipitation and the pressure range over which the particles continue to precipitate is investigated. The effectiveness of the different ligands and solvents on the monodispersity of particles obtained after fractionation is also scrutinized with the belief that understanding the ligand-solvent interactions will help further our fundamental understanding of the phenomenon of nanoparticle precipitation from GXLs. The applicability of these relatively monodisperse particles in areas like catalysis and emulsion stabilization will also be discussed.

(393w) Exploring the Dispersibility of Gold Nanoparticles in DMSO Using Gas-Expanded Liquid Systems

Jennifer N. Duggan[1], Pranav S. Vengsarkar[2] and Christopher B. Roberts[2], (1)Chemical Engineering, Auburn University, Auburn University, AL, (2)Department of Chemical Engineering, Auburn University, Auburn, AL

Nanoparticles exhibit very unique and highly size-dependent magnetic, mechanical, chemical, electronic, and optical properties. Thus, it is necessary to develop technologies to produce monodisperse samples of nanoparticles in order to exploit these size-dependent properties in various applications. Our group has developed a technique that enables the separation of nanoparticles into different size ranges by manipulating the thermophysical properties of a solvent/antisolvent mixture in a gas expanded liquid system (GXL). GXLs have solvent strength and transport properties that are intermediate to those of liquids and gases which can be controlled by pressure variations, and these properties can be utilized for nanoparticle separation applications. In our lab, GXLs have been successfully used for this purpose to size-selectively separate alkyl-capped metal nanoparticles and semiconductor quantum dots dispersed in organic solvents. Ongoing research in our lab is centered on using this GXL separation technique to size-selectively separate metal nanoparticles that have been stabilized by non-polar ligands with different structures (linear and branched) and dispersed in a variety of non-polar alkyl solvents (linear and branched). However, it is desired to understand the versatility of the GXL nanoparticle size-selective separation process using other molecules, such as polar and organosulfur compounds, in order to gain a better understanding of general nanoparticle dispersibility behavior in these solvents. Thus, an alternative means of nanoparticle stabilization and subsequent size-selective separation has been developed by dispersing the nanoparticles in a functional solvent, such as dimethyl sulfoxide (DMSO). Interestingly in some cases, DMSO can adequately serve as both the ligand and solvent for nanoparticle systems. It has been reported in the literature and observed in our lab that a sufficient amount of CO_2 can diffuse into DMSO to create a GXL for nanoparticle precipitation from DMSO. Preliminary precipitation experiments have been undertaken to determine whether a CO_2 gas expanded DMSO system may be used to size-selectively separate gold nanoparticles by tuning the applied CO_2 pressure. Initial results from UV-vis spectroscopy analysis indicate that the DMSO molecule alone may not be able to sufficiently stabilize the gold nanoparticles long enough to successfully perform precipitation experiments using the GXL technique. To gain more control over gold nanoparticle dispersibility in these unique DMSO-GXL systems, alternative stabilizing agents, such as fatty acids found naturally in animals and vegetables, have been added to the DMSO nanoparticle system. In each case, UV-vis spectroscopy is used to understand the gold nanoparticle/DMSO/fatty acid ligand system by measuring the absorbance of the gold nanoparticles as a function of the applied CO_2 pressure. Furthermore, a cascaded vessel apparatus can be used to perform size-selective separations of the gold nanoparticle dispersions from DMSO using GXL. Comparisons will be made to gold nanoparticles that have been size-selectively separated from the traditional non-polar solvent/ligand systems in order to demonstrate the versatility of using DMSO as an alternative solvent for nanoparticle systems.

(393x) Microfluidic in-Channel Growth of 3-D Nanostructures

Joseph Parisi Jr., Chemical, Materials, and Biomolecular Engineering, University of Connecticut, Storrs, CT and Yu Lei, Chemical, Materials & Biomolecular Engineering, University of Connecticut, Storrs, CT

3-D Nanostructure Sensors in Microfluidic Channels

Joseph Parisi, Yu Lei

Department of Chemical, Materials, & Biomolecular Engineering, University of Connecticut, 191 Auditorium Road, Unit 3222, Storrs, Connecticut 06269

Research for the detection of chemical and biological species is consistently moving forward and creating more accurate and sensitive devices. However, many of these devices require multiple steps for not only fabrication of the sensor material but also for preparation of the sample for testing. In many of these examples there is no simple approach to economically fabricate novel 3-D sensing materials and package them for future application. Our work focuses on *in-situ* preparation of 3-D nanostructured electrodes within a microfluidic channel. These nanostructures are fabricated in a fraction of the time of traditional lithography, CVD and hydrothermal techniques, require less material and limit any complicated fabrication steps. Once the nanostructures are grown they are immediately available for species detection. Since the sensor is directly fabricated and confined to the microfluidic channel no additional fabrication steps are needed. The nanostructures in microfluidic channel based sensing platform exhibit ultrasensitive detection limits as they combine the advantages of nanomaterials and microfluidics.

(393y) Approximating the Solution to the Master Equation to Simulate Directed Self Assembly of Nanostructures

Sivaraman Ramaswamy[1], Richard Lakerveld[1,2], George Stephanopoulos[3] and Paul I. Barton[1], (1)Chemical Engineering, Massachusetts Institute of Technology, Cambridge, MA, (2)TU Delft, Netherlands, (3)Massachusetts Institute of Technology, Cambridge, MA

Reliable self assembly of nanoscale systems into a structure of desired geometry will enable various future technological applications such as nanoelectronic circuits and molecular factories [1]. Significant progress has already been made on the self assembly of periodic nanostructures (for example, self-assembled monolayers [2]). However, less progress has been achieved in fabricating systems with non-periodic nanostructures. Self assembly, being spontaneous, fast and having high spatial resolution, provides a choice for fabricating non-periodic nanostructures. Various interaction forces (electrostatic, Van der Waals, hydrogen bonding, etc.) play a role in self assembly at the nanometer scale [3]. The nanoparticles can interact with each other and with an external force field to form a desired structure [1]. Such a force field can be the result of nanoscale electrodes (or actuators) that are placed on a scaffold, which can be fabricated using various top-down methods such as lithography [4]. The key question is to design the nanoparticles and utilize the external fields to guide self assembly towards a desired structure rapidly and with high probability.

Lakerveld et al. [5] demonstrated efficient algorithms to solve simultaneously a very large number of master equations that describe the probability of finding the system in a certain configuration at a given time during the course of directed self assembly including parametric sensitivities. However, instances of their model eventually become prohibitively large upon increasing the size of the physical domain as each possible configuration is being considered explicitly.

The aim of this contribution is to demonstrate the application of the Finite State Projection (FSP) method [6] to simulate directed self assembly. The FSP method is based on simulation of only a subset of the total configuration space (called a projection space). In this work, the projection space is reduced at equal intervals of time. The projection space is also adjusted if the error (caused by simulating only a subset of the total configuration space) exceeds a bound. This adaptation of the projection space is guided by the probability and flux values, and hence it brings the error back within the bound. The adjustment of the projection space is continued until the maximum allowable size of the projection space is reached. This method helps in identifying the time scale over which the system can be simulated in detail and this will depend on the computational resources and the user specified error bounds. It also provides critical information about the dominant configurations and the transition flux between those configurations. A couple of case studies are chosen to demonstrate the method and useful information is revealed about them.

References

[1] N. Stephanopoulos, E. O. P. Solis, and G. Stephanopoulos, "Nanoscale process systems engineering: Toward molecular factories, synthetic cells, and adaptive devices," *AIChE Journal*, vol. 51, no. 7, pp. 1858-1869, 2005.

[2] J. C. Love, L. A. Estroff, J. K. Kriebel, R. G. Nuzzo, and G. M. Whitesides, "Self-assembled monolayers of thiolates on metals as a form of nanotechnology," *Chemical Reviews*, vol. 105, no. 4, pp. 1103-1170, 2005.

[3] K. J. M. Bishop, C. E. Wilmer, S. Soh, and B. A. Grzybowski, "Nanoscale forces and their uses in self-assembly," *Small*, vol. 5, no. 14, pp. 1600-1630, 2009.

[4] B. D. Gates, Q. Xu, M. Stewart, D. Ryan, C. G. Willson, and G. M. Whitesides, "New approaches to nanofabrication: molding, printing, and other techniques," *Chemical Reviews*, vol. 105, no. 4, pp. 1171-1196, 2005.

[5] R. Lakerveld, G. Stephanopoulos, and P. I. Barton, "A master-equation approach to simulate kinetic traps during directed self assembly," *The Journal of Chemical Physics*, In press, 2012.

[6] B. Munsky and M. Khammash, "The finite state projection algorithm for the solution of the chemical master equation," *The Journal of Chemical Physics*, vol. 124, no. 4, p. 044104, 2006.

(393z) Simulating the Mechanical Properties of Silica Aerogels by Means of a Coarse-Grained Flexible Model

Carlos Ferreiro, Materials Science and Engineering, University of Texas at Dallas, Richardson, TX and Lev Gelb, University of Texas at Dallas, Richardson, TX

Aerogels are materials with very low densities and high surface areas prepared through sol-gel processing. The unique properties of aerogels make them very attractive for many applications,[1] but their highly compliant nature hinders their applicability. The bulk moduli of aerogels are related to their density by a power-law relation [2] with exponent between 3 and 4 [3]. Some previous attempts at modeling the mechanical properties of these systems have been made, using finite-element method (FEM) modeling of structures generated using a diffusion-limited cluster-cluster aggregation (DCLA) algorithm [4,5]. More recently, Gelb [6] introduced a coarse-grained flexible model where Langevin dynamics are used to mimic the gelation process of the aerogel primary particles. This model also naturally includes the effects of relaxation and fluctuation on the gel structure. Using this model, we study the mechanical properties of aerogels as a function of both the primary particle size and the material density. These simulations provide a better understanding of the nature of the dependence of moduli on structural properties, as well as of the roles of gel network relaxation and plastic deformation.

(1) J. L. Gurav, et al., J. Nanomat. 2010 (2010).
(2) G. W. Scherer, Adv. Coll. Int. Sci. 76-77 (1998) 321.
(3) G. W. Scherer, et al., J. Non-Cryst. Solids 186 (1995) 316.
(4) H.-S. Ma, et al., J. Non-Cryst. Solids 277 (2000) 127.
(5) H.-S. Ma, et al., J. Non-Cryst. Solids 285 (2001) 216.
(6) L. D. Gelb, J. Phys. Chem. C 111 (2007) 15792.

The Role of the Intermolecular Potential on the Dynamics of C_2H_4 Confined in Cylindrical Nanopores

Fernando J.A.L. Cruz,[1] Erich A. Muller,[2] Jose P.B. Mota[1]
[1] Requimte/CQFB, Dept. Chemistry, Universidade Nova de Lisboa, 2829-516 Caparica, Portugal. [2] Dept. Chem. Eng., Imperial College London, South Kensington Campus, London SW7 2AZ, UK.
fj.cruz@fct.unl.pt

I. Introduction

The confinement of fluids in nanoporous solids can be accompanied by striking effects on the molecular dynamics,[1] which do not necessarily possess an analogy in the bulk phase. Upon confinement, molecules interact with the solid walls to an extent that markedly depends on the chemical nature of the solid and its corresponding pore size. Bhatia *et al.*[2] studied the transport of CH_4 adsorbed into armchair single-walled carbon nanotubes (SWCNTs), and concluded that the diffusion coefficient decreases with adsorbate density. The dynamics of CH_4 and C_2H_6 inside a zig-zag SWCNT have been studied by Krishna *et al.*[3] The diffusivity of gases and liquids confined in carbon nanotubes is far from being thoroughly understood and, furthermore, results are sometimes conflicting. We aim herein to understand the effect of the level of detail used in the description of the fluid's energetics, upon the effective observed dynamics. Much can be gained or lost in terms of computational effort and physical detail by coarse-graining (CG) molecular models; Depa *et al.* showed that by CG the intermolecular potential of polymers,[4] the bulk fluid would exhibit faster dynamics than the corresponding atomistic representations, and related this to the enhanced softness of the CG potential. It is the purpose of this work to study the influence of the intermolecular potential on the dynamical behavior of C_2H_4, in a bulk phase, and also confined inside a (16,0) SWCNT (D_{eff} = 9.1 Å). This issue has been partially addressed by Mao and Sinnott[5] and Cruz and Muller.[6] We employ molecular dynamics (MD) simulations with five distinct intermolecular potential models (Fig.1), whose level of CG corresponds to different degrees of detailing the C_2H_4 molecule; including a fully atomistic representation of ethylene (AA-OPLS), validated by comparison against bulk experimental data,[7] but also simpler models with varying degrees of CG ranging down to a single isotropic Lennard-Jones sphere (1CLJ). The capabilities and drawbacks of each of these different molecular descriptors are explored in a wide range of densities (< 15.751 mol/L) and temperatures ($220 < T\,(K) < 340$).

II. Results and Discussion

The effect of confinement on the self-diffusion coefficient, D, of ethylene, is observed to be dependent on the particular potential model employed (Fig.2), in a manner which is different than in bulk, where the calculated D values are rather independent of the potential used. As C_2H_4 is confined, differences between molecular potentials are enhanced resulting in discrepancies in the observed self-diffusivities. It is generally argued that coarse-graining intermolecular potentials results in increased self-diffusivities. In the present case, if the AA-OPLS model is used as benchmark, it becomes clear that the effect of CG is unpredictable, and an appropriate and accurate parameterization may faithfully reproduce self-diffusion data. Furthermore, the existence of explicit electrostatic details in the potential seems to induce short-range order in the confined fluid (Fig.3), resulting in a slowing down of molecular mobility. The radial distribution function (Fig.3) for the simple 1CLJ potential exhibits a sharp peak at r = 4.73 Å and a second (less intense) long-range peak located at r = 8.73 Å, indicating that molecules are densely packed together with almost no free volume between them (*Sigma* = 4.6 Å). When the fluid model is increasingly refined, a different picture is obtained. If the C_2H_4 molecule is described with the TraPPE potential, not only does the initial *g(r)* peak becomes much broader, with an half-width length of *ca.* 1.95 Å, but it also appears closer at r = 4.48 Å; the second peak observed in the 1CLJ model is no longer clearly observable. If electrostatics are now explicitly added to the fluid molecules, via atomic partial charges (AA-OPLS), a further refinement of the initial peak takes place (magnification inset of Fig.3). The r = 4.48 Å peak of the TraPPE potential essentially gets split into two contributions, r = 4.03 Å and r = 4.78 Å, suggesting the insurgence of a very short-range ordered arrangement of molecules in the AA-OPLS model, that is absent in the other potentials.

In spite of the marked anisotropy introduced by the solid, both the bulk fluid and anisotropic systems data are observed to roughly collapse onto a same master curve (Fig.2), in this work described as a simple power-law dependency of the self-diffusion coefficient with molecular density, $D = D_0 \rho^{\lambda}$ ($-1.269 < \lambda < -0.943$). The isochors in Fig.4 show that the confined fluid response to temperature is similar, either above or bellow the critical value, T_c=282.35 K, and in all cases obeys the Arrhenius law, $D = A\exp(-E_a/RT)$. At constant density, it was observed[7] that the self-diffusion coefficients are remarkably dependent of the particular potential model, D (1CLJ) < D (AA-OPLS) < D (TraPPE). From the Arrhenius fits, it becomes clear that for each potential studied, regardless of its own specificities and degree of molecular detail, the activation energy (E_a/R) decreases monotonically with increasing molecular density. Because molecules in the dense fluid are highly packed, with little free volume between them, exchange of momentum via molecular collisions is highly efficient, thus dramatically increasing the self-diffusional process. We are unaware of experimental work reporting self-diffusion data for confined ethylene. Nonetheless, in order to probe the simultaneous effect of confinement and T on the fluid dynamics, we compare in Fig.5 our results with bulk experimental data[8] using the AA-OPLS model as benchmark. It is clear from Fig. 5 that, when the solid is saturated with fluid (15.024 mol/L), the D coefficients show a similar response to applied temperature, both in bulk and under confinement. The activation energies obtained from the Arrhenius plots, $(E_a/R)^{bulk}$ = 214.12 K and $(E_a/R)^{SWCNT}$ = 242.85 K, suggest that confinement exerts a major influence on the absolute values of D, but not so much on their intrinsic dependence on temperature. To a reasonable approximation, the self-diffusivity differences between the bulk and confined phases are roughly constant over the entire temperature domain, $(D^{bulk} / D^{SWCNT})^{[220K\text{-}340K]} = 2.52 \pm 0.13$. A totally different picture results when density is lowered, and the existence of free volume becomes a relevant parameter. For the isochore at 5.331 mol/L (Fig.5), free volume inside the solid is more than 50 % of total available volume, and therefore a sharp discrepancy can be observed in the slope of the corresponding bulk system (5 mol/L), leading to activation energies that are quite different between the bulk and confined fluid, $(E_a/R)^{bulk}$ = 301.64 K and $(E_a/R)^{SWCNT}$ = 533.96 K. For those lower densities, the self-diffusivities ratio increases dramatically to $(D^{bulk} / D^{SWCNT})^{[220K\text{-}340K]} = 10.2 \pm 0.5$. One can extrapolate the previous observations to a very dilute fluid, and postulate that as the free volume inside the solid increases, so do the differences in the corresponding activation energies between the bulk and confined fluid.

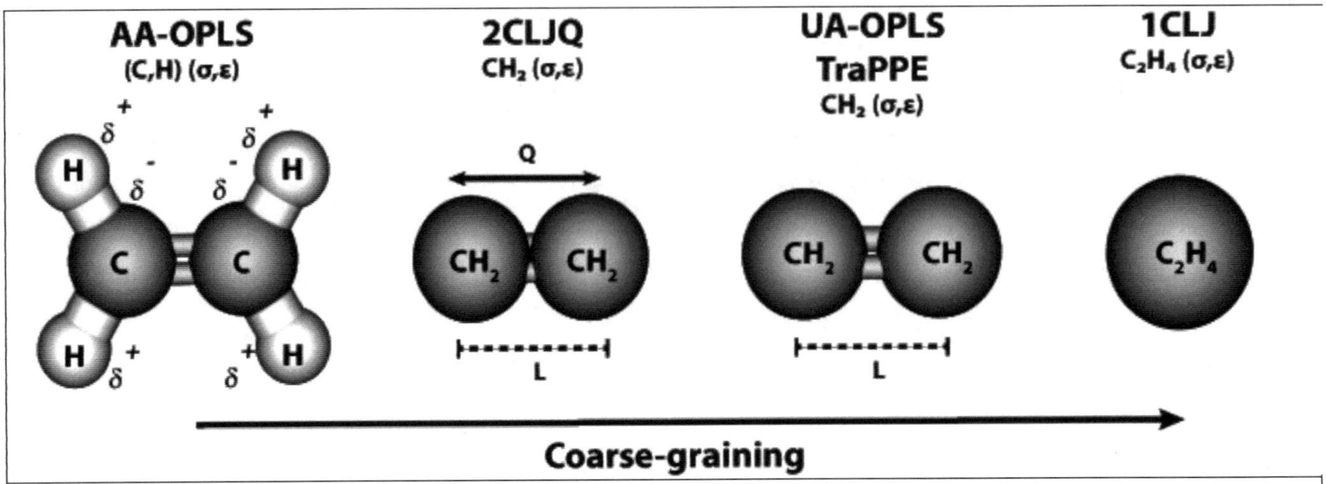

FIG. 1 The five different intermolecular potentials used in the MD simulations; each sphere represents a pseudo-atom with a characteristic Lennard-Jones pair (σ, ε), δ^i is the partial electrostatic charge on particle *i*, *Q* is the molecular quadrupole corresponding to the 2CLJQ force-field, and *L* is the chemical bond length between two CH_2 pseudo-atoms.

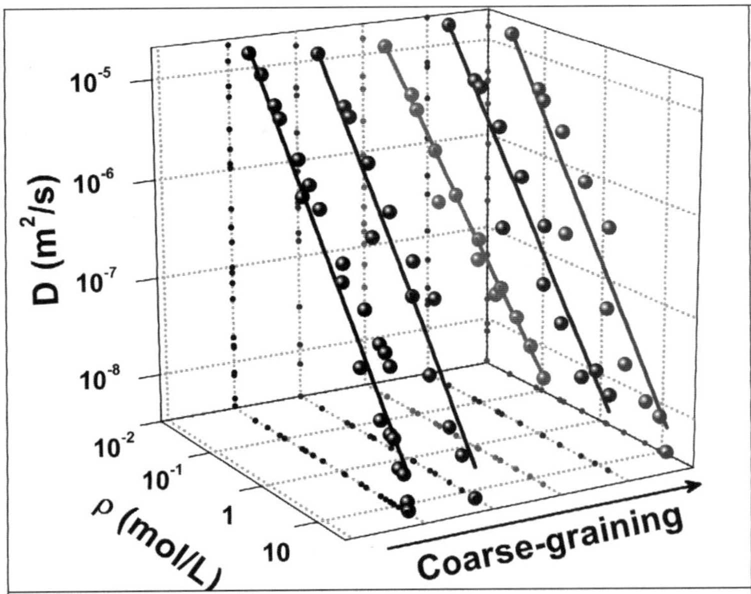

FIG.2 Influence of potential CG upon bulk and confined C_2H_4 (T = 300K). Symbols are MD data and lines correspond to a power-law, $D=D_0\rho^{\lambda}$($-1.269 \leq \lambda \leq -0.943$); AA-OPLS (black), 2CLJQ (blue), TraPPE (red), UA-OPLS (purple) and 1CLJ (green).

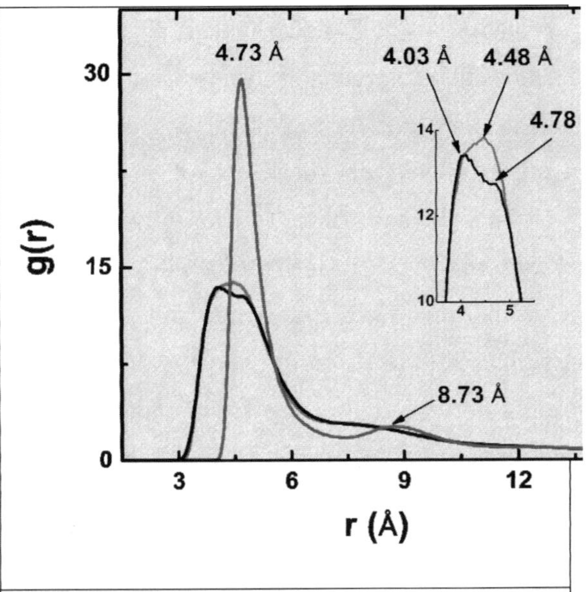

FIG.3 Radial distribution functions, *g(r)*, of confined C_2H_4 (T=300 K): AA-OPLS (black, 15.024 mol/L), TraPPE (red, 15.024 mol/L) and 1CLJ (green, 11.632 mol/L).

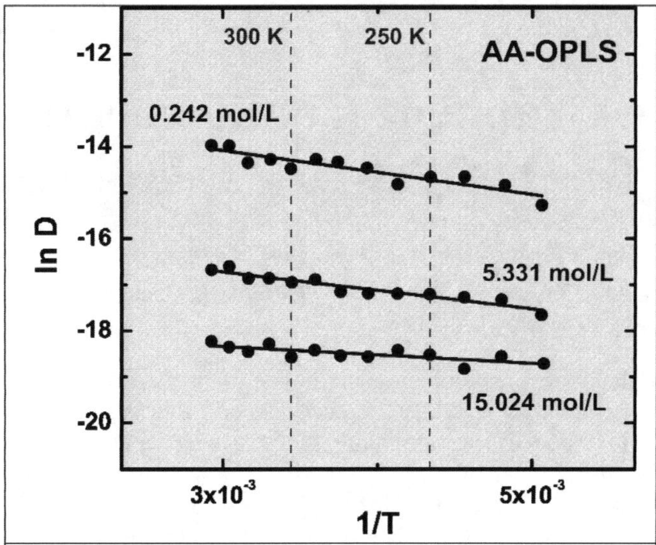

FIG.4 Temperature dependence of the confined fluid self-diffusion coefficient (220 ≤ T (K) ≤ 340). Symbols represent MD results and lines correspond to Arrhenius fits, $D=A\exp(-E_a/RT)$.

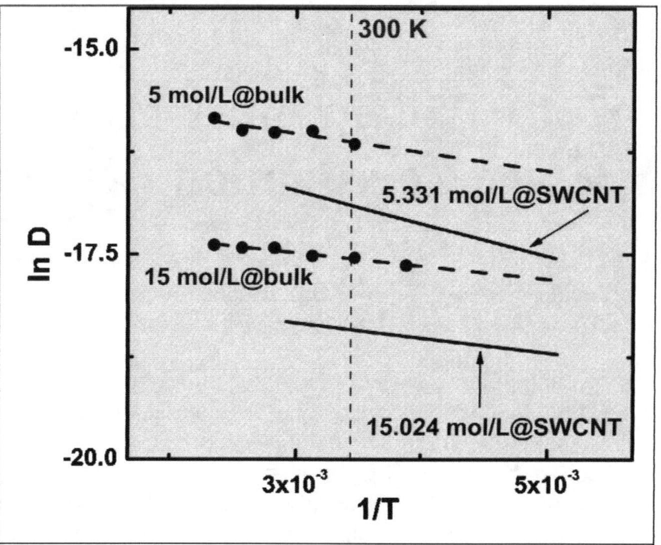

FIG.5 Temperature dependence of the fluid self-diffusion coefficient (220 ≤ T (K) ≤ 398.15). Symbols represent bulk experimental data[8] and lines correspond to the Arrhenius equation for the bulk (dashed) and confined (full) phases using the AA-OPLS potential.

References

[1] Gelb *et al.*, Rep. Prog. Phys. **62**, 1573 (1999). Bhatia, Ads. Sci. Technol. **24**, 101 (2006). Coasne *et al.* Mol. Phys. **104**, 3491 (2006). Bhatia *et al.* Phys. Rev. Let. **100**, 236103 (2008). Gubbins *et al.*, Phys. Chem. Chem. Phys. **13**, 58 (2011).
[2] S. K. Bhatia, H. Chen, D. S. Sholl, Mol. Sim. **31**, 643 (2005).
[3] R. Krishna, J. M. van Baten, Fluid Phase Eq. **45**, 2084 (2006).
[4] P. Depa, C. Chen, J. K. Maranas, J. Chem. Phys. **134**, 014903 (2011).
[5] Z. Mao, S. B. Sinnott, Phys. Rev. Lett. **89**, 278301 (2002).
[6] F. J. A. L. Cruz, E. A. Müller, Adsorption **15**, 13 (2009).
[7] F. J. A. L. Cruz, E. A. Müller, J. P. B. Mota, RSC Advances **1**, 270 (2011).
[8] B. Arends, K. O. Prins, N. J. Trappeniers, Physica A **107**, 307 (1981). E. S. Baker, D. R. Brown, J. Jonas, J. Phys. Chem. **88**, 5425 (1984).

(424b) Water Under Extreme Graphene Confinement: Surface Corrugation Effects On the Wet-Dry Transition

Ariel A. Chialvo, Chemical Science Division, Oak Ridge National Laboratory, Oak Ridge, TN, Lukas Vlcek, Chemical Sciences Division, Oak Ridge National Laboratory, Oak Ridge, TN and Peter T. Cummings, Center of Nanophase Materials Sciences, Oak Ridge National Laboratory, Oak Ridge, TN

The wetting of solid surfaces by water (or any other fluid for that matter) is a phenomenon of relevance involving natural biochemical and geochemical processes as well as energy-related industrial applications, and current evidence indicates that the surface wettability becomes strongly affected by the surface topography (corrugation). In fact, it has been known for some time that surface roughness can significantly modify the contact angle of a fluid, a phenomenon frequently described in terms of the meso-scale models of Wenzel [1] and Cassie[2]. However, theory and simulation of solid-water interfacial phenomena involving nanoscopically-flat surfaces indicate that the interfacial region extends only a few molecular diameters; consequently, the effect of surface roughness on interfacial phenomena might become more significant when the surface roughness is within the same characteristic length-scale, *i.e.*, of a few angstroms [3].

More importantly, it is not just the effect of surface roughness on the solid-water interfacial behavior ¾ on the inhomogeneous solid-fluid interfacial (SFI) region when a fluid becomes in contact with a solid surface ¾ but also, the compounding effect resulting from the overlapping of approaching SFI regions with the formation of confined-fluid environments. Because experimental nano-scale studies of SFI's are significantly challenging and their outcome must be interpreted in terms of pre-defined models, the molecular simulation of precisely defined realistic model systems can provide direct molecular-based microscopic information to aid the understanding of microscopic processes and the relationships between wetting and surface structure/roughness. In this context molecular simulation becomes a versatile tool to link every microscopic details of the intermolecular forces describing the system and the resulting structural and dynamical behavior of water at and under confinement of SFI's.

Therefore, the overarching goal of this effort is the molecular-based investigation of the surface-corrugation effect of finite-size graphene plates on the interfacial and confinement behavior of water at ambient conditions, to address key issues regarding the microscopic links between the strength of the graphene-water interactions, the evolution of their hydro-philic/phobic nature with plate corrugation, and the impact on the local microstructural and thermodynamics properties of interest. For that purpose, we performed isothermal-isobaric molecular dynamics simulations to study the behavior of water under confinement between uncharged graphene plates, to simultaneously characterize the behavioral differences between water at interfacial and under confinement, while in equilibrium with its own bulk [4]. Using the slit-pore (graphene flat-plates) configuration as a reference system, we introduce four precisely defined corrugation patterns that preserve the original pore volume, a feature that facilitate the interpretation of simulation results. The characterization involves the axial profiles of the potential of mean force (PMF), local isothermal compressibility, diffusivity, and thermal expansivity, as well as interpreted in terms of the corresponding profiles for hydrogen bonding structure and water coordination. Moreover, the profiles of the solvent contribution to the PMF in conjunction with that of the local isothermal compressibility are then used to assess the surface-corrugation effects on the hydro-philic/phobic nature of the water-graphene interactions and on the (oscillatory) onset of wet-dry transition. We also make contact with quasi-elastic neutron scattering (QENS) experiments by calculating the simulated scattering spectra and interpreting them in terms of microscopic information from the corresponding molecular dynamics simulation [5].

[1] Wenzel, R. N. *Industrial & Engineering Chemistry* **1936**, *28*, 988-994.

[2] Cassie, A. B. D.; Baxter, S. *Transactions of the Faraday Society* **1944**, *40*, 546-551.

[3] Striolo, A. *Adsorption Science & Technology* **2011**, *29*, 211-258.

[4] Chialvo, A. A.; Cummings, P. T. *The Journal of Physical Chemistry A* **2011**, *115*, 5918-5927.

[5] Mamontov, E.; Wesolowski, D. J.; Vlcek, L.; Cummings, P. T.; Rosenqvist, J.; Wang, W.; Cole, D. R. *Journal of Physical Chemistry C* **2008**, *112*, 12334-12341.

(424c) Dynamics of Ion Transport in Single-Walled Carbon Nanotubes As a Function of Diameter and Temperature

Wonjoon Choi[1], **Steven Shimizu**[2], Darin Bellisario[3], Zack Ulissi[4] and Michael S. Strano[2], (1)Mechanical Engineering, Massachusetts Institute of Technology, Cambridge, MA, (2)Department of Chemical Engineering, Massachusetts Institute of Technology, Cambridge, MA, (3)Chemistry, MIT, Cambridge, MA, (4)Chemical Engineering, Massachusetts Institute of Technology, Cambridge, MA

The study of transport through nanopores is a growing field of research due to the possibility of inexpensive DNA sequencing, single-molecule detection, and incorporation into efficient membranes. Our previous studies[1,2] demonstrated the presence of the Coulter effect when individual ions block an otherwise stable proton current flowing through the interior of the carbon nanotube. In this study, we demonstrated further evidence for this mechanism by showing the inverse relationship between dwell times and applied voltage, which indicate the presence of a charged blocker. Furthermore, temperature experiments demonstrate the possibility of an activated process involved in the ion translocation. Finally, Raman spectroscopy was used to characterize the diameter of individual single-walled carbon nanotubes before they were incorporated into devices. Stochastic ion pore-blocking was observed for several devices containing SWNT of diameters ranging from 0.9 and 2.0 nm. Interestingly, we observe a maximum blocked proton current around a nanotube diameter of 1.6 nm, with decreased pore-blocking currents at smaller and larger diameters.

1. Lee, C. Y., Choi, W., Han, J.-H., & Strano, M. S. (2010). Coherence Resonance in a Single-Walled Carbon Nanotube Ion Channel. *Science*, *329*(5997), 1320-1324.
2. Choi, W., Lee, C. Y., Ham, M.-H., Shimizu, S., & Strano, M. S. (2010). Dynamics of Simultaneous, Single Ion Transport through Two Single-Walled Carbon Nanotubes: Observation of a Three-State System. *Journal of the American Chemical Society*, *133*(2), 203-205.

(424d) Role of Adsorbed Surfactant in the Reaction of Aryl Diazonium Salts with Single-Walled Carbon Nanotubes

Andrew J. Hilmer[1], Tom P. McNicholas[1], Shangchao Lin[2], Jingqing Zhang[1], Qing Hua Wang[1], Jonathan D. Mendenhall[1], Changsik Song[3], Daniel A. Heller[1], Paul W. Barone[1], Daniel Blankschtein[1] and Michael S. Strano[1], (1)Department of Chemical Engineering, Massachusetts Institute of Technology, Cambridge, MA, (2)Department of Chemical Engineering and Department of Mechanical Engineering, Massachusetts Institute of Technology, Cambridge, MA, (3)Department of Chemistry, Sungkyunkwan University, Suwon, South Korea

Because covalent chemistry can diminish the optical and electronic properties of single-walled carbon nanotubes (SWCNTs), there is significant interest in developing methods of controllably functionalizing the nanotube sidewall. To date, most attempts at obtaining such control have focused on reaction stoichiometry or strength of oxidative treatment. Here, we examine the role of surfactants in the chemical modification of single-walled carbon nanotubes with aryl diazonium salts. The adsorbed surfactant layer is shown to influence the diazonium derivatization of carbon nanotubes in several ways, including electrostatic attraction or repulsion, steric exclusion, and direct chemical modification of the diazonium reactant. Electrostatic effects are most pronounced in the cases of anionic sodium dodecyl sulfate and cationic cetyltrimethylammonium bromide, where differences in surfactant charge can significantly affect the ability of the diazonium ion to access the SWCNT surface. For bile salt surfactants, with the exception of sodium cholate, we find that the surfactant wraps tightly enough such that exclusion effects are dominant. Here, sodium taurocholate exhibits almost no reactivity under the explored reaction conditions, while for sodium deoxycholate and sodium taurodeoxycholate, we show that the greatest extent of reaction is observed among a small population of nanotube species, with diameters between 0.88 and 0.92 nm. The anomalous reaction of nanotubes in this diameter range seems to imply that the surfactant is less effective at coating these species, resulting in a reduced surface coverage on the nanotube. Contrary to the other bile salts studied, sodium cholate enables high selectivity toward metallic species and small band gap semiconductors, which is attributed to surfactant-diazonium coupling to form highly reactive diazoesters. Further, it is found that the rigidity of anionic surfactants can significantly influence the ability of the surfactant layer to stabilize the diazonium ion near the nanotube surface. Such Coulombic and surfactant packing effects offer promise toward employing surfactants to controllably functionalize carbon nanotubes.

(424e) Charge Transfer in Junctions of Single Layer Graphene and Metallic Single Walled Carbon Nanotubes

Geraldine LC Paulus[1], Qing Hua Wang[1], Zach Ulissi[2], Tom McNicholas[2], Chih-Jen Shih[1], Zhong Jin[2] and Michael S. Strano[1], (1)Department of Chemical Engineering, Massachusetts Institute of Technology, Cambridge, MA, (2)Chemical Engineering, Massachusetts Institute of Technology, Cambridge, MA

We fabricate and study junctions between single walled carbon nanotubes (SWNTs) and monolayer graphene for the first time. A single layer graphene (SLG) sheet grown by chemical vapor deposition (CVD) was transferred onto a Si/SiO_2 wafer with aligned CVD-grown SWNTs. Raman spectroscopic mapping is used to identify metallic-SWNT/SLG junctions, and we report a method for spectroscopic deconvolution of the overlapping G peaks of SWNT and SLG by making use of the polarization dependence of the SWNT. A comparison of the Raman peak positions and intensities of the individual SWNT and graphene to the SWNT-graphene junction indicates an electron transfer of 1.12 x 10^{13} cm^{-2} to the SLG at the junction. We claim part of these electrons come directly from the SWNT where the direction of transfer is in agreement with the work functions of each. It is likely another part of the additional electrons observed in the graphene at the junction results from the fact that the latter is no longer in direct contact with the p-doping SiO_2 substrate. We find that the compression of the SWNT by the graphene increases the broadening of the radial breathing mode (RBM) peak from 3.6+/-0.3 to 4.6+/-0.5 cm^{-1} and the G peak from 13+/- 1 to 18 +/- 1 cm^{-1}, in reasonable agreement with a molecular dynamics simulation we performed of the system. However we conclude that RBM and G peak position shifts are due primarily to charge transfer with minimal contributions from strain. With this method, we are able to dope the graphene on a nanometer lengthscale.

(424f) Characterizing the Structural and Electronic Properties of Metallocene-Functionalized, Single-Walled Carbon Nanotubes: A DFT Study

Zhongtao Zhang[1], Wei An[2] and C. Heath Turner[1], (1)Chemical and Biological Engineering, University of Alabama, Tuscaloosa, AL, (2)Department of Chemical and Biological Engineering, University of Alabama, Tuscaloosa, AL

Characterizing the Structural and Electronic Properties of Cyclopentene -Metal- Singlewalled Carbon Nanotube Complexes: a DFT Study

The complexes of metallocene(Cp_2M) between single walled carbon nanotubes (SWCNT) on the surface or inside,

graphene are studied thoroughly, and shown potential applications as sensors. CpM and SWCNT/ graphene complexes drawed our attentions due to the similarity with the tunable electronic structures of the complexes of SWCNT/ graphene and organic molecules/ metal atoms. Meanwhile, our previous research showed significantly stabilization of transition metals on B-doped nanocarbon, indicates that CpM may aslo be stabilized. Therefore, we perform a theoretical study on the structural and electronic properties of CpM (M= Fe, Co, Cr, V, Ni) and pristine or B-doped SWCNTs. A significant increasing of binding energies comparing with our previous work on metal atoms was observed (increased about 2eV). Also, density of states (DOS), and band strucutres of CpM (M= Fe, Co, Cr, V, Ni) ligands on the outer surface of 4 different SWCNTs: prinstine (6, 6), pristine (8, 0), B-doped (6, 6), B-doped (8, 0) SWCNTs were calculated using Density Functional Theory (DFT) to explore the potential application of CpM SWCNT complexes for sensors as well as the possibilities of immobilization of CpM based catalyst by B-doped SWCNTs.

Thermal Transport Properties of Energetic Composites with Graphene and Carbon Nanotube Additives: An Experimental Study

*Keerti Kappagantula** and *Michelle L. Pantoya*
Mechanical Engineering Department,
Texas Tech University,
Lubbock, TX 79409-1021
Email: keerti.kappagantula@ttu.edu,
michelle.pantoya@ttu.edu

ABSTRACT

Reactive materials such as aluminum and polytetrafluoroethylene (Teflon) are used for energy generation applications and specifically in ordnance technologies. With the advent of nanotechnology various nano-scale additives have become incorporated into reactive material formulations with the hope of enhanced performance. An important component to the study of energy generation is an examination of energy transport through a reactant matrix. This study examines an experimental approach to quantifying thermal properties of an Al/Teflon nanocomposite reactant matrix that has been impregnated with carbon additives. Various structures of carbon are investigated and include amorphous nanoscale carbon spheres (nano C), graphene flakes and unaligned multiwalled carbon nanotubes (CNTs). The additives were selected based on their completely different structures with the hypothesis that the structure of the additive will influence the thermal transport properties of the matrix. Results show graphene has the greatest influence on the thermophysical properties. For example, thermal conductivity of the composites containing graphene increased by 98%. Graphene similarly enhanced the thermal diffusivity and specific heat of the Al/Teflon matrix. Conversely, nano C and CNTs decreased the thermal conductivity and thermal diffusivity of the samples significantly.

INTRODUCTION

Energetic composite materials consisting of metals and fluoropolymers are referred to here as reactive materials. Compared to monomolecular energetic materials, the fuel-oxidizer ratios in reactive materials may be varied to tailor flame front propagation speed, final product composition, gas production and energy density requirements. A number of variables such as fuel or oxidizer particle size and distribution, metal passivation layer variations, bulk density, synthesis technique and stoichiometry may be altered to tune the reaction characteristics of a reactive material towards a specific application. Ever since the synthesis of CNTs and the isolation of free-standing single layer graphene, both the allotropes of carbon have been subject to a lot of attention due to their exceptional thermal, mechanical and electrical properties. Literature shows a lot of work where graphene and CNTs are incorporated into polymers and composites to enhance the thermophysical properties of the mixture (1-5). However, only recently did graphene find its way into pyrotechnic and propellant applications as an additive (6). The objective of this study is to examine the effects of carbon additives on the thermal properties of Al/Teflon nanocomposite reactive material.

EXPERIMENTAL

Nanometric aluminum (nano Al) was used as the fuel and obtained from Novacentrix (USA). The nano Al particles were encapsulated in an alumina (Al_2O_3) passivating shell with a spherical shape, an average diameter of 80 nm and a shell thickness of 4 nm. Polytetrafluoroethylene, produced by Dupont, under the trade name of Teflon (Zonal MP 1150) was used as the oxidizer. The Teflon particles have an average diameter of 200 nm and are a good source of fluorine (about 75% by weight). Graphene, carbon nanotubes (CNT) and amorphous nanometric carbon spheres (nano C) were used as the additives. The graphene used was procured from Graphene Supermarket (USA). These particles have an average thickness of 8 nm (consisting of 20-30 monolayers) and a lateral size of averaging 550 nm. The CNTs were obtained from Alfa Aesar (USA) and are multiwalled tubes with 3 – 20 nm outer diameter, 1 – 3 nm inner diameters and 0.1 to 10 μm long. The samples obtained were 95% pure. The nano C was purchased from Sigma Aldrich and have an average size of 30 nm. They are amorphous in their structure and were made from a laser decomposition process.

A slightly fuel rich composition with an equivalence ratio of 1.4 was maintained for all Al/Teflon baseline nanocomposites. To this base composition 1, 5 and 10 percent by mass of additives were added to understand how their effect varies with increasing concentration. Detailed procedure for sample preparation may be found elsewhere [4]. Measured quantities of samples prepared were pressed into cylindrical pellets having a TMD of 75-80%. The pellets were all maintained at a diameter of 12.7 mm and a thickness of 0.23 mm. These were then coated with a dry graphite film lubricant.

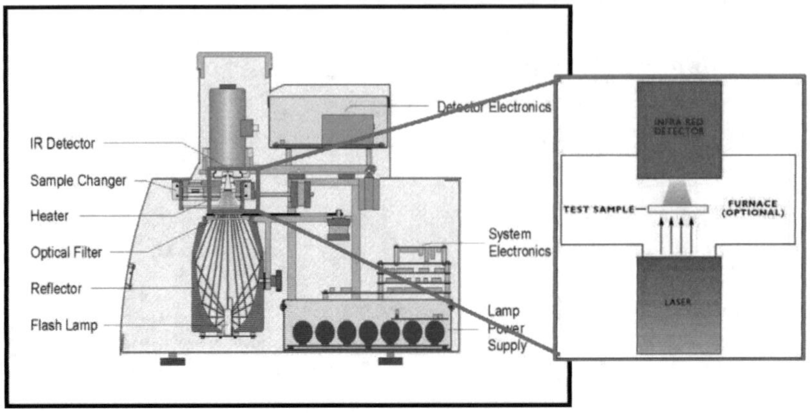

Fig. 1: Schematic showing the cross sectional view and the principle on which the Netzsch LFA 447 Nanoflash LFA diagnostic works.

For each combination of Al/Teflon/Additive, thermal diffusivity, specific heat and heat capacity were measured using a Netzsch LFA 447 Nanoflash laser flash analysis (LFA) diagnostic. The LFA is a direct measurement method for determining thermal diffusivity and an indirect method for thermal conductivity. Radiation from a xenon flash lamp is incident on the bottom surface of the sample. The temperature response on the top surface of the sample due to the propagation of heat from the bottom surface is measured using an IR detector and plotted as a function of time. By analysis of the resulting temperature history, the thermal diffusivity is determined by the software using the following equation:

$$\alpha = 13.88\frac{l^2}{t_{50}} \qquad \text{Equation [1]}$$

where α (mm^2/s) is thermal diffusivity, l (cm) is the thickness of the test piece and t_{50} is the "half-rise-time", corresponding to the time taken by the top surface to get to the 50% of its maximum temperature. Specific heat is then measured by comparing the actual temperature rise of the sample to the temperature rise of a reference sample of a known specific heat. The equation used is given below:

$$C_p = \frac{Q}{m.\Delta T} \qquad \text{Equation [2]}$$

where C_p (kJ/kg.K) is the specific heat capacity of the sample, m (kg) is the mass of the sample and ΔT is the temperature rise of the sample. The instrument can measure thermal diffusivity (α) and specific heat (C_p) simultaneously. The software then uses these values and the bulk density (ρ) supplied prior to the commencement of the experiment to calculate thermal conductivity (k) from the Laplace equation shown below.

$$k = \alpha * \rho * C_p \qquad \text{Equation [3]}$$

Since Teflon has a very low melting point of 300 – 310°C, lowest energy pulses at 190 V and 0.18 ms durations were fired at the samples to test them at temperatures starting from 25°C to 200°C with 8 steps of 25°C increase. The optical filter was kept at 100% and a single layer measurement technique was employed using the software. The Cowan-plus pulse method was used to account for the radiation losses from the pellet and measure the thermophysical properties desired. A cylindrical Pyrex pellet supplied by the manufacturer was used as the reference sample for the calculation of specific heat.

RESULTS AND DISCUSSION

The thermal diffusivity, specific heat and thermal conductivity are plotted as a function of pellet temperature in figures 2-4. Two pellets were made for every composition and the thermophysical properties were evaluated for seven tests from each pellet, each data point in figures 2-4 is an average of 14 measurements. There is a 7% associated with the diagnostic.

The most interesting observation from this analysis is the uniquely opposite behavior graphene introduces as compared to the other carbon additives: nano C and CNTs. Specifically, small amounts of graphene were shown to increase thermal diffusivity properties up to 75%. In contrast, nano C decreases thermal diffusivity by 80% and CNTs exhibit a similar decreasing trend. Heat capacity is similarly affected. Graphene tends to increase heat capacity properties of Al/Teflon, while other carbon additives show a decrease. The thermal energy transport properties shown in figures 2-4 for graphene suggest superior energy propagation behaviors ideal for energetic material applications.

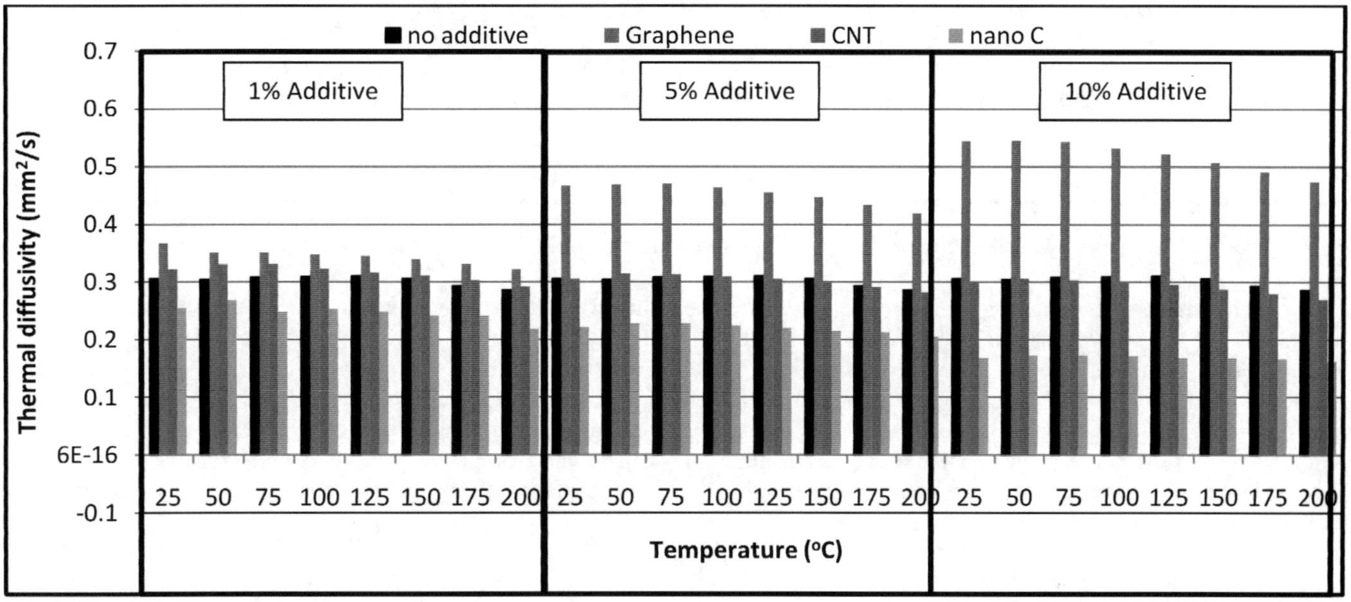

Fig. 2: Thermal diffusivity of nanocomposites as a function of temperature.

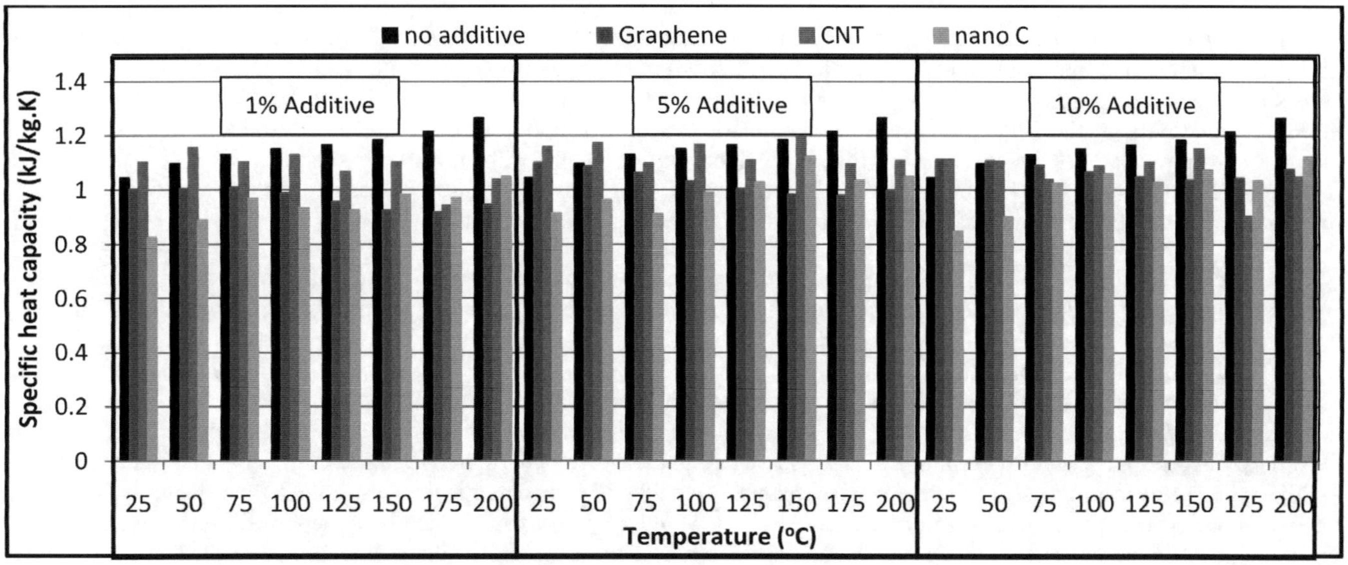

Fig. 3: Specific heat capacity of nanocomposites as a function of temperature.

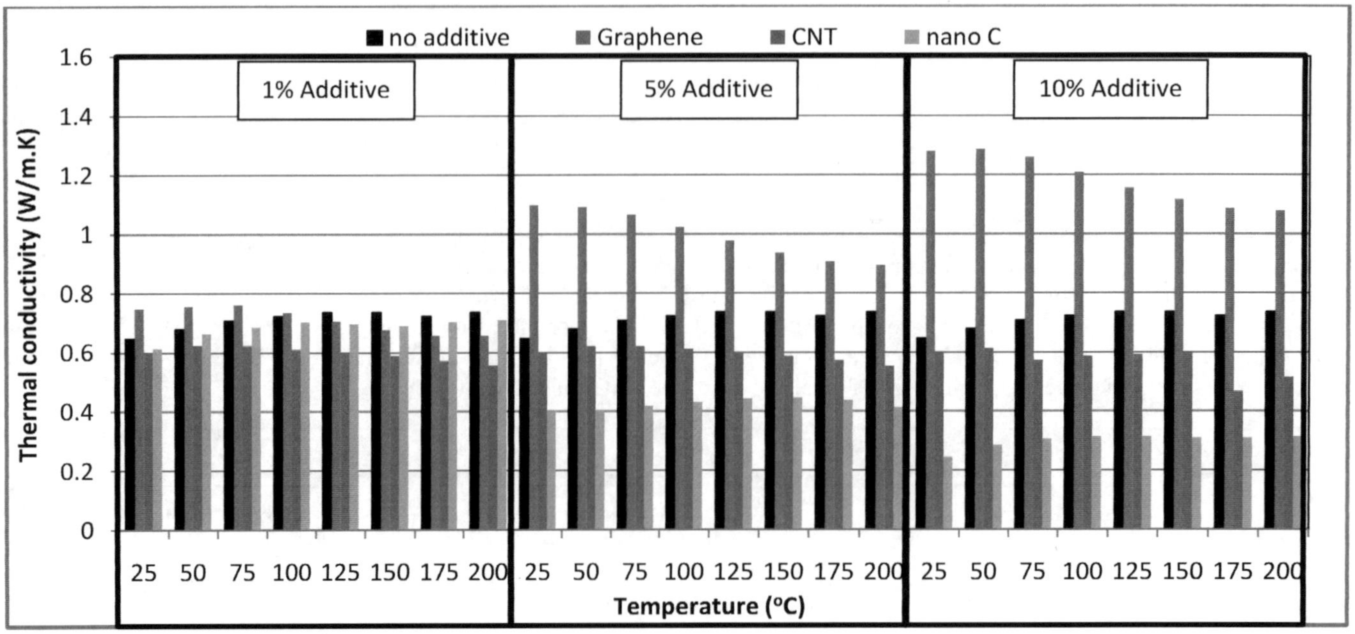

Fig. 4: Thermal conductivity of nanocomposites as a function of temperature.

Thermal conductivity of a composite is very closely linked to the phenomenon of heat transport between the individual components of the composite (7, 8). Presence of additives with thermal conductivities different from that of the matrix causes a qualitative change in the mechanism of heat transport, leading to a meaningful increase or decrease of thermal conductivity of the nanocomposite as a whole. Nano C has a thermal conductivity lower than CNTs and graphene. It seems almost intuitive to postulate that presence of nano C with an amorphous structure in the Al/Teflon matrix (k = 0.65-0.74 W/m.K) would serve to decrease the thermal conductivity of the nanocomposite. This is exactly reflected in figure 3. This argument also supports the experimental observation that as the concentration of amorphous nano C increases in the Al/Teflon matrix, the thermal conductivity decreases.

A similar trend is displayed in figure 4, where the addition of CNTs to Al/Teflon decreases the thermal conductivity of the nanocomposite. Thermal transport is closely related to the intrinsic properties of the grain boundaries between the individual components in a composite. Although it has been reported that addition of CNTs to nanocomposites has enhanced the thermal properties of the grain boundaries and subsequently the thermal conductivity of the composite (9), the results from the present work suggest the contrary. This may be because the CNTs used are not aligned. Also, CNTs agglomerate, forming sporadic particle cliques and bundles in the matrix. Increasing the amount of CNTs in Al/Teflon matrix may lead to their uneven, unaligned dispersion, thus distribution the thermal transport network by creating interfacial thermal barriers and increasing contact resistance (10). Fig. 4 also shows that introducing graphene into Al/Teflon leads to an increase in the thermal conductivity of the nanocomposite. Similar results have been reported (11-13) where introducing graphene into a polymer substrate lead to the increase of the thermal properties of the substrate, both theoretically (13) and through experimental results (11, 12). Klemmens et al. showed that when graphene's volume fraction reached a critical value, the improvement of the thermal conductivity of the composite above that of the substrate became significant (13).

Specific heat is the thermophysical property that describes a material's ability to store thermal energy. Specific heat capacity is defined here as the amount of heat energy required to raise the temperature of a specific mass of a substance through a defined temperature (without phase change). Specific heat capacity is an intensive property of the material, based on the energy stored in the form of kinetic and potential energy, depending on the degrees of freedom available for the atoms inside it. The specific heat capacity of the nanocomposites with nano C, CNT and graphene additives are presented in figure 3. It may be seen from these results that the C_p of the nanocomposites with additives decreases when compared to the C_p of Al/Teflon. Incorporating additives into a matrix serves to increase the thermal boundaries between the matrix and the additives. Consequently, the C_p of the composite (matrix + additive) is expected to increase rather than decrease. However, the results obtained oppose this reasoning. Various reasons were explored explain this phenomenon. Initially, packing factor was thought to be the cause for this. However, an examination of the densities of the pellets used revealed that all of them had more or less the same TMD (about 75 – 80%). Also, if the packing density was a factor contributing to the decreased C_p, the trend should also have been reflected in the other thermal transport properties measured. In addition, since C_p is an intensive property, mass of the pellet should not really affect it. Next, it was reasoned that the counterintuitive results may be because of some error in the LFA diagnostic's measuring technique. To cross-check the results, two calibration runs were performed with samples whose thermophysical properties were already known: Pyroceram 9606 and Pyrex 7740. However, both the samples yielded results well within the uncertainty limits of the LFA and which concurred with the thermal property tables published. The reason for the decreased C_p of the nanocomposites with additives is not yet understood and is still being explored.

Thermal diffusivity of a material measures the ability of a material to conduct thermal energy relative to its ability to store thermal energy. Consequently materials with a larger thermal diffusivity will respond quickly to changes in their thermal environment. Figure 2 shows the variation of thermal diffusivity of Al/Teflon and the nanocomposites with nano C, CNTs and graphene. Adding nano C and CNTs to Al/Teflon decreases the thermal diffusivity of the nanocomposite, where as the graphene additive increases it. Since thermal diffusivity is the ratio of the thermal conductivity to the volumetric specific heat of a material, it is closely affected by both these properties. The reason for the decrease of the thermal conductivity of the nanocomposites with nano C and CNTs was explained earlier in this section. Also, it is seen that the additives have almost no effect on the specific heat capacity of the nanocomposite. It naturally follows there would be a corresponding decrease in the thermal diffusivity for graphene containing nanocomposites. The thermal conductivity increases with increasing graphene as does the volumetric specific heat capacity. However, the increase in the thermal conductivity is much greater than that in the volumetric heat capacity and consequently, this leads to increased thermal diffusivity.

The additives were selected to study the influence of their structure on the thermal properties of the reactant matrix. It is clearly shown from the experimental results that particles with amorphous structure (nano C) decrease the thermophysical properties of the nanocomposites. On the other hand, CNTs and graphene, although consisting of sp^2 bonded carbon atoms, have very different structures. If both graphene and CNTs are viewed as spheroidal with principal dimensions $a_1 = a_2$ and a_3, then under ideal circumstances, graphene sheets may be seen as oblate spheroids with $a_3/a_1 \rightarrow 0$ and CNTs may be considered to be prolate spheroids with $a_3/a_1 \rightarrow \infty$. It has been shown that composite properties greatly depend on the aspect ratios and geometries of the additives introduced into their matrices (14-15). In terms of thermal conductivity, Xie et al. (16) showed theoretically that graphene reinforced composites showed greater values compared to the CNT reinforced composites. When the aim is to obtain enhanced thermal properties in the final nanocomposites, graphene sheets enjoy certain advantages over the CNTs. Whereas CNTs might be susceptible to irregular arrangement, point defects and poor wetting properties, graphene is immune to these deficiencies. On the other hand, graphene sheets too show some surface and edge defects (17-20). Despite this, however, the thermal transport in graphene is not affected to a great extent because of a large lattice area available for heat transfer. In addition, CNTs are stiff in nature whereas graphene is more flexible which is advantageous when the substrate nanocomposites are pressed to form pellets. These arguments suggest that graphene composites would outperform CNT composites in terms of thermophysical properties and this is reflected in the experimental results.

REFERENCES

1. W. Hong and N. Tai, *Diam. Relat. Mater.*, 2008, 17, 1577.
2. K. Mackenzie and M. Bolton, *J. Mat. Sci.*, 2009, 44, 2851.
3. S. Stankovich, *Nature*, 2006, 442, 282.
4. S. Watcharotne et al., *Nano Lett.*, 2007, 7, 1888.
5. D. Wang et al., *ACS Nano,* 2009, 3, 1745.
6. J. Sabourin et al., *ACS Nano,* 2009, 3, 3945.
7. M. Kaviany, *Principles of Heat Transfer in Porous Media,* 1991.
8. R. Carbonell and S. Whitaker, *Fundamentals of Transport Phenomena in Porous Media*, 1984, 121.
9. B. Li et al., *Mater. Sci. Forum*, 2009, 610-613, 559.
10. K. Ahmad and W. Pan, *Ceram. Eng. Sci. Proc.*, 2009, 29, 49.
11. K. Jagannadham, *J. Electron. Mater.*, 2011, 40, 25.
12. A. Yu et al., *J. Phys. Rev. C*, 2007, 111, 7565.
13. P. Klemmens, *Int. J. Thermophys.*, 2001, 22, 265.
14. F. Deng et al., *Appl. Phys. Lett.*, 90 (2) (2007) 021914.
15. S. Yao et al., *Appl. Phys. Lett.,* 2007, 91, 212901.
16. S. Xie, Y. Liu and J. Yi, *Appl. Phys. Lett.*, 2008, 92, 243121.
17. M. Morooka, T. Yamamoto and K. Watanabe, *Phys. Rev. B*, 2008, 77, 033412.
18. K. Saito, J. Nakamura and A. Natori, *Phys. Rev. B,* 2007, 76, 115409.
19. M. Trushin and J. Schliemann, *Phys. Rev. Lett.*, 2007, 99, 216602.
20. M. Lusk and L. Carr, Phys. Rev. Lett., 2008, 100, 175503.

(424h) Study of Single-Walled Carbon Nanotube Aerogel Thermal Conductivity in the Gas Environment with Varying Aerogel Density

Scott N. Schiffres[1], Lin Hu[1], Kyu Hun Kim[2], Mohammad F. Islam[2], Alan J. H. McGaughey[1] and Jonathan A. Malen[1], (1)Mechanical Engineering, Carnegie Mellon University, Pittsburgh, PA, (2)Chemical Engineering and Materials Science & Engineering, Carnegie Mellon University, Pittsburgh, PA

We will report on the thermal conductivity of a gas permeated single-walled carbon nanotube (SWCNT) aerogel as a function of density. The SWCNT aerogel is a nano-porous material that is solution-formed and contains only pure SWCNTs. The SWCNTs of the aerogel interact via van der Waals forces. Densities in the range of 8-15 kg/m^3 were considered. By using kinetic theory to analyze the gas contribution to thermal conductivity in the SWCNT aerogel , a mean energy transport length scale was extracted. This mean energy length scale was compared to the structurally limited mean free path of the gas molecules derived from mesoscale simulations. The difference between these lengthscales was used to determine the thermal energy accommodation of various gases with the SWCNTs. Measurements at a single aerogel density indicate low energy accommodation. The SWCNT's effective thermal conductivity versus the volumetric density of tube-tube junctions will be correlated from the density dependent measurements

(439a) Enzyme Microreactors and Their Prospective Application in CO2 Bioconversion

Man Bock Gu, College of Life Sciences and Biotechnology, Korea University, Seoul, South Korea

Enzyme Microreactors and Their Prospective Application in CO_2 Bioconversion

Ee Taek Hwang, Haemin Gang, Jinyang Chung, Ji Hoon Kim, Hyun Lee, Rameshwar Tatavarty, and Man Bock Gu*

College of Life Sciences and Biotechnology, Korea University, Republic of Korea

E-mail: mbgu@korea.ac.kr

TEL: +82-2-3290-3417, FAX: +82-2-928-6050

Enzymes immobilized on nano/micro sized organic/inorganic hybrid materials have been known to be stabilized. For meeting the requirement of enzymatic processes, such as highly stabilized activity, enhanced specific activity with high enzyme loading, and easy scale up, the enzymes have been subjected to be aggregated on the surfaces of quantum dots-embedded nanofibers, entraped in calcium phosphate nanoparticles, or organic-inorganic hybrid microspheres as porous support materials for good accessibility of substrate. The successful immobilization of enzymes on these enzyme-nano/micro sized organic/inorganic hybrid materials have well been evidenced by using Fourier transform infrared (FTIR) spectroscopy, scanning electron microscopy (SEM), Transmission electron microscopy (TEM), while physical characterizations such as like Dynamic light scattering (DLS), Elemental analysis (EDAX), particle size etc were done. As results, the immobilized enzymes have shown the good storage and recyclable stabilities retained in terms of its enzyme activity, in repeated usages for more than ten times. Therefore, considering CO_2 enzymatic bioconversion into useful materials, this unique enzyme stabilization technology developed here is strongly considered to be a key tool for realizing enzyme-based CO_2 bioconversion. An example of enzyme-based CO_2 bioconversion will be presented in this talk with clear perspectives on its enzyme microreactor-based CO_2 bioconversion.

REFERENCES

1. E.T. Hwang, H. Lee, J.H Kim, R. Tatavarty, M.B. Gu (2011), *Journal of Materials Chemistry*,21 (18), 6491 – 6493

2. E.T. Hwang[+], R.Tatavarty[+], H. Lee, J Kim, M.B. Gu (2011), *Journal of Materials Chemistry*, 21 (14), 5215-5218, (Selected on the cover page) ([+]equally contributed authors)

(439b) Nano Carbon Materials for Efficient Cofactor Regeneration – Toward Solar Driven Biosynthesis

Ping Wang, Department of Bioproducts and Biosystems Engineering, University of Minnesota, St Paul, MN

One unique feature of nanostructured carbon materials for electrochemical applications is that they can substantially reduce overpotentials of electrobiochemical reactions. Cofactors such as NADH have been shown greatly valuable in developing cell-free multistep biosynthetic reaction pathways, including the direct enzymatic reduction of carbon dioxide to value-added chemicals. While such reactions are greatly appealing, regeneration of cofactors using either chemical or electrochemical means can be expensive, often presents a major hurdle in large scale applications. Solar driven cofactor regeneration may afford a new avenue for energy-efficient biosynthesis. Toward that, we have demonstrated that regeneration of NADH from NAD^+ can be achieved at a potential ~0.5 V, very close to the theoretical redox potential (0.33V) of NADH, and much lower than those realized with regular carbon electrodes (~1.0 V). Similar to biological energetic pathways where cofactors serve as the major energy storage agents, this work also promises new strategies in harvesting solar energy for both biosynthesis and other bioelectrochemical applications.

(439c) In-Situ Study of Native Cellulose Structure in Intact Lignocellulose Biomass and Its Influence On Enzymatic Hydrolysis Process

Kabindra Kafle[1], Christopher Lee[2], Heenae Shin[3], Sunkyu Park[3] and **Seong H. Kim**[2], (1)Chemistry, Pennsylvania State University, University Park, PA, (2)Chemical Engineering, Pennsylvania State University, University Park, PA, (3)Forest Biomaterials, North Carolina State University, Raleigh, NC

In order to efficiently convert biomass to other useful chemicals, it is necessary to understand and overcome its recalcitrance to enzymatic deconstruction processes. The first key question involved in the biomass study is the correlation between the enzymatic activity and the cellulose structure. Although many experimental studies have been attempted previously to answer this fundamental question, it still remains elusive. The difficulty is that it requires molecular analysis of crystalline carbohydrate polymers (cellulose) in amorphous matrix containing the same organic functional groups. Current analytical methods cannot provide sufficient details about crystalline cellulose in lignocellulose biomass. In this talk, we will present the recent results from sum-frequency generation (SFG) vibration spectroscopy study of the structure of cellulose in intact lignocellulose biomass and its correlation with enzymatic deconstruction process. The main advantage of the SFG spectroscopy technique is the detection of crystalline cellulose in various biomass samples without spectral interferences from amorphous matrix carbohydrate. Thus, it reveals new molecular insights that cannot be obtainable by other analytical methods.

(439d) Engineering Biomaterials for Bioremediation Applications

Eduardo Reategui[1], Erik Reynolds[2], Lisa Kasinkas[1], Michael Sadowsky[2], Lawrence Wackett[2] and **Alptekin Aksan**[1,2], (1)Mechanical Engineering Department, University of Minnesota, Minneapolis, MN, (2)Biotechnology Institute, University of Minnesota, Minneapolis, MN

To be useful in a water-treatment applications, a bioremediation method should: (i) have stable, long-term degradation activity; (ii) be mechanically stable and sturdy; (iii) be conducive to high water flow; (iv) maintain active enzymes or cells in the matrix without significant release, and (v) be inexpensive. We developed hybrid biomaterials to be used for atrazine (2-chloro-4-ethylamine-6-isopropylamino-s-triazine) biodegradation. The herbicide atrazine is currently used in 70 countries at an estimated annual rate of 111,000 tons for control of broadleaf weeds, principally in corn, sorghum, and sugarcane. The hybrid biomaterial consisted of recombinant *E. coli* cells overexpressing atrazine chlorohydrolase (AtzA), which converts atrazine into hydroxyatrazine. The cells were encapsulated in a polymer/silicon oxide matrix by a sol-gel process (Figure 1).

Figure 1: SEM image of E-coli encapsulated in silica gels

The silica base gel used to encapsulate the microorganisms consisted of a combination of silicon oxide precursors (e.g., silica nanoparticles, alkoxides) and a biocompatible organic polymer (e.g., polyethylene glycol, PEG). The porous material enabled diffusion of water and atrazine into the gel and diffusion of hydroxyatrazine out of the gel. The gel additionally adsorbed atrazine, a property that contributed to removal of atrazine from the solution in the process. Moreover, high temperature exposure was utilized to ensure that the encapsulated cells were non-viable but remained fully active in degrading atrazine over a long time scale.

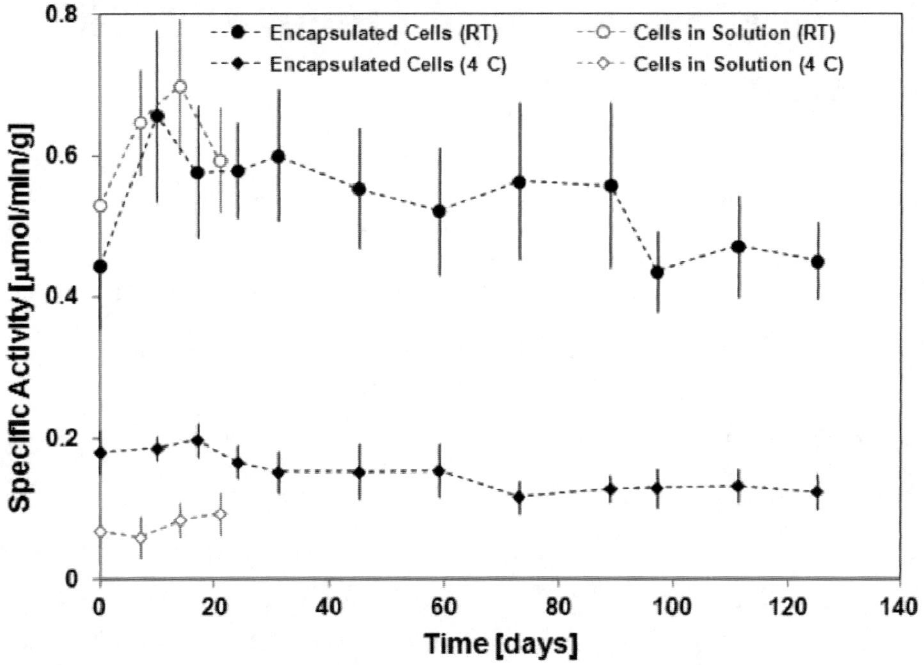

Figure 2: Specific activity of encapsulated (vs. free) cells measured over 4 months.

We measured the activities of the free and silica encapsulated cells over 4 months (Figure 2). When the activity was measured at room temperature, free cells showed an average of (0.61 ± 0.04) mmol/g/min of activity over 21 days. After 21 days, significant cell lysis was observed in the free cells; this was likely due to long-term hypo-osmotic stress induced by water. Therefore, the experiments on the free cells were stopped at that time point. On the other hand, cells encapsulated in porous silica gels showed stable activity between (0.44 ± 0.06) mmol/g/min to (0.66 ± 0.12) mmol/g/min for up to 4 months. This showed that even though the encapsulated cells were non-viable and had lost some membrane integrity, AtzA was protected and active in the silica matrix. The activities of the free and encapsulated cells were found to be temperature dependent. At 4°C, activity dropped by 45% and 30% for the free and encapsulated cells, respectively. The activity of encapsulated cells at 4°C was 33.3% higher than the cells in solution.

This showed that bioremediation using encapsulated microorganisms is a safe and feasible technique. Currently, we are working towards using similar techniques to remove chemicals from wastewaters generated by hydraulic fracturing for shale gas and oil.

(439e) Highly Efficient Enzyme Immobilization and Stabilization within Meso-Structured Onion-Like Silica for Biodiesel Production

Seung-Hyun Jun, Department of Chemical & Biological Engineering, Korea University, Seoul, South Korea, Jinwoo Lee, Department of Chemical Engineering and School of Environmental Science & Engineering, Pohang University of Science and Engineering, Pohang, South Korea, Taeghwan Hyeon, National Creative Research Center for Oxide Nanocrystalline Materials and School of Chemical and Biological Engineering, Seoul National University, Seoul, South Korea and Jungbae Kim, Chemical and Biological Engineering, Korea University, Seoul, South Korea

Highly Efficient Enzyme Immobilization and Stabilization within Meso-Structured Onion-like Silica for Biodiesel Production

Seung-Hyun Jun, Jinwoo Lee, Taeghwan Hyeon, and Jungbae Kim*

Department of Chemical and Biological Engineering, Korea University, Seoul, Korea

Advanced Functional Nanomaterials Laboratory, Department of Chemical Engineering, Pohang University of Science and Technology, Pohang, Kyungbuk 790-784, Korea

National Creative Research Initiative Center for Oxide Nanocrystalline Materials and School of Chemical Engineering, Seoul National University, Seoul 151-744, Korea

Abstract

Enzyme was immobilized and stabilized in meso-structured onion-like silica (Meso-Onion-S), which has highly-curved meso-pores of 10 nm diameter in a multi-shell structure in a form of nanoscale enzyme reactors (NERs). NERs were prepared via a two-step process of enzyme adsorption and subsequent enzyme crosslinking, which effectively prevents the leaching of crosslinked enzyme aggregates from highly-curved mesopores of Meso-Onion-S. As a result, NERs in Meso-Onion-S significantly improved the enzyme stability as well as the enzyme loading. For example, NER of lipase (NER-LP) was stable under rigorous shaking for 40 days, while the control sample of adsorbed LP (ADS-LP) with no enzyme crosslinking showed a rapid inactivation due to rigorous enzyme leaching under shaking. Stable NER-LP was successfully employed to produce biodiesels, fatty acid methyl esters, from the LP-catalyzed transesterification of soybean oil with methanol. Interestingly, the specific activity of NER-LP was 23 and 10 times higher than those of free LP and ADS-LP, respectively, revealing the importance of LP stabilization in the form of NER-LP in the presence of organic solvents.

(440a) High Throughput Sequencing of Antibody Heavy and Light Chain Pairings

Brandon DeKosky[1], Ryan P. Deschner[1], Brandon Rawlings[1], Gregory Ippolito[1], C. Grant Willson[2], Andrew D. Ellington[3] and George Georgiou[4], (1)Department of Chemical Engineering, The University of Texas at Austin, Austin, TX, (2)Departments of Chemical Engineering and Chemistry, The University of Texas at Austin, Austin, TX, (3)Department of Biochemistry, University of Texas at Austin, Austin, TX, (4)Departments of Chemical Engineering and Biomedical Engineering, The University of Texas at Austin, Austin, TX

Next generation DNA sequencing (NGS) is transforming the field of immunology by giving researchers the potential to sequence immune repertoires at unprecedented depth. However, the utility of NGS for repertoire analysis is severely curtailed by its inability to resolve the pairing between antibody heavy and light chains and T-cell receptor α and β strands. To directly address this issue, we have developed an emulsion-based method to sequence natively paired antibody heavy and light chains at high-throughput. V_H-V_L pairings have been validated using mixtures of various immortalized B-cell clones with known antibody sequences to confirm accuracy. High-throughput sequencing of the heavy and light chain pairing is a significant step toward greater understanding of the immune repertoire, and potential applications include immune response analysis and antibody discovery.

This work was supported by the University of Texas at Austin Cockrell School of Engineering, the Harrington Foundation, the National Science Foundation, and The Hertz Foundation.

(440b) Characterization and Optimization of Inhalable Pegylated Phospholipid Microparticles and Nanoparticles Containing Paclitaxel for Targeted Pulmonary Nanomedicine in Lung Cancer

Samantha A. Meenach[1,2], Kimberly W. Anderson[2], J. Zach Hilt[2], Ronald C. McGarry[3] and Heidi M. Mansour[1], (1)Pharmaceutical Sciences - Drug Development Division, University of Kentucky, Lexington, KY, (2)Chemical and Materials Engineering, University of Kentucky, Lexington, KY, (3)Radiation Medicine, University of Kentucky, Lexington, KY

Despite the significant advances in the treatment of lung cancer, it is a disease that still has signifies poor prognosis and challenges in implementation of treatment. Targeted pulmonary inhalation drug delivery offers many advantages for lung cancer patients in comparison to conventional systemic chemotherapy. These include the potential to deliver local therapeutically effective concentrations of drug directly to the lung, minimized side effects due to limited systemic delivery, and ease of use for the patient. Inhalable dry powder formulations of nanoparticles and microparticles containing a chemotherapeutic are advantageous in their ability to deliver drug deep in the lung via optimally sized particles, higher local drug dose delivery, and long-term storage capability. In this work, novel advanced spray-dried inhalable PEGylated phospholipid microparticulate/nanoparticulate powders containing the chemotherapeutic paclitaxel were successfully designed and produced via dilute organic solution advanced spray drying under various conditions. Fixed ratios of dipalmatoylphosphatidylcholine (DPPC) and dipalmatoylphosphatidylethanolamine poly(ethylene glycol) (DPPE-PEG) at three different polymeric chain lengths were mixed with various ratios of paclitaxel in a dilute methanol solution. Upon optimization of the spray drying conditions (e.g. pump rate), the physicochemical characterization of the particles was completed. Scanning electron microscopy (SEM) images showed the spherical particle morphology of the inhalable particles. The size of the particles was statistically analyzed using these images SigmaScan software, and these particles were determined to be 600 nm – 1.2 μm in diameter, which is optimal for efficient targeting of the deep lung alveolar and small airway regions for enhanced local deposition. Differential scanning calorimetry (DSC) and X-ray powder diffraction (XRPD) were performed to analyze solid-state transitions and long-range molecular order, respectively, and allowed for the confirmation of the presence of phospholipid bilayers and/or paclitaxel and their phase transition behavior. The water content of the particles was very low as quantified analytically via Karl Fisher titration. The composition of the particles was confirmed using attenuated total reflectance Fourier transform infrared (ATR-FTIR) spectroscopy. Confocal Raman microspectroscopy was employed in chemical imaging to assess particle composition and miscibility. The amount of paclitaxel loaded into the particles was analyzed via high performance liquid chromatography (HPLC), and their aerosol performance was evaluated using the Next Generation Impactor (NGI) and an approved dry powder inhaler (DPI) device for human use to determine the emitted dose, respirable dose, fine particle fraction and mass median aerodynamic diameter. Overall, these results demonstrate this novel therapeutic nanomedicine platform as one capable of effectively delivering paclitaxel directly to the lung in high local concentration for the treatment of lung cancer.

(440c) Stability and Immunogenicity of H5N1 Hemagglutinin Antigens Upon Sustained Release From Polyanhydride Nanoparticles

Kathleen A. Ross[1], Lucas Huntimer[2], Wuwei Wu[3], Susan Carpenter[3], Amanda Ramer-Tait[2], Michael J. Wannemuehler[2] and Balaji Narasimhan[1], (1)Chemical and Biological Engineering, Iowa State University, Ames, IA, (2)Veterinary Microbiology and Preventive Medicine, Iowa State University, Ames, IA, (3)Animal Science, Iowa State University, Ames, IA

Respiratory pathogens, including influenza, have largely increased in occurrence over the last decade. Specifically, highly pathogenic avian influenza, or H5N1 avian influenza, is becoming a contributor to these incidents with a reported 60% mortality rate among confirmed cases and growing resistance to antiviral treatments. With the potential of the virus to mutate to allow human to human transmission and develop into a global pandemic and/or a biodefense agent, it is necessary to design novel vaccines for H5N1 prevention.

Polyanhydride nanoparticle vaccines have shown efficacy in eliciting protection against intranasal pathogens. These copolymers, consisting of sebacic acid (SA), 1,6-bis-(*p*-carboxyphenoxy) hexane (CPH), and 1,8-bis-(*p*-carboxyphenoxy)-3,6-dioxaoctane (CPTEG), promote protein stability by reducing aggregation and exposure to acidic polymer degradation products. Amphiphilic polyanhydride nanoparticles also provide the ability to tailor the release kinetics of stable antigens through their amphiphilic monomers and a combination of surface and bulk erosion. Finally, polyanhydrides have been observed to have immunomodulatory capabilities, which can be harnessed to design vaccines against viral pathogens.

Previously, we have studied the stability of the H5N1 HA antigen under slightly acidic environments, which correlates strongly with the pH of polyanhydride copolymers. In this current work, we have determined the stability and release of both monomeric and trimeric forms of encapsulated HA antigens from polyanhydride nanoparticles of different chemistries. During the course of one month, released protein was quantified and examined for structural stability and functional capacity, such as the ability of HA to bind sialic acid residues and inhibit virus neutralizing antibodies. These studies identified several polyanhydride nanoparticle formulations that stabilize HA antigens. Currently, *in vivo* studies to examine the efficacy of these nanoparticle formulations as compared to traditional adjuvants or immunostimulatory excipients are ongoing, as well as observing the differences between monomeric HA or HA stabilized in the native trimeric state.

(440d) Fucoidan Loaded PCL-PEO Nanofibrous Scaffold for Small Diameter Vascular Graft - a Solution to Intimal Hyperplasia

S. Uday Kumar, Centre for nanotechnology, Indian Institute of Technology Roorkee, Roorkee,India, Roorkee, India, R. Jayaganthan, Centre of Nanotechnology, Indian Institute of Technology Roorkee, Roorkee,India, Roorkee, India and **Narayan C. Mishra**, Paper Technology, Indian Institute of Technology Roorkee, Roorkee,India, Saharanpur, India

Intimal hyperplasia is the thickening of the Tunica intima of a blood vessel as a complication of a reconstruction procedure, and it is the universal response of a vessel to injury and is an important reason of late bypass graft failure, particularly in vein and synthetic vascular grafts. Fucoidan, the drug under this investigation, has been proved to be a prospective drug to overcome intimal hyperplasia in vitro, and it remains yet to be monitored in vivo. With this as the basis, in this study, fucoidan loaded nanofibrous SDVG (small diameter vascular graft, < 5mm dimater) is fabricated from a blend of PEO and PCL polymer by using electrospinning method. Fucoidan is extracted from brown seaweed *Sargassum weitti* and then is blended with PEO/PCL polymer to form a homogeneous electrospinnable solution. It was optimized the electrospinning parameter for producing uniform fibers from PEO/PCL blend. The vascular conduit fabricated in this work is a bi-layered conduit, in which fucoidan loaded PEO/PCL intermediate layer is sandwiched between two layers of PEO/PCL nanofibers. The study establishes the PEO/PCL bilayered vascular graft as a reservoir of fucoidan and thereby gave an improved drug delivery in the later phase of graft implantation. The drug release profile showed a phase of drug release with an initial burst phase accounting for release of 30% of the loaded fucoidan. The dimensional uniformity of the vascular graft is confirmed by examining the cross section at different lengths along the graft. The study of interface under SEM showed a good overlap of the intermediate layer with the first and the second layer of PEO/PCL nanofiber. The Fucoidan loaded PEO/PCL graft fabricated can be a potential solution for long term patency of SDVG with a controlled and extended drug delivery.

(440e) Engineered Nanoparticles That Mimic Bacterial Pathogens for the Treatment of Breast Cancer

Lina Herrera Estrada and Julie Champion, Chemical and Biomolecular Engineering, Georgia Institute of Technology, Atlanta, GA

Bacterial pathogens have evolved mechanisms by which they can control the fate of host cells during infection. Some Gram-negative bacteria use needle-like structures, known as the type III secretion system (T3SS), through which they can inject effector proteins directly into the cytoplasm of host cells. These effector proteins have high potential as therapeutic agents because they are capable of efficiently subverting a variety of eukaryotic signaling pathways by regulating protein factors involved in signal transduction. The NF-κB and MAPK pathways are common targets of these proteins and are fundamental in the development and progression of inflammatory diseases and cancer. However, bacterial delivery of effector proteins for therapeutic purposes is not a viable option.

In order to mimic the mechanism used by bacterial pathogens we have selected one of these effector proteins, acetyltransferase-H (ATH), which simultaneously down-regulates the NF-κB and MAPK pathways. We have engineered ATH as a fusion protein with glutathione S-transferase (GST) that assembles into nanoparticles with 100nm average hydrodynamic diameters. *In vitro* studies with two breast cancer cell-lines show that within 72 hours the ATH-GST fusion protein induces 91% cell death of MCF-7 cells, which are hormone receptor positive and don't overexpress MAPK or NF-κB. Strikingly, we have observed statistically higher cytotoxicity levels (98% cell death) in SK-BR-3 cells over the same period of time. Because SK-BR-3 cells have overactive MAPK and NF-κB pathways they are more prone to develop chemotherapy-resistance and become metastatic.

We have demonstrated that ATH retains its normal activity as a fusion protein by confirming its ability to down-regulate the MAPK pathway. Consequently, this fusion protein nanosystem is a promising novel delivery alternative for the replacement of the T3SS of bacterial pathogens for efficient delivery of effector proteins for therapeutic purposes.

(440f) Pharmaceutical Bio-Applications of Nanocomposite Layers On Drug Carriers

Brandon Tracy, Brandon Byrnes and **Kalyana Pingali**, Paper Engineering, Chemical Engineering and Imaging, Western Michigan University, Kalamazoo, MI

Interactions of nanocomposite layers on active pharmaceutical ingredients and their carrier materials during pharmaceutical formulation and mixing process have been investigated with a focus on bio-applications related to drug release and bioavailability. The influence of loading of nanocoating related to contact hardness, adhesion force and nanoporosity was studied in order to understand the bio-availability of drug component. Field Emission Scanning Electron Microscopy (FESEM), Transmission Electron Microscopy (TEM), Inductively Coupled Plasma Mass Spectroscopy (ICPMS) and Atomic Force Microscopy (AFM) were used to characterize the nanocomposite layers deposited on particle surfaces. Nanosurface studies performed using atomic force microscopy (AFM) showed that the topography of nanostructures changed considerably based not only on the type of contact (low contact, medium contact or high contact), but also on the % contact area. In addition, it was found that the thermal interaction at nanoscale played a crucial role in controlling the nanoporosity of nanolayers, thereby increasing or decreasing the adhesion force between two drug or excipient particles. ICPMS studies confirm that the nanocomposite structures formed on particle surfaces control the mixing at nanoscale thereby resulting in the formation of new nanostructures at the inter-particle interface. Force distributions on homogeneous and non-homogenous layers showed that the adhesion force on the particle surface is non-uniform depending upon the topography of nanostructure. We expect that these variations in uniformity in loading which controls the drug release pattern have wide applications in biomedicine.

(512a) A Specific Non-Covalent Enzyme Immobilization Approach with Nanomaterials to Enable High Activity Retention

Rongrong Jiang, School of Chemical and Biomedical Engineering, Nanyang Technological University, Singapore, Singapore

Nanotechnology-inspired biocatalyst systems have attracted a lot of attention in enzyme immobilization recently. Theoretically, nanomaterials are ideal supporting materials because they can provide the upper limits on enzyme-efficiency-determining factors such as surface area/volume ratio, enzyme loading capacity and mass transfer resistance. However, common immobilization methods have limited the applicability of these biocatalysts owing to enzyme leaching, 3D structure loss, and strong diffusion resistance. Expensive enzyme purification step is also required for these methods before immobilization. Here, we want to show an efficient immobilization method based on specific interaction between His-tagged NADH oxidase and functionalized nanomaterials without requiring enzyme purification for immobilization. The first target enzyme was NADH oxidase. We cloned the annotated NADH oxidase gene from *Bacillus cereus* genome and overexpressed with pET30 vector encoding N-terminal 6 × His tag. The His-tagged NADH oxidase was then immobilized onto single-walled carbon nanotubes (SWCNTs) functionalized with N_{α},N_{α}-bis(carboxymethyl)-L-lysine hydrate. The resulting nanoscale biocatalyst had overcome the foresaid limitations, and demonstrated good loading capacity and stability while maintaining 92% maximum activity of the native enzyme. We further demonstrate that the immobilization is reversible and can retain ca. 92% activity for a couple of loading cycles.

Glycerol dehydrogenase (GlyDH) was also immobilized via the same method and the resulting SWCNT-GlyDH could retain 100% free enzyme activity. We further tested this method with other carbon nanomaterials such as multi-walled carbon nanotubes and nanospheres, and the resulting nanomaterial-enzyme conjugates all showed over 80% activity retention.

(512b) Multi-Enzyme Co-Localization with Nanoscale Pluronic–QD Micelles

Feng Jia, Surya Mallapragada and Balaji Narasimhan, Department of Chemical and Biological Engineering, Iowa State University, Ames, IA

To mimic the structure and functionality of multi-enzyme complexes, which are widely present in Nature, Pluronic®-based nanomicelles were designed to co-localize multiple enzymes. To stabilize the micelles as well as to enable characterization of single enzyme immobilization and multi-enzyme co-localization using Förster Resonance Energy Transfer (FRET), quantum dots (QDs) were encapsulated into the polymeric micelles to form Pluronic®-QD nanomicelles using a novel micro-reactor. Model enzymes, glucose oxidase (GOX) and horseradish peroxidase (HRP), were respectively labeled with fluorescent dyes. The results showed that FRET occurred between the QDs and dyes that labeled each type of enzyme in single enzyme immobilization as well as between the different dyes that served to label the enzymes. These observations were consistent with dynamic light scattering data. The enzymatic activity of single GOX or HRP was retained after immobilization. The optimized sequential co-localization of GOX and HRP was found to improve the overall conversion rate by approximately 100% compared to the equivalent amount of free enzymes in solution or a physical mixture of individual immobilized enzymes on nanoparticles. This study demonstrates a simple platform for multi-enzyme co-localization and a facile and effective strategy to characterize multi-enzyme immobilization and co-localization.

(512c) Surface Functionalization of Magnetic Nanoparticles for Covalent Immobilization of Candida Antarctica Lipase B

Maria Cristiane Martins de Souza[1], Rafael Melo Freire[2], Antonio César Honorato Barreto[2], Pierre Basílio Almeida Fechine[2] and Luciana Rocha Barros Goncalves[1], (1)Departament of Chemical Engineering, Universidade Federal do Ceará, Fortaleza, Brazil, (2)Department of Analytical and Physical Chemistry, Universidade Federal do Ceará, Fortaleza, Brazil

The Lipase B from *Candida Antarctica* (Calb, EC 3.1.1.3), an enzyme capable of performing a wide variety of reactions, such as hydrolysis, transesterification, and aldol reactions, can be immobilized onto functionalized magnetic nanoparticles, preserving its enzymatic activity. The immobilized enzyme can be easily recovered from the reactional media with a magnet, allowing its reuse. Beyond these properties, it has environmental compatibilities, being considered a green chemistry, allowing high recovering and long catalytic useful life. Iron magnetic nanoparticles (Fe_3O_4) were produced by co-precipitation method. The size (31.1nm) and density (5.24 g/cm^3) of the nanoparticles were measured using the technique of X-ray diffraction (XRD) with subsequent refinement of the phases obtained by the Rietveld method. Modifications were carried out on the nanoparticles surface with g-aminopropyltriethoxysilane (APTS). Modification with APTS in the presence of nitrogen was also analyzed. After this modification, the structures were activated with 25% glutaraldehyde solution. The experimental conditions of Calb immobilization in functionalized magnetic nanoparticles, stirring speed (20-45 rpm), enzyme load (50-650 UpNPB/$g_{support}$), incubation time (0,5-5 hours) and reuse of the biocatalyst (five cycles) were optimized. The immobilization was performed in presence of 100mM bicarbonate buffer, pH 10, at 25 ºC. After immobilization, Calb exhibited improved thermal and operational stabilities. The best result (Immobilization yield: 53% and immobilized enzyme activity: 29.1 UpNPB/$g_{support}$) was obtained at 45 rpm, using 200 UpNPB/$g_{support}$ and 1h of incubation. Furthermore, immobilized Calb maintained approximately 53% of initial activity after five cycles of hydrolysis.

(512d) Assembly of Multi-Enzyme Cascades Using Synthetic Scaffoldins Displayed On the Yeast Surface

Fang Liu, Department of Chemical and Biomolecular Engineering, University of Delaware, Newark, DE, Scott A. Banta, Columbia University, New York, NY and Wilfred Chen, Chemical and Biomolecular Engineering, University of Delaware, Newark, DE

Substrate channeling is an important process involved in the transfer of intermediates between multiple enzymes. For most multi-enzyme cascades, enzymes are randomly distributed within the complexes. An optimal system with the appropriate stoichiometric ratio and the right ordering of enzymes will likely result in the highest reaction rate because of the more efficient utilization of the substrate. Cellulosome is a naturally occurring enzyme complex that has been shown to enhance the hydrolysis rate of cellulose by more than 10-fold. The major component of the cellulosome is a structural scaffoldin that consists of repeating cohesin domains, which are docked individually with an enzyme tagged with the corresponding dockerin domain. Because the interaction of the cohesin and dockerin modules is species-specific, artificial "designer cellulosomes" have been created containing the cohesin/dockerin pairs from different species. In this talk, a multi-enzyme cascade was established by assembly three different enzymes onto a synthetic scaffoldin displayed on the yeast surface. The effect of enzyme ordering and spacing on the conversion of methanol to carbon dioxide will be discussed.

(512e) Nanobiocatalytic Enzyme Stabilization

Jungbae Kim, Chemical and Biological Engineering, Korea University, Seoul, South Korea

The specificities of enzyme catalysts can be used in various applications such as fine-chemical synthesis, pharmaceuticals, bioremediation, biobleaching, polymerase chain reaction, protein digestion in proteomic analysis, biosensors and biofuel cells. However, the short lifetimes of enzymes often hampers their usefulness. Nanobiocatalysis, using nanostructured materials for enzyme immobilization, has gathered growing attention due to its recent successes in stabilizing the enzyme activity. This presentation will discuss recent developments of nanobiocatalysis to improve the enzyme stability by using crosslinked enzymes in mesoporous media and on nanofibers/nanotubes/nanoparticles. The approach of nanometer-scale enzyme reactors (NERs) stabilized enzymes in mesoporous media via a ship-in-a-bottle approach, which employs adsorption of enzymes followed by enzyme crosslinking via glutaraldehyde treatment. A similar approach resulted in the enzyme coating on the surface of electrospun nanofibers, which also stabilized the enzyme activity in a vivid way. Nanobiocatalytic systems with stabilized enzyme activity have been employed in various applications, including biofuel cells, and enzyme-linked immunosorbent assay, and trypsin digestion in proteomic analysis. Several successful examples of nanobiocatalytic enzyme stabilization in biosensor and biofuel cell applications will be introduced in details.

NANOSTRUCTURING BY ELECTRONANOPATTERNING AND COLLOIDAL TEMPLATING

Rigoberto C. Advincula, Department of Macromolecular Science and Engineering, Case Western Reserve University, Cleveland, OH

Abstract

Nanostructuring involves the application of materials and processing methods to achieve unique dimensional structures at the nanoscale. To this respect, a number of strategies have been developed by our group to achieve patterned materials based on electropolymerization and electrodeposition. In particular, we have used the precursor polymer approach and the electronanopatterning method using conducting AFM. Recently, a number of strategies have been based on colloidally templated arrays. The fabrication scheme combines the LB-like technique and electropolymerization process called colloidal template-assisted electropolymerization or template-directed electrosynthesis followed by particle removal. The process is done under ambient conditions pivotal to further surface alteration of the unmodified areas including: layer-by-layer assembly, self-assembled monolayer, growth of polymer brush, and another electrodeposition. In this study, the modification of the unpatterned surface is elaborated through the use of self-assembled monolayers (SAM)s which can then undergo a variety of chemistries or polymerization protocols. Until now, colloidal template 2D electropolymerization remains largely an unexplored method, and there are only a few accounts on colloidal template electropolymerization technique for micropatterning polymer films. The same technique is amenable for making highly ordered macroporous structures by electrodeposition of metals (e.g. gold, platinum, nickel, palladium, and cobalt) and metal oxides (e.g. iridium oxide, zinc oxide) in between the void spaces of the sacrificial template colloidal particles.

Introduction

Electrically conducting or π-conjugated organic polymers are of high interest for metallic-like behavior, semiconductor and photochemical properties. They are useful in fabricating thin film energy conversion devices due to their ease of processability, high energy quantum efficiency, and absorption in the visible spectrum. Unlike metals, they can be accessed from a variety of organic synthetic routes or molecular/macromolecular architectures and compositions. They can be prepared by chemical oxidative, electrochemical reactions, metathesis, condensation, and metal mediated coupling or cross-coupling reactions. Optical and electronic applications of organic and polymer materials, such as organic light-emitting devices (OLEDs), organic field effect transistors (OFETs) and solar cell devices, are well-reported and can revile solid-state device technologies in performance. They can also have advantages in terms of thermo-mechanical stability as well as intrinsically low band-gap (metallic conduction) properties.

Lithographic and non-lithographic patterning is of high interest towards device technologies and applications including sensors, biochips, displays, and microfluidic systems. Pattering involves various resolutions and size features, sometimes in hierarchy. Recently, there has been a lot of interest in combining chemistry and polymerization methodologies towards patterning in polymer films. There are a number of strategies for end-grafting of polymers to form "polymer brush" systems. Polymer brushes have drawn significant attention because of the wide range of mechanisms suitable for surface initiated polymerization (SIP) and the physical/mechanical properties of the grafted films in confined regions. Patterned brush surfaces are normally produced using complicated or time consuming lithographic techniques such as soft-lithography, electron-beam lithography, scanning probe lithography, and imprint lithography, which require a master pattern or mask.

In template patterning, a technique that has been under-utilized for two dimensional (2-D) surface patterning is the use of ordered colloidal crystals and templates. Object formation or porogenic templating has been done on ordered colloidal crystals that serve as sacrificial templates (colloidal templating) for structuring inverse three-dimensional (3-D) colloidal arrays. Other 3-D sacrificial templates can come from a variety of materials including anodized alumina, block copolymers, and inorganic or organic colloidal spheres. Among them, the organic colloidal spheres i.e. polystyrene (PS) microspheres, are of interest, because they are both easy to handle and are commercially available. Following the formation of the ordered colloidal crystals, a variety of materials have been used to fill-in the interstitial void spaces of the arrays. For example, metals, inorganic oxides, diamond and glassy carbon, or conducting polymers have all been templated.

Results and Discussions

In a series of recent works and publications, the aspects of the combined use of conducting polymers, in electropatterning and colloidal templating has been highlighted. It is essential to see the surface properties as a continuum of these elements where the electropolymerizability is related to patternability.

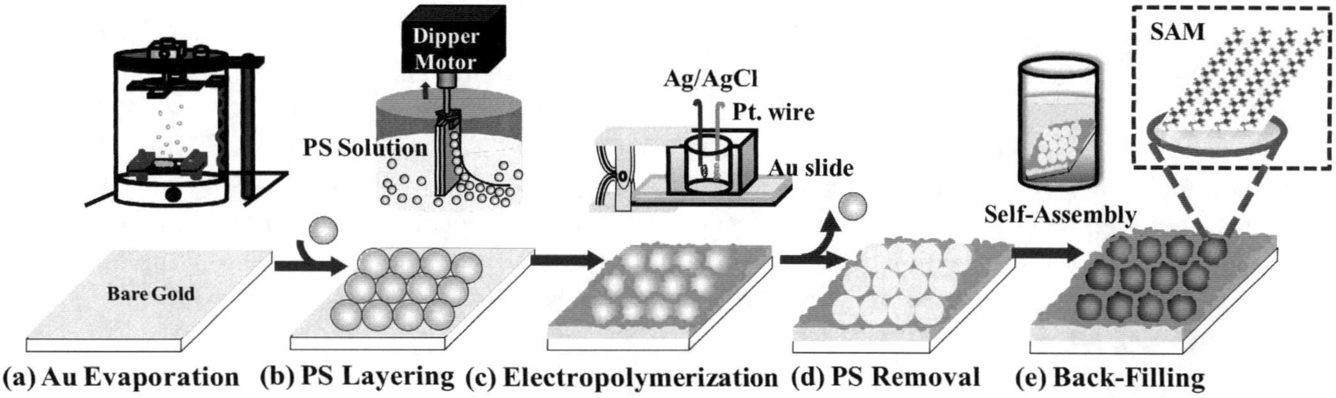

Figure 1. Fabrication of highly ordered monolayer colloidal crystals and inverse patterned colloidal crystals of conducting polymer film.

We have reported on a facile approach to a binary compostion of highly ordered 2-D conducting polymer pores and/or array objects by template-directed electropolymerization with self-assembled monolayer (SAM) patterning in thin films (Figure 1). The work illustrated the generalization of a colloidal template methodology for the production of high surface area monolayer nano/macroporous structures of polymer objects. The polymer network electrosynthesis displayed a smooth CV deposition as observed by EC-QCM and verified by using AFM, XPS, and contact angle measurements. The back-filling of the inside cavities by the SAM approach resulted in a "2-D binary patterned chemistry". Thus this procedure provides an alternative route towards the fabrication of dual patterned surfaces, which are usually accomplished by more complicated and expensive lithographic and non-lithographic methods. Because of the versatility of the fabrication scheme, the holes can also be used for other electropolymerization and electrografting procedures. A number of patterning protocols combining the deposition of polymer brushes by surface initiated polymerization (SIP) and the use of molecularly imprinted polymers (MIP) are being pursued by our group.

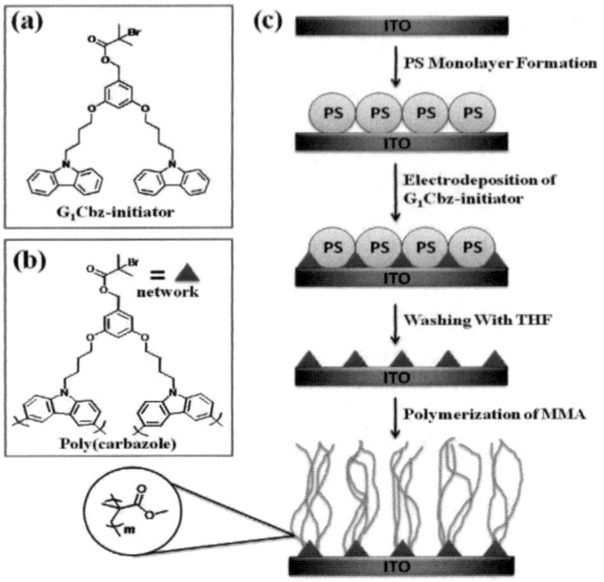

Figure 2. Structure of the G1Cbz-initiator (a), poly(carbazole) net-work after electrodeposition (b). Fabrication of a highly ordered monolayer of colloidal crystals (500 nm diameter PS microspheres), inverse colloidal G1Cbz-initiator arrays, and patterned polymer brushes via SIP of methyl methacrylate (c).

We have also reported on the fabrication of an electro-patterned polymer brush using polystyrene (PS) microsphere templates combined with electrodeposition of an electroactive carbazole dendron - atom transfer radical polymerization (ATRP) initiator. From the hexagonally packed template, surface-initiated polymerization (SIP) via ATRP of methyl methacrylate (MMA) was done to produce novel non-lithographically patterned polymer brush surfaces (Figure 2).

References

1. Pernites, R.; Foster, E.; Felipe, J.; Robinson, M.; Advincula, R. (2011), "Patterned Surfaces Combining Polymer Brushes and Conducting Polymer via Colloidal Template Electropolymerization" *Adv. Mater.*, 23 (10) 1287-1292.
2. Pernites, R.; Ponnapati, R.; Felipe, J.; Foster, E.; Advincula, R. (2011) "Colloidally Templated Two-Dimensional Conducting Polymer Arrays and SAMs: Binary Composition Patterning and Chemistry" *ACS Appl. Mater. Interfaces,* 3 (3), 817–827.

(513b) Collective Osmotic Shock (COS) a New Way to Create Polymeric Templates for Inorganic Nanomaterials

Paul Zavala-Rivera[1], Qilei Song[1], Sanna K. Nataraj[1,2], Shaheen A. Al-Muhtaseb[2] and Easan Sivaniah[1], (1)Cavendish Laboratory, Physics, University of Cambridge, Cambridge, United Kingdom, (2)Chemical Engineering, Qatar University, Doha, Qatar

Collective osmotic shock (COS) a new way to create polymeric templates for inorganic nanomaterials.

Paul Zavala-Rivera[a],

Sanna K. Nataraj[a,c], Shaheen A. Al-Muhtaseb[c], *Easan Sivaniah*[a,*]

[a]Cavendish Laboratory, Cambridge University, Cambridge, CB3 0HE, U.K.

[c] Department of Chemical Engineering, Qatar University, P.O. Box 2713 , Doha, Qatar

Collective Osmotic Shock (COS)[1] is a phenomenon responsible for generating template structures that could lead to the development of new materials and their applications in polymer technology. This process consists of creating bicontinuous polymeric membranes from block copolymers (BCP) such as PS-b-PMMA (polystyrene-*b*-polymethyl methacrylate). To produce the COS effect, self-assemble spherical micellar BCPs nanostructures are treated with UV under air, inducing the cross-linking of the PS block and the degradation of the PMMA part. The final stage consists of treatment of the membrane with a specific solvent that dissolves the minor block and diffuses through the mayor matrix. The dissolution of the degrade PMMA micelles induces an osmotically derived mechanical stress strong enough to deform and evacuate the nanostructures leading to the formation of mesoporous and macro-mesoporous layered architectures. This process is robust in creating bicontinuous polymeric nanostructures. These self-assembled nanostructures are used as templates for creating inorganic matrices with promising applications in optoelectronics, nanophotonics, energy production and storage devices. This is demonstrated with transformation of the COS scaffolds using sol-gel, electrochemistry and biomimetic deposition of various oxides and carbonates.

[1] Zavala-Rivera, P., Channon, K., Nguyen, V., Sivaniah, E., Kabra, D., Friend, R.H., Nataraj, S.K., Al-muhtaseb S.A., Hexemer, A., Calvo, M.E., Miguez, H., 2012. Collective osmotic shock in ordered materials. *Nature Materials*, 2012, **11**(1), pp 53.

(513c) Effect of Chain Architecture On the Viscoelasticity and Shock Response of Block Copolymers

Bedri Arman[1], Srinivas Reddy[2] and Gaurav Arya[1], (1)Department of Nanoengineering, University of California, San Diego, La Jolla, CA, (2)La Jolla Bioengineering Institute, La Jolla, CA

A clear understanding of the molecular mechanisms responsible for energy and shock dissipation in elastomeric materials is essential for the design of next-generation materials for blast mitigation. Even though current high-end, protective gear typically rely on polyurea and polyurea-based materials for protection against blasts, the molecular basis for the dissipative properties of polyurea remains unknown. It may be hypothesized that the multiblock chain architecture of polyurea--repeating units of hard and soft segments--might be partly responsible for its superior dissipative properties. As a first step towards addressing this question, we carry out a detailed comparison of the microstructure, viscoelastic properties, and shock response of coarse-grained models of multiblock and diblock copolymers using molecular dynamics simulations. We find that the multiblock copolymer forms small, interconnected, rod-shaped, hard domains surrounded by a soft matrix, whereas the diblock copolymer forms larger, unconnected, hard domains. Viscoelastic analyses indicate that multiblock copolymer is not only more elastic but also more dissipative at low to intermediate frequencies. Shock simulations reveal that shock waves propagate slower in the multiblock copolymer compared to the diblock copolymer, which may be due to the more deformable hard domains in the former system. Our results thus suggest that the multiblock architecture of polyurea might endow polyurea with smaller, more deformable, and interconnected hard domains that could lead to improved energy dissipation and lower shock speeds.

(513d) The Effect of Polymer Structure and Solvent Selectivity On Ordering Kinetics in Solution Cast Block Copolymer Films

Stephen M. Martin[1], D. G. Baird[1], Michael J. Heinzer[1], Eugene Joseph[2] and John A. Pople[3], (1)Chemical Engineering, Virginia Tech, Blacksburg, VA, (2)ICTAS, Virginia Tech, Blacksburg, VA, (3)Stanford Synchrotron Radiation Laboratory, Stanford Linear Accelerator Center, Stanford, CA

The kinetics of ordering in solution cast block copolymer films has been studied using a novel *in situ* small-angle X-ray scattering technique. By tracking changes in the scattering intensity as solvent is continually removed from the film, information on the ordering rate was obtained as the film progresses from the dilute to the concentrated regime. The ordering rate has been found to be controlled by thermodynamics below a critical concentration and by limited chain mobility at higher concentrations. Previous work in our group has considered the effects of temperature, sweep gas and solvent saturation on the ordering kinetics of a polystyrene-b-polybutadiene (SB) diblock copolymer that forms hexagonally packed cylinders in a neutral solvent (toluene.) The effect on ordering kinetics of changes in solvent selectivity and drying rate in the same polymer system and a closely related triblock copolymer system (SBS) are now reported. In addition, the effect of adding butadiene-selective and styrene-selective tackifying agents will be discussed. Finally, the ordering kinetics of a polymethylmethacrylate-b-polybutylacrylate (PMMA-PBA) triblock copolymer system with varying PMMA to PBA ratios, resulting in different ordered phase morphologies, is presented.

(513e) Nature Inspired Functional Polymer Composites with Novel Properties

Jian Zhu, Christine M. Andres, Huanan Zhang, Ming Yang, Bongjun Yeom and Nicholas A. Kotov, Chemical Engineering, University of Michigan, Ann Arbor, MI

Advanced technologies require advanced material platforms with unique combinations of material properties. Our motivation is to translate the perfect nanoscale electrical, thermal and mechanical properties of carbon nanotubes (CNT) and graphene to tangible macroscale composite materials with tunable combinations of properties by taking advantage of Nature's intriguing design. Layer-by-layer (LBL) assembly of CNT composites, with its ubiquitous nanometer resolution, can be used to tune the energy gap of CNT in a way that optimizes both the optoelectric and mechanical properties of the resulting composite. The universality of layer-by-layer assembly is further demonstrated to give both CNT and graphene composites a highly desirable balance of strength and toughness. Alternatives to LBL assembly have been developed to speed up the film production process and allow for non-aqueous systems to be used. Here we show an evaporation assisted spray technique that can quickly deposit polymers and nanofillers with controlled volume and thickness while preserving the uniformity of the coatings. We also introduce vacuum assisted assembly and template infiltration techniques which are capable of producing materials with similar mechanical properties to LBL assembled graphene nanocomposites and can be used to prepare hydrophobic polyurethane and epoxy composites. In addition, these techniques are further exploited to be used in other nanofiller systems, such as with aramid (Kevlar) fibers to obtain nanocomposites with excellent mechanical properties. Along with the unique electrical and mechanical properties of hierarchical nanocomposites, our graphene oxide assembly demonstrates negative thermal expansion properties, which have potential applications in areas such as sensors, actuators and other MEMS/NEMS applications. Coefficient of thermal expansion of our composite can also be easily tuned allowing for a composite by design to be created.

(513f) Synthesis and Crystallization of All-Conjugated Block Copolymers

Rafael Verduzco[1], Kendall Smith[1], Yen-Hao Lin[1], Chloe Kempf[1], Dana Dement[1], Jim Howe[2], Seth B. Darling[3], Deanna Pickel[4], Enrique D. Gomez[5] and Changhe Guo[5], (1)Chemical and Biomolecular Engineering, Rice University, Houston, TX, (2)Rice University, Houston, TX, (3)Center for Nanoscale Materials, Argonne National Laboratory, Argonne, IL, (4)Oak Ridge National Laboratory, Oak Ridge, TN, (5)Chemical Engineering, Pennsylvania State University, University Park, PA

The nanoscale structure of the active layer plays a key role in determining the efficiency of photon-to-electricity conversion in bulk heterojunction (BHJ) organic photovoltaics (OPVs), which are typically comprised of a blend of p-type (hole-conductive) and n-type (electron-conductive) organic semiconductors. While some degree of phase separation is desired for creating continuous charge transport pathways, large-scale phase separated domains are unfavorable due to reduced interfacial area for efficient charge separation. This presents significant challenges for all-polymer OPVs, which are made up of a blend of a p-type and n-type conjugated polymers; large-scale phase separation is commonly observed due to the reduced entropy of mixing for polymer blends. While the best performance in all-polymer OPVs (~ 2%) is significantly lower compared with state-of-the art polymer-fullerene BHJs, reducing or eliminating large-scale phase separation in all-polymer OPVs may dramatically improve performance. Advantages of all-polymer OPVs over polymer-fullerene OPVs include a typically higher V_{oc} and broader absorbance due to the presence of two polymeric semiconductors in the active layer.

All-conjugated block copolymers with p- and n-type blocks represent a promising approach to improving the performance of all-polymer OPVs. Phase separation can be avoided and block copolymer self-assembly may lead to ideal structures for charge dissociation and transport. However, the synthesis of well-defined, high molecular weight all-conjugated block copolymers is challenging, and a broad understanding of crystallization and micro-phase segregation in all-conjugated block copolymers is lacking. In this work, we present new synthetic approaches which enable the preparation of all-conjugated block copolymers with little or no homopolymer impurities. Additionally, these approaches enable the preparation of a systematic series of all-conjugated block copolymers with molecular weights greater than 50,000 g/mol in some cases. In one approach, a poly(alkyl thiophene) polymer is coupled to a poly(9,9 dioctyl fluorene) polymer or copolymer through copper-catalyzed azide-alkyne "click" coupling. End-group control in high molecular weight polymers is achieved through the use of an functionalized, external initiator. In a second approach, the Suzuki polymerization of various poly(fluorene) copolymers in the presence of a poly(alkyl thiophene) macroreagent enables the preparation of high molecular weight (> 50,000 g/mol) all-conjugated block copolymers in two steps. This latter approach is particularly useful for synthesizing all-conjugated block copolymers with both p- and n-type polymer blocks.

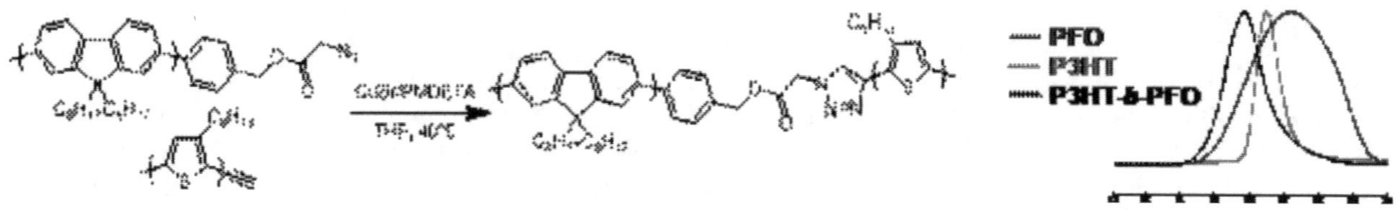

Figue 1. Reaction scheme for the "click" coupling of two conjugated polymer to make a block copolymer and SEC chromatograph of the starting homopolymers and final block copolymer.

Crystallization and micro-phase segregation in all-conjugated block copolymers is analyzed through a combination of atomic force microscopy (AFM) and grazing-incidence x-ray scattering (GIXS). While previous work with poly(3-hexylthiophene) (P3HT) block copolymers has found that crystallization of P3HT typically dominates the final morphology, GIXS measurements show that P3HT crystalllinity can be suppressed for a large second block (see Figure below). Further evidence is provided by AFM and DSC measurements.

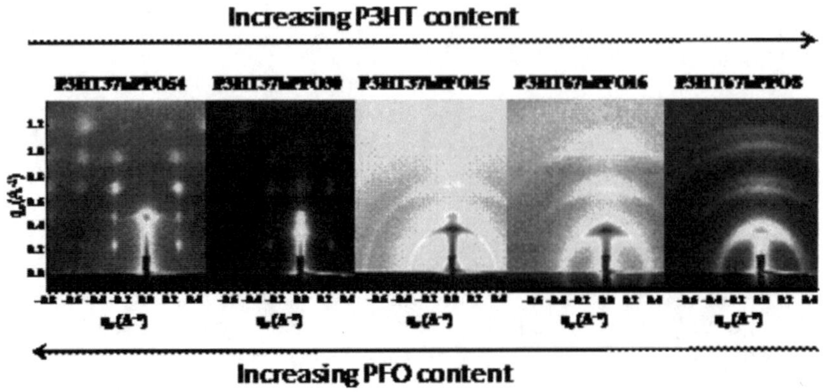

As systems relevant to organic photovoltaics, we investigate block copolymers with P3HT and poly((9,9-dioctylfluorene)-2,7-diyl-alt-[4,7-bis(thiophen-5-yl)-2,1,3-benzothiadiazole]-2',2"-diyl) (PFOTBT). The limited solubility of PFOTBT precludes the preparation of high molecular weight block copolymers, but we find that the addition of P3HT-b-PFOTBT block copolymers can improve the performance of P3HT/PFOTBT all-polymer photovoltaics. A PCE of 1.5 % is achieved in an all polymer system with 20 wt % added block copolymer, compared with a PCE of just 0.5 % for an all-polymer blend.

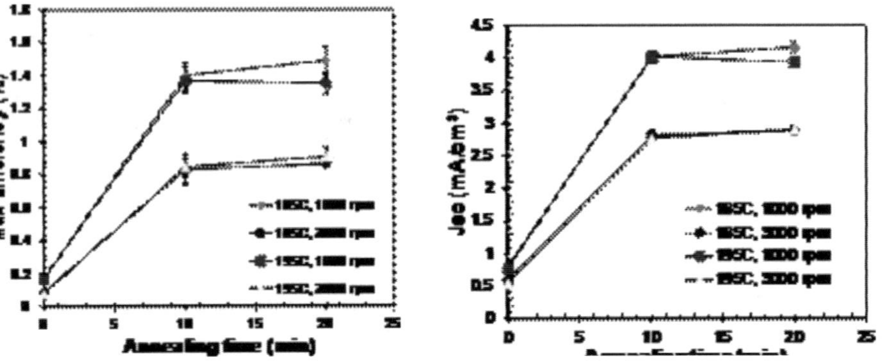

This work demonstrates new methods for synthesizing all-conjugated block copolymers and provides strong evidence for improved performance in block copolymer OPVs.

Acknowledgements

This work was carried out with support from the Welch Foundation for Chemical Research (Grant #C-1750), the Shell Center for Sustainability, and Louis and Peaches Owen. Use of the Center for Nanoscale Materials and Advanced Photon Source at Argonne National Laboratory was supported by the U. S. Department of Energy, Office of Science, Office of Basic Energy Sciences, under Contract No. DE-AC02-06CH11357. A portion of this research was conducted at the Center for Nanophase Materials Sciences, which is sponsored at Oak Ridge National Laboratory by the Scientific User Facilities Division, Office of Basic Energy Sciences, U.S. Department of Energy. J. A. H. acknowledges support from the Century Scholars program.

(524a) Microwave-Assisted Templated Synthesis of II-VI Semiconductor Nanocrystals

Ryan Reeves, Ying Qi, Jun Wang, Geoffrey Tompsett, W. Curtis Conner and T. J. Mountziaris, Department of Chemical Engineering, University of Massachusetts, Amherst, MA

Semiconductor nanocrystals are interesting nanomaterials whose tunable optical and electronic properties make them ideal for applications in biological sensing and imaging, light-emitting devices, displays, and solar cells. [1] Nanocrystals of II-VI semiconductors, such as CdSe and ZnSe, and core/shell structures based on them, such as CdSe/ZnS and ZnSe/ZnS, have attracted significant attention because of their unique optoelectronic properties. [2] The commercial exploitation of these materials requires the development of synthesis techniques that are scalable, economical, and environmentally friendly, while enabling precise control of the size, shape and size distribution of the nanocrystals. The most common synthesis technique for II-VI nanocrystals employs small batch reactors in which nanocrystals grow as a function of time following a rapid injection of organometallic precursors into a hot coordinating solvent. [2] The limitations of this process for large-scale commercial exploitation stem from the incomplete mixing of the precursors in large batches that can lead to non-uniform nucleation and broad particle size distributions, from the high cost, flammability, and toxicity of the organometallic precursors, and from its operator-intensive nature.

Templated synthesis techniques for II-VI nanocrystals have distinct advantages over other methods, including more precise control of particle size, shape, and size distribution and easier scalability for commercial applications. Stable microemulsions and liquid crystals, formed by self-assembly of an amphiphilic block copolymer in the presence of a polar and a less polar solvent, have been successfully employed as templates for the synthesis of II-VI nanocrystals [3-6]. The nanocrystals are formed by reacting a group II precursor dissolved in the dispersed phase of the template with a group VI hydride gas. Water-in-oil microemulsions and liquid crystals have emerged as the most promising templates, because they employ inexpensive water-soluble metal salts and eliminate the need for using more expensive and pyrophoric organometallic compounds. [5] To eliminate the toxic and flammable group VI hydride gases, a microwave-assisted templated synthesis route is being developed. The technique employs bursts of microwaves to selectively heat the aqueous dispersed droplets of water-in-oil microemulsions that contain the water-soluble precursors of the group II and VI elements, thus leading to nucleation and formation of a single nanocrystal inside each droplet. In proof-of-principle experiments, luminescent CdS nanocrystals were synthesized by exposing water-in-oil microemulsions containing cadmium acetate and thiourea in their aqueous dispersed phase to bursts of microwaves. Nanocrystals grown by the microwave-assisted route exhibited higher quantum efficiency when compared to nanocrystals grown by the original templated technique. Results from experiments aiming at optimizing the microwave-assisted templated synthesis technique and expanding it to other II-VI nanocrystals will be presented.

[1] A.P. Alivisatos, "Semiconductor Clusters, Nanocrystals, and Quantum Dots", *Science* 271(5251), 933-937 (1996).

[2] C.B. Murray, C.R. Kagan and M.G. Bawendi, "Synthesis and Characterization of Monodisperse Nanocrystals and Close-Packed Nanocrystal Assemblies", *Annual Review of Materials Science,*30, 545–610 (2000).

[3] G.N. Karanikolos, P. Alexandridis, G. Itskos, A. Petrou and T.J. Mountziaris, "Synthesis and Size Control of Luminescent ZnSe Nanocrystals by a Microemulsion-Gas Contacting Technique", *Langmuir*, 20(3), 550-553 (2004).

[4] G.N. Karanikolos, P. Alexandridis, R. Mallory, A. Petrou and T.J. Mountziaris, "Templated synthesis of ZnSe nanostructures using lyotropic liquid crystals", *Nanotechnology,*16 (10), 2372-2380 (2005).

[5] G.N. Karanikolos, N.L. Law, R. Mallory, A. Petrou, P. Alexandridis and T.J. Mountziaris, "Water-based synthesis of ZnSe nanostructures using amphiphilic block copolymer stabilized lyotropic liquid crystals as templates", *Nanotechnology,*17 (13), 3121-3128 (2006).

[6] G.N. Karanikolos, P. Alexandridis and T.J. Mountziaris, "Growth of ZnSe and CdSe nanostructures in self-assembled block copolymer-stabilized templates," *Materials Science and Engineering B: Advanced Functional Solid-State Materials*, 152(1-3), 66-71 (2008).

(524b) Layer-by-Layer Silica Reactive Assembly On Nanoscale Chemical Templates

Juan Pablo Hinestrosa, Center for Nanophase Material Sciences, Oak Ridge National Laboratory, Oak Ridge, TN and Scott Retterer, Cnms, Oak Ridge National Laboratory, Oak Ridge, TN

Layer-by-Layer Silica Reactive Assembly on Nanoscale Chemical Templates

Juan Pablo Hinestrosa[*] & Scott Retterer[*,µ]

[*]Center for Nanophase Material Sciences, Oak Ridge National Laboratory, Oak Ridge, TN 37834

[µ]Biosciences Division, Oak Ridge National Laboratory, Oak Ridge, TN 37834

Reactive assembly of silica onto nano- and micro-scale organic templates was achieved using a combination of chemical lift-off and solution chemistry performed under mild conditions of pH and temperature. Well-defined gratings, honeycomb-lattices, and dot arrays were created by electron beam lithography and used as templates for the generation of silica nanostructures. Initially, an amine-silane monolayer was vapor deposited onto the patterned substrates. They were subsequently reacted with silicic acid solutions made in buffers of varying pH and salt content at room temperature. The process of vapor treatment followed by the reactive assembly of silica in solution could be repeated multiple times to perform layer-by-layer deposition while keeping the nanoscale organic templates intact, as seen in the Figure. It was observed that the conditions for the silicic acid deposition, namely pH, salts and solution age, had a strong effect on thickness of each layer and the morphology of the amorphous silica formed.

Following silica reactive assembly, the templates can be easily removed by sonication in acetone leaving only the silica nanostructures on the substrates. 'Defects' in the arrays of silica nanostructures were relatively minor (~ 10% after three depositions) and do not affect the overall organization of the layers. These defects were attributed to difficulties associated with completely removing the PMMA mask from underneath the deposited silica. The number of defects increased with the addition of each silica layer. The biomimetic method described here allows for the generation of well-defined, complex nanostructures starting from small molecules under mild conditions; thus, we believe it provides a robust platform for high-throughput reactive assembly process and growth of organic-inorganic hybrid nanostructures.

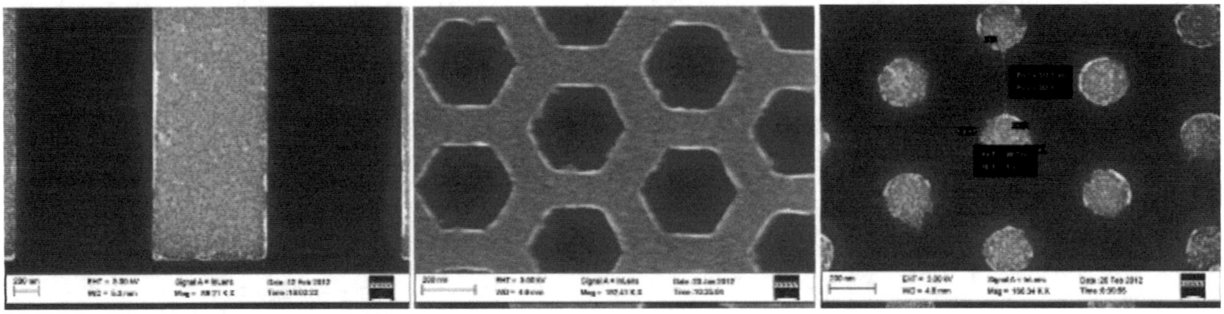

Figure. SEM images of silica gratings, honeycomb lattices and dot arrays after reactive assembly from silicic acid solution.

(524c) Modulating Electron Transport in PbSe Inorganic Nanocomposites: Doping Through Surface Engineering

Richa Sharma[1], April M. Sawvel[1], Raffaella Buonsanti[1], Jeffrey Urban[2] and Delia J. Milliron[1], (1)Lawrence Berkeley National Laboratory, Berkeley, CA, (2)Lawrence Berkeley National Laboratories, Berkeley, CA

In this work we report the use of inorganic matrices to engineer the carrier transport and interparticle spacing for lead selenide (PbSe) nanocrystal array. Electrically insulating PbSe nanocrystal superlattices are converted into conductive nanocomposites by post assembly ligand exchange of the organic ligands with inorganic chalcogenidometallate clusters (CHAMs) of $(N_2H_5)_4Sn_2S_6$ / $(N_2H_4)_x(N_2H_5)_4Ge_2S_6$ or neutral polymorphs of $(N_2H_4)_2ZnTe$. Thermal annealing or chemical treatment further eliminates the hydrazine/hydrazinium cation and excess chalcogen to give the final inorganic nanocomposites. This work will discuss the fabrication method of defect-free and highly ordered nanocomposite films of PbSe nanocrystals. Our synthesis procedure entails independent processing of nanocrystals and inorganic matrix materials. Thus, it enables more control over the composition of resulting nanocomposites. Interestingly, we demonstrate it is possible to create either n- or p- type thin film transistors of PbSe nanocomposites based upon the choice of the inorganic matrix material (SnS_2 / GeS_2 / $(N_2H_4)ZnTe$). Our findings suggest a practical route to control doping in semiconductor nanocrystal array of PbSe.

(524d) Facile Synthesis of Silica Micro- and Nano-Rods

Marco Furlan and Marco Lattuada, Adolphe Merkle Institute, University of Fribourg, Marly, Switzerland

The use of nano-building blocks as fillers to enhance the property of composites is a topic of great interest. The most used nano-blocks are spherical nanoparticles, which can be prepared using a variety of methods. Nevertheless, there is a growing interest in the use of anisotropic building blocks, such as nanotubes, platelets, rods, ellipsoids, etc. In this work we propose a new and reliable method for the production of large quantities of magnetic silica micro and nano-rods. This method is based on the spinodal decomposition of a silica precursor leading to a bicontinuous phase, the structure of which is controlled by the incorporation of polymer-magnetite nano-spheres with the aid of an external magnetic field. Our system is based on the silica precursor TMOS and the polymer PEG, which undergoes a phase separation following a spinodal decomposition due to the chemical cooling induced by the condensation polymerization of the silica precursor. Polymer-magnetite nanoparticles are incorporated in the silica rich phase due to the strong affinity of the magnetite to the silica precursor. The accumulation of the nanocolloids in the silica rich phase during the spinodal decomposition leads to the formation of long rod-like structures in the presence of an external magnetic field, which form a monolith. At the end of the reaction, the application of strong ultrasonication leads to the disintegration of the monolith and to the recovery of the rods. Different parameter such as rods diameter and length can be tuned by varying reactant concentration, the applied magnetic field and the sonication time. The applied magnetic field has an enormous influence on the length of the produced rods and depending on the kind of field applied, homogeneous or alternating, the length can be varied by more than two orders of magnitude.

(524e) Larger-Scale Production of Janus Silica-On-Silica Nanoparticles Via Adsorption On Charged Surface

Dieu Huong Pham, Ayae Sugawara-Narutaki, Atsushi Shimojima and Tatsuya Okubo, Department of Chemical System Engineering, The University of Tokyo, Tokyo, Japan

Recently, Janus particles whose surface structure or function differs between two sides of the particles have attracted great attention. Owing to the anisotropy in their structure and properties, these particles are expected to become an important element of functional nanomaterials if they can be assembled into useful structures. However, the preparation of anisotropic particles less than 100 nm in diameter has been a difficult challenge. To make these particles practical, they should be produced in large scale. In this study, a simple method to synthesize Janus nanoparticles in large quantity by utilizing their adsorption onto a charged surface is demonstrated. First, spherical silica nanospheres (SNSs) less than 100 nm in diameter were mixed together with hydrotalcite (HT), a plate-like layered crystals, in water at pH 11.5. Under this pH condition, SNSs are adsorbed on the surface of HT by electrostatic interaction because the surface of SNSs is negatively-charged while that of HT is positively-charged. Next, the exposed surfaces in the immobilized particles are modified with aminopropylsilanes so that electrostatic adsorption of additional silica nanospheres can be operative. Finally, HT substrates were removed by being dissolved in acidic water to give anisotropically surface-modified SNSs. Gram-scale synthesis of Janus particles can be achieved by this approach.

ONE STEP SYNTHESIS OF RASPBERRY-LIKE GOLD NANOPARTICLES

Chunrong Wang[1], Xianzai Yan[1], Xunpeng Xu[1] and Yun Fang[2]*

[1]*School of Food Science and Engineering, Jiangxi Agricultural University, Nanchang, Jiangxi, P. R. China*

[2] *School of Chemical and Material Engineering, Jiangnan University, Wuxi, Jiangsu, P. R. China*

Abstract. Raspberry-like gold nanoparticles were prepared by the reduction of $HAuCl_4$ in the solution containing soybean peptides and sodium dodecyl benzene sulfonate (SDBS). Soybean peptide and SDBS together acted as soft template in reacted solutions. Soybean peptide also played another role of reducing agent. The resulting gold nanostructures were characterized by transmission electron microscopy (TEM), electron diffraction pattern (ED) and UV-vis spectra. It was found that anisotropic gold nanostructures with various shapes were prepared by changing the concentration of SDBS. With a fixed concentration of 0.2 mM $HAuCl_4$ and 0.4% Soybean peptide, as there were SDBS less than 0.1 mM, spherical gold nanoparticles were prepared; as 0.3 mM, some multipodlike gold nanoparticles formed; as 0.5 mM, raspberry-like gold nanoparticles synthesized. Electron diffraction pattern showed that the crystal structures of raspberry-like gold nanoparticles scattering points corresponding to different gold crystalline facets. It is confirmed that raspberry-like gold polycrystalline nanoparticles grew from many small particles that have random and independent orientation.

Keywords: gold; nanoparticles; raspberry; soybean peptide

Introduction

In the past few years, metal nanostuctures have been the focus of intensive research due to their novel electronic, optical, magnetic, thermal, catalytic, and other properties that are distinctly different from their bulk counterparts, and considerable attention from both fundamental and applied research has been paid to synthesizing and characterizing metal nanostructures are closely related to their size and shape.[1-3] In these metal nanostructures, special anisotropic noble metal nanoparticles has attracted lots of science researchers because of their dramatic effect on the electric, optical, and catalytic application[3-4].

To prepare structurally isotropic nanostructures is a challenging task. There have been several recent reports describing differently shaped gold nanostructures including nanowire, nanocubes, nanoplates, nanobelts and nanorings made through various chemical approaches[5]. For preparing these nanostructures, macromolecules method has been proved to be effective[6-7]. soybean peptide is a nontoxic macromolecule can be used to synthesizing nanostructures. In this paper, we attempted to use the soybean peptide to synthesize Raspberry-like gold nanoparticles.

Experimental section

Chemicals

Hydrogen tetracholoaurate ($HAuCl_4 \cdot 4H_2O$) and soybean peptide were purchased from China Medicine (Group) Shanghai Chemical Reagent Corporation. All reagents were of analytical grade and used as received without further purification. The water used was deionized water.

Preparation of gold nanoparticles

The water solution containing the reagents ($HAuCl_4$ and soybean peptide) was placed in a 100 mL round-bottomed flask in a water bath maintained at a constant temperature at 70±1 °C for 15 h. The products were separated by centrifugation (TGL-16G, Anting Scientific Instrument Factory), washed with deionized water several times. Optical spectra were recorded using a SPECORD200 spectrophotometer (Analytik Jena AG, Germany). TEM images and electron diffraction patterns (ED) were recorded on JEM-2010 microscopy operated at 200 kV.

Results and Discuss

In this work, Au nanostructures were synthesized by reducing $HAuCl_4$ with soybean peptide in water at 70 °C. Compared to nonaqueous systems, water-based synthesis can offer an environmentally benign route to the production of metal nanoparticles because it does not involve toxic organic solvents.

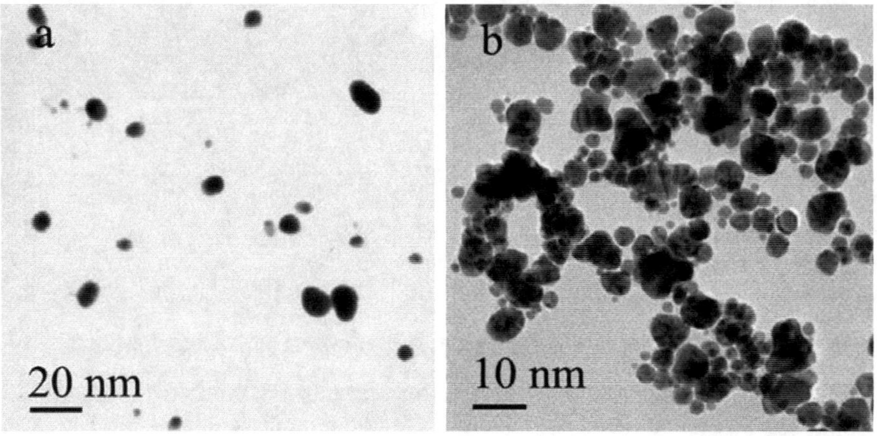

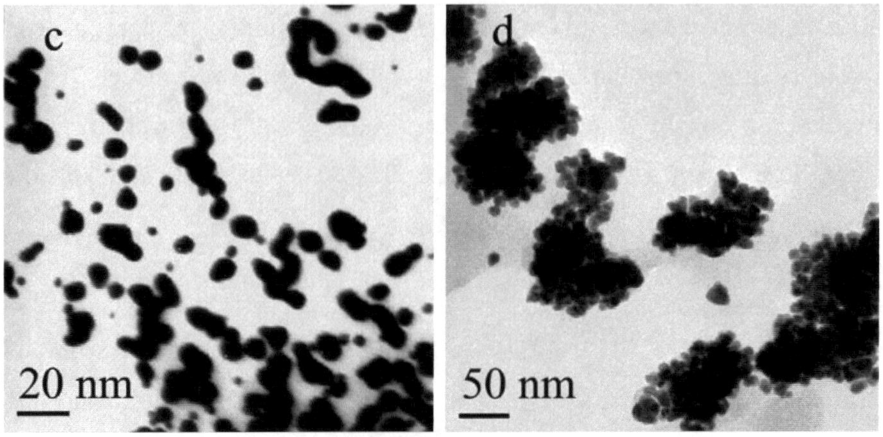

Figure 1 TEM images of gold nanostrurctures

The reacted solutions contain fixed concentration of 0.2 mM $HAuCl_4$ and 0.4% Soybean peptide, with different concentration of SDBS: (a) 0, (b) 0.1 mM, (c) 0.3mM and (d) 0.5mM.

Figure 1 shows TEM images of a sample that was prepared in a solution containing 0.2 mM $HAuCl_4$ and 0.4% Soybean peptide with different SDBS at 70 °C. Under various reaction condition, the resulting nanostructures changed from spherical nanoparticles (Figure 1a), branched particles (Figure1b and c), to raspberry-like nanoparticles (Figure1d) as the reaction proceeding.

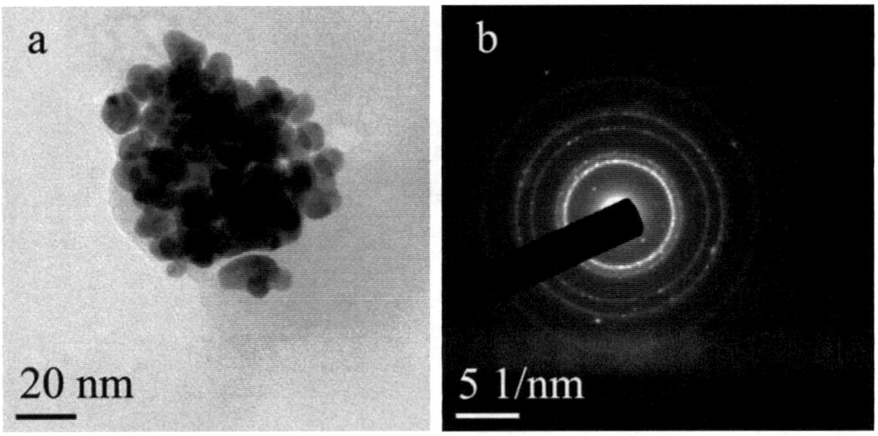

Figure 2 Typical gold nanoparticles and ED pattern

Figure 2a shows the typical raspberry-like gold nanoparticles and Figure2b shows the electron diffraction (ED) corresponded. ED pattern shows the diffraction rings.

Gold nanoparticles dispersed in a liquid medium usually display a very UV-vis absorption spectrum. Figure3 shows the UV-vis absorption spectrum of gold nanopartices. It is clearly that the spectrum above 600 nm has stronger absorption indicating there are large size gold structures formed.

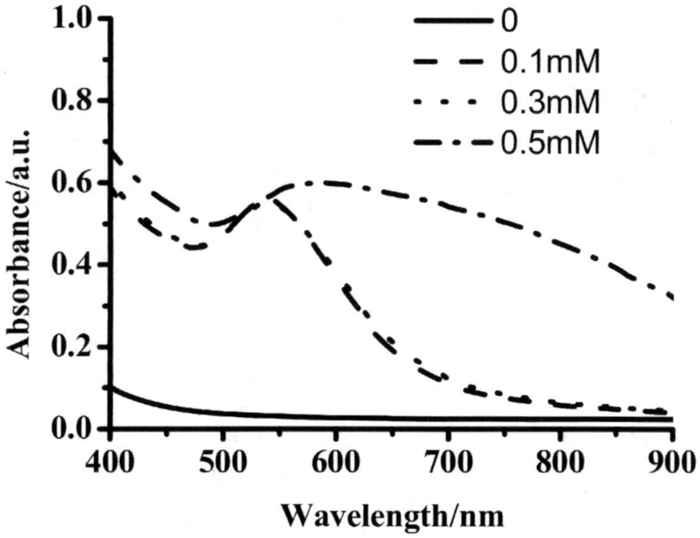

Conclusions

Raspberry-like gold nanoparticles were prepared by the reduction of $HAuCl_4$ in the solution containing soybean peptides and SDBS. Raspberry-like gold polycrystalline nanoparticles grew from many small particles that have random and independent orientation.

Acknowledgement

We thank the National Science Foundation (31000820) and Jiangxi Agricultural University Doctor Start Foundation (2313) for funding this research.

References

1 Tan, Y. N.; Su, X.; Liu, E. T.; Thomsen, J. S.(2010), "Gold-Nanoparticle-Based Assay for Instantaneous Detection of Nuclear Hormone Receptor-Response Elements Interactions", *Analytical Chemistry*, 82, (7), pp2759-2765.

2. Qu, S.; Li, H.; Peng, T.; Gao, Y.; Qiu, J.; Zhu, C.(2009), "Optical nonlinearities from transverse plasmon resonance in gold nano-rods", *Materials Letters*, 58, (9), pp1427-1430.

3. Ko, K. D.; Kumar, A.; Fung, K. H.; Ambekar, R.; Liu, G. L.; Fang, N. X.; Toussaint, K. C.(2011), "Nonlinear Optical Response from Arrays of Au Bowtie Nanoantennas", *Nano Letters*, 11, (1), pp61-65.

4. Wu, B.; Boland, J. J.(2006), "Synthesis and dispersion of isolated high aspect ratio gold nanowires", *Journal of Colloid and Interface Science*, 303, (2), pp611-616.

5 R. Rodríguez-Oliveros J A S. Gold nanostars as thermoplasmonic nanoparticles for optical heating,[J]. Opt. Express. 2012, 20(1): 621-626.

6. Pardinas-Blanco, I.; Hoppe, C. E.; Pineiro-Redondo, Y.; Lopez-Quintela, M. A.; Rivas, J.(2008), "Formation of Gold Branched Plates in Diluted Solutions of Poly(vinylpyrrolidone) and Their Use for the Fabrication of Near-Infrared-Absorbing Films and Coatings", *Langmuir*, 24, (3), pp983-990.

7. Park, J.; Atobe, M.; Fuchigami, T.(2006), "Synthesis of multiple shapes of gold nanoparticles with controlled sizes in aqueous solution using ultrasound", *Ultrason. Sonochem.*, 13,, pp237-241.

(577a) High-Rate NADH Electrocatalysis At Nanoscale Carbon Electrodes

Hanzi Li, Rui Li, R. M. Worden and Scott A. Calabrese Barton, Chemical Engineering & Materials Science, Michigan State University, East Lansing, MI

Implementation of NADH-dependent enzymes in electrochemical bioreactors, biosensors, and biofuel cells is currently limited by poor kinetics of electrochemical NADH regeneration to form NAD^+. However, we have shown that high-rate NADH oxidation may be achieved by combining nanomaterials with deposited azines. For example, high-rate electrodes may be fabricated by adsorbing methylene green on electroactivated carbon surfaces. A glassy carbon electrode was activated by cyclic voltammetry at 2.5 V to -1.5 V in phosphate buffer, leading to the formation of redox active quinones on the carbon surface. The electroactivated glassy carbon electrode possesses high capacitance and achieves NADH oxidation current density comparable to a 0.21 mg cm^{-2} CNT electrode, a more than 200-fold increase compared to untreated glassy carbon. Deposition of methylene green (MG) on activated glassy carbon electrode and pretreated CNTs was successfully achieved by facile adsorption. The MG-CNT electrode demonstrates a nearly twofold increase compared to electropolymeried MG on CNT. We have shown that the electrogenerated NAD^+ is nearly 100% bioactive, and have implemented this electrode in a bioreactor designed to oxidize glycerol to dihydroxyacetone.

(577b) The Three-Pronged Approach to Investigate the Performance of Various Enzyme Immobilization Techniques

Youngho Wee, Chemical & Biological Engineering, Korea University, Seoul, South Korea, Jungbae Kim, Chemical and Biological Engineering, Korea University, Seoul, South Korea and **Su Ha**, Washington State University, Pullman, WA

We will explore various enzyme immobilization techniques with different nanomaterial supports to study which mechanism shows the most potential for further use, what the shortcomings of each method are, and to suggest and best ways in which these methods might be improved upon. To achieve this goal, we will apply "the three-pronged approach." This approach has the advantage of providing us with information about the performance of the complexes, morphology of the complexes, and where the bottlenecks lie for each of the complexes tested. For example, the potentiometric experiments performed in an electrochemical cell will provide information about the onset potential of the complexes (the potential at which current first is able to be generated), the activity of the complexes (how much current they can produce as a function of applied potential), and about the stability of the complexes (current output as a function of time at a fixed potential). Confocal or TEM imaging of the tagged enzymes incorporated into the complexes will provide us with information on the quantity of enzymes immobilized, the structure and morphology of the immobilized enzymes, and the porosity of the complexes. Finally, impedance analysis performed in an electrochemical cell will indicate what the major resistances to efficient power output arise from. For example, impedance analysis can be used to compare fuel mass transfer resistance, mediator mass transfer resistances, enzyme kinetics, and electron transfer resistances, allowing their relative magnitudes to be weighed thus pinpointing the bottlenecks of each complex. Once the relative advantages of various immobilization methods and electron transfer mechanisms are known, we can proceed toward developing further improved enzymatic electrode materials

(577c) Nanostructured Photosystem Complexes As Biological Fuel Cell Catalysts

Ramaraja P. Ramasamy, The University of Georgia, Athens, GA

Manipulating photosynthesis for electricity generation is an interesting concept. Here we report for the first time, the demonstration of light to electrical power conversion in photosynthetic biological fuel cell using plant photosystems complexes and enzyme biocatalysts. The system works by combining the photo-induced water oxidation reaction on nanostructured photosystem protein based anode and oxygen reduction reaction onbio-cathode. Both the photosystem complex and cathode biocatalyst molecules were immobilized onto multiwalled carbon nanotubes through molecular tethers on gold electrodes. The nanostrcutured electrodes possess high surface area for high biomolecule loading and the carbon nanotubes served as excellent immobilization support for the biocatalysts. The photosystem-carbon nanotube composite exhibited high photo-electrochemical activity upon light illumination. For the first time ever, direct electrochemical communication was observed between plant proteins and the electrode. The photo-induced water splitting reaction has been confirmed to be the main source of electrons for photo-current generation. The favorable electron transfer pathways in the photosynthetic process for photo-electrochemical activity have been identified. The direct redox electrochemistry observed has significant impact in artificial photosynthesis research.

(577d) Functional Hybrid Nanomaterials for Biosensors

Jinwoo Lee, Department of Chemical Engineering and School of Environmental Science & Engineering, Pohang University of Science and Engineering, Pohang, South Korea

Self-assembled nanostructured materials attracted much attention due to well-defined porous structures. Various types of mesostructured materials have been synthesized via self-assembly method employing surfactants and block copolymers. Here, we present synthesis of self-assembled nanostructured materials such as mesoporous carbons and silicates and their applications to biosensors. Large cellular porous silica materials were synthesized by the self-assembly of commercially available block copolymers, P123. One-pot multi-catalyst system, so called "nanofactory", was developed entrapping magnetic nanoparticles and oxidases in large cellular mesoporous silica with high loadings of simultaneously above 40 wt% MNPs and 20 wt% enzymes. Our approach provided highly loaded MNP system and any highly loaded enzymes with superior activity, stability, and reusability, thereby making further applications as versatile sensors for detecting DNA, protein, and cell highly promising. A conductive nanocomposite consisting of Fe_3O_4 magnetic nanoparticles (MNPs) and oxidative enzyme co-entrapped in the pores of mesoporous carbon is fabricated for highly efficient and robust electrochemical biosensing platform.

(577e) A High-Power Enzymatic Fuel Cell That Can Completely Oxidize Glucose

Zhiguang Zhu and Y-H. Percival Zhang, Biological Systems Engineering, Virginia Polytechnic Institute and State University, Blacksburg, VA

An enzymatic fuel cell (EFC) is believed as a promising technology to meet the growing need to power miniaturized microelectronic, micromechanical and microfluidic devices, such as environmental sensors, battery chargers, and medical implants. However, it is still on its early stage, because of high cost of manufacturing, low power output, instabilities of biocatalysts and cofactors, limited fuel utilization efficiency, and so on. To develop more stable and powerful enzymatic fuels, cell-free synthetic pathway biotransformation has been applied for enzymatic fuel cells. Our previous studies have shown two key dehydrogenases in the pathway – glucose-6-phosphate dehydrogenase and 6-phosphogluconate dehydrogenase – can work well and the EFCs can generate electricity from glucose-6-phosphate, maltodextrin and glucose with power densities around $10^{-2} \sim 10^{-1}$ mW/cm^2. In this work, we have developed a novel carbon nanotube electrode disk that generates several mW/cm^2 power density. With more enzymes that are responsible for complete oxidation of glucose and no enzyme immobilization, glucose fuel can be completely oxidized and is projected to produce 24 electrons per glucose unit ideally.

(578a) Dynamic Processes in Diblock Copolymer Micelles with a Semi-Crystalline Core

Megan Robertson, Avantika Singh and Maria Marquez, Chemical and Biomolecular Engineering, University of Houston, Houston, TX

Diblock copolymers, which form micelle structures in selective solvents, offer advantages of robustness and tunability of micelle characteristics as compared to small molecule surfactants. Diblock copolymer micelles in water have been a subject of great interest in drug delivery applications based on their high loading capacity and targeted drug delivery. The aim of this work is to understand the dynamic processes which underly the self-assembly of diblock copolymer micelle systems which have a semi-crystalline core. Due to the large size of the molecules, the self-assembly of block copolymer micelles occurs on significantly longer time scales than small molecule analogues. The present work focuses on amphiphilic diblock copolymers containing blocks of poly(ethylene oxide) (a hydrophilic polymer) and polycaprolactone (a hydrophobic, semi-crystalline polymer), which spontaneously self-assemble into spherical micelles in water. A variety of experimental techniques are used to probe the kinetic processes relevant to micelle self-assembly, including time-resolved neutron scattering, dynamic light scattering, pulsed field gradient nuclear magnetic resonance, and fluorescence resonance energy transfer experiments.

(578b) The Effects of Copolymerized Blocks On the Self-Assembly of Acrlyic Terpolymers in Solution

James A. Bergman, Chemical and Biological Engineering, Iowa State University, Ames, IA and Jennifer O'Donnell, Chemical and Biological Engineeirng, Iowa State University, Ames, IA

Obtaining the optimal polymer microstructure with the optimal chemical domains for specific applications is difficult. Because of this, there has been much research into how to manipulate the microstructures of self-assembling polymers such that desired chemical functionalities and microstructures can be simultaneously achieved. The specific microstructure formed in a phase-separated system is determined by minimizing the free energy of the polymer chain confirmations and the interfacial energy between the two microdomains. To date, the thermodynamics that define the interfacial curvature, and therefore, the microstructure have largely been tuned by changing polymer composition. Other techniques for tuning the thermodynamics include additives (solvents, surfactants or counter ions), and manipulation of architectures (miktoarm star, janus particles and graft copolymers). Terpolymers offer another variable for changing the microstructure at a specific composition, namely the sequencing of the blocks. For example, it has been shown that two triblock terpolymers, ABC and ACB, both with the same block sizes, do not self-assemble into the same microstructures. Over the past twenty years the interest in polymers with three chemically unique repeat units has increased. Compared to diblock copolymers, triblock terpolymers have been shown to self-assemble into a greater number of microstructures. One of the exciting promises of terpolymers is the wider availability of multi-continuous phases.

Here, two of the three monomers in an acrylic terpolymer consisting of hydroxyethyl acrylate (HEA), methyl acrylate (MA), and octyl acrylate (OA) are copolymerized, and the effects of the architecture on the self-assembled microstructures in water and tetrahydrofuran are examined. The copolymerization of the HEA and MA or OA and MA is expected to dampen the interfacial energy between domains, and the entropic penalties of self-assembling, allowing for greater segregation. All of the polymers were synthesized using reversible addition chain transfer (RAFT) polymerization techniques. The entirety of this work has studied four different terpolymer compositions with a constant molecular volume, and four different architectures. Here, we will limit the discussion to two of the compositions. The first composition had 69 repeat units of HEA, 20 repeat units of MA and 9 repeat units of OA; the second composition had 54 repeat units of HEA, 20 repeat units of MA and 17 repeat units of OA. The first two architectures were triblocks, HEA – b – OA – b – MA and HEA – b – MA – b – OA. The third and fourth architectures consisted of one homoblock and one copolymerized block, HEA – b – (OA – co – MA) and (HEA – co – MA) – b – OA. The monomer conversion was verified gravimetrically, and molecular weight was determined by gel permeation chromatography after each monomer addition. All of the polymers were equilibrated at four concentrations in water and four concentrations in tetrahydrofuran. Small angle x-ray scattering was used to characterize morphologies of the neat polymers and equilibrated samples.

Lamellar and hexagonal microstructures have been identified, and differences in self-assembled microstructures based on architectures were observed. The results indicate a strong segregation for the terpolymers with copolymerized blocks. Further characterization of these systems using cryo-TEM is being conducted.

(578c) Dielectric Relaxation of Tethered Polyisoprene in Nanoscale

Sung A Kim, Praveen Agarwal and Lynden A. Archer, School of Chemical and Biomolecular Engineering, Cornell University, Ithaca, NY

Most studies about relaxation of tethered polymers have been performed on the systems where polymers are attached to a platelet structure or adsorbed onto a surface of other geometries. This talk presents the relaxation dynamics of polymers which are confined in nanoscale by tethering to spherical nanoparticles. In our system, all polymer chains are grafted to nanoparticles and as-synthesized materials are well dispersed and suspended without any additional solvents. Therefore, understanding of nanoscale dynamics of polymer in the system is straightforward from bulk measurement.

In this study, synthesized cis 1-4 Polyisoprene (PI) is tethered to the surface of silica nanoparticles, and dielectric relaxation behavior is probed by broadband dielectric spectrometer. Global chain relaxation is conveniently explored since the net dipole moment of an entire chain of cis 1-4 PI is parallel to the end to end vector of the repeating unit, which is the characteristics of type A dielectric polymers. Segmental mode relaxation provides the information about the local relaxation from the components perpendicular to the chain backbones. The effects of grafting densities and molecular weights in the spectrum from unentangled to highly entangled regime have been studied. For example, when PI molecular weight is in unentangled range, we observe that global chain relaxation time ~ 1.65 E-01 s of tethered 3K PI is ~10^5 times slower than the relaxation time ~ 2.27 E-06 s of untethered 3K PI at 20 °C, indicative of highly entangled polymer dynamics for tethered 3K PI. Segmental mode relaxation time also becomes ~10^3 times longer when 3K polymers are tethered. We propose the model to understand the effect of molecular weight and grafting density on global chain and local relaxation when all polymers are attached to nanoscale spherical particles.

(578d) Structure and Properties of Polyelectrolyte Multilayers in the Exponential and Linear Growth Regimes

Biswa P. Das and Marina Tsianou, Department of Chemical and Biological Engineering, University at Buffalo - The State University of New York (SUNY), Buffalo, NY

Multilayered thin films have been the focus of intense research because of their vast array of applications. We investigate the growth and disassembly under different pH conditions of layer-by-layer (LbL) films composed of oppositely charged polyelectrolytes (including poly(diallyldimethylammonium chloride) (PDDA), poly(styrene sulfonate) (PSS), and poly(acrylic acid) (PAA)) and synthetic clays. We monitor the assembly growth by UV-vis spectroscopy, estimate the assembly thickness by scanning electron microscopy (SEM) cross-section images and the surface roughness by atomic force microscopy (AFM). We further examine the charge density of films by Fourier transform infrared (FTIR) spectroscopy, we determine the layer spacing by X-Ray diffraction (XRD), and we evaluate the organic-inorganic content by thermogravimetric analysis (TGA). Our results demonstrate that PDDA-PAA-PDDA-Laponite multilayers display exponential growth aided by internal diffusion of PAA molecules, whereas PDDA-PSS-PDDA-Laponite and PDDALaponite multilayers display linear growth by sequential molecular level deposition. We also illustrate that different assembly regimes are attained by manipulating the deposition pH of PAA solution. The results on the structural characteristics and properties of multilayers formed under different conditions can provide insights into their behavior and sensitivity to external stimuli during various applications.

(578e) Self-Assembly of Polystyrene-Polyisoprene STAR Copolymers

Juan Pablo Hinestrosa, Center for Nanophase Material Sciences, Oak Ridge National Laboratory, Oak Ridge, TN, S. Michael Kilbey, Departments of Chemistry and of Chemical and Biomolecular Engineering, University of Tennessee, Knoxville, TN and Jamie Messman, Oak ridge National Laboratory, Oak Ridge, TN

SELF-ASSEMBLY OF POLYSTYRENE-POLY (N-ISOPROPYLACRYLAMIDE) STAR COPOLYMERS

Juan Pablo Hinestrosa,[1] Jason P. LeJeune,[1] S. Michael Kilbey II,[1,2] Jamie M. Messman[1]

[1]Center for Nanophase Material Sciences, Oak Ridge National Laboratory, Oak Ridge, TN 37834; [2]Department of Chemistry, University of Tennessee, Knoxville, TN 37996

A modular synthetic process combining the utility of RAFT polymerization with 'click' chemistry yields well-defined star copolymers suitable for fundamental studies of self-assembly in selective solvents and for bio-related applications, such as drug delivery, due to their temperature responsiveness at physiological conditions. Copolymers composed of polystyrene (PS)-poly (N-isopropylacrylamide) (PNIPAM) are synthesized by this modular approach yielding three-arm stars as probed by FTIR, GPC and NMR measurements. Self-assembly studies of these copolymers in H_2O, a selective solvent for PNIPAM, show micellization into sphere-like structures with sizes ~ 20 nm. For a diblock copolymer of PS (P)-*block*-PNIPAM (N), large aggregates, i.e. micelle clusters, with sizes ~ 100 nm are detected in coexistence with micelles, as can seen in the Figure. This behavior is accentuated as temperature is increased closer to the coil-to-globule transition for PNIPAM where this block becomes more hydrophobic.. Measurements at ~ 34 °C showed the formation of clusters for the P_2N_1 star, i.e. two PS arm and one PNIPAM arm, but not for star P_1N_2.

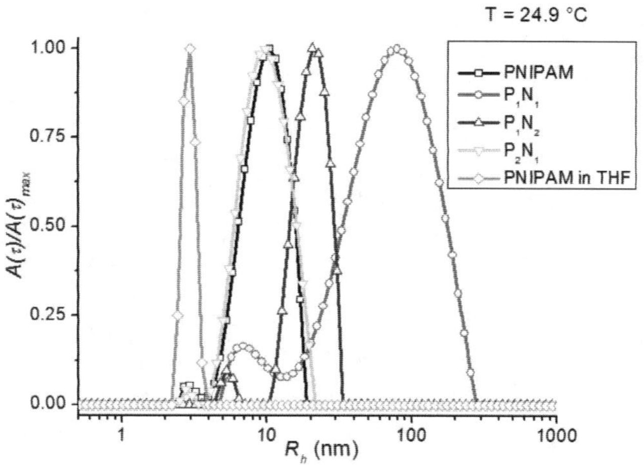

Figure. Normalized hydrodynamic radius, R_h, distributions for PS (P)-PNIPAM (N) star copolymers in H_2O at 1 mg/mL and for the PNIPAM arm in THF at 5 mg/mL. The subscripts indicate the number of PS or PNIPAM arms in each copolymer.

(578f) Super Gas Barrier of All-Polymer Layer-by-Layer Assemblies

Laura Bolling[1], You-Hao Yang[1], Merid Haile[2], Frank A. Malek[1] and Jaime C. Grunlan[3], (1)Chemical Engineering, Texas A&M University, College Station, TX, (2)Chemistry, Texas A&M University, College Station, TX, (3)Department of Mechanical Engineering and Department of Chemical Engineering, Texas A&M University, College Station, TX

Super Gas Barrier of All-Polymer Layer-by-Layer Assemblies

Layer-by-layer assembly of cationic branched polyethylenimine (PEI) and anionic poly (acrylic acid) (PAA) was investigated with regard to pH of each polymer solution during deposition. Thickness and surface roughness can be altered dramatically due to the pH-responsive morphology of PEI and PAA (from 90 nm to 4.74 mm for 30-bilayer films). Crosslinking these films with glutaraldehyde (GA) creates an inhibition of polymer interdiffusion, causing exponential film growth to be reset. The oxygen permeability was improved by increasing the number of PEI/PAA bilayers. P_{O2} of an 8-bilayer, glutaraldehyde-crosslinked PEI/PAA film (305 nm thick) is below 3.2×10^{-21} $cm^3 \cdot cm/cm^2 \cdot s \cdot Pa$, which is believed to be the lowest permeability of an all-polymer system ever reported. Tremendous PEI/PAA interfaces result in a scrambled egg structure with a high gas barrier, so this relatively simple recipe may be of use for a variety of applications, including polymeric membranes, flexible electronics, and food packaging.

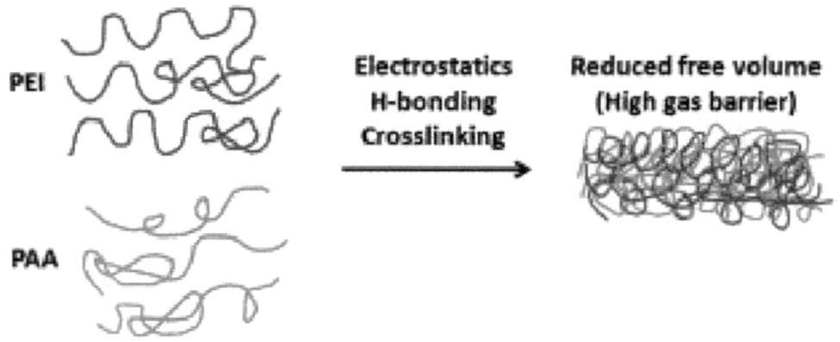

(578g) Mesocellular Phenol Formaldehyde Foams: Synthesis, Characterization and Lysozyme Adsorption Study

Manasa Sridhar[1], Krishna Reddy Gunugunuri[1], Naiping Hu[2], Dale W. Schaefer[3], Stephen W. Thiel[4] and Panagiotis Smirniotis[1], (1)Chemical Engineering Program, School of Energy, Environmental, Biological and Medicinal Engineering, University of Cincinnati, Cincinnati, OH, (2)Electrical Engineering Program, University of Cincinnati, cincinnati, OH, (3)Chemical and Materials Engineering, University of Cincinnati, Cincinnati, OH, (4)Chemical & Materials Engineering, University of Cincinnati, Cincinnati, OH

This study aims at systematic investigation of nanocasting strategy to produce organic mesoporous polymers and exploration of their potential as sorbents for bio-separations. Siliceous mesocellular foam (MCF) templates of regular and modified mesostructures and morphologies have been synthesized by symbiotic interplay between TMB and TEOS concentration in the seminal microemulsion. MCFs with thickened walls, polyhedral foam-like and ultra large macroporous textures and interconnected rod-like morphologies have been fabricated. These interesting templates have been employed to produce mesocellular phenol formaldehyde foams (MPFFs). TEM, SEM images in conjunction with N_2 physisorption analysis suggest disordered but nearly faithful replication of cavity sizes and morphologies. Strikingly, hierarchical trimodal MPFFs have been synthesized when high TMB/P123 and TEOS/P123 mass ratios are utilized to make the MCF templates. Ultra-large 100 nm diameter nearly spherical cavities were witnessed in the TEM images of MPFF@$R_1$2.5$R_2$4.4 which may evince the relatively large Lysozyme adsorption capacities in comparison to the MPFFs templated using conventional MCFs. These MPFFs posses narrow PSDs, high surface areas of above 900 m^2/g and large pore volumes up to 1.37 cm^3/g. Micropore volumes constituting more than 20% of the total pore volume are expected to play a vital role in mass transport of biomolecules further augmenting the dynamics of diffusion. Lysozyme adsorption capacities of more than 230 mg/g have been achieved, which holds enough promise to further enhance their capacities using post-synthesis functionalization techniques.

(579a) Self-Assembling DNA Nanotubes From Programmable Seeds

Abdul M. Mohammed and Rebecca Schulman, chemical and biomolecular engineering, Johns Hopkins University, Baltimore, MD

Self-Assembling DNA Nanotubes from Programmable Seeds

One of the fundamental challenges in the bottom-up assembly of nanoscale devices is the limited control over the organization of molecular assemblies. For example, it is not well understood how to assemble a molecular link between two points in 3D space. To address this problem, we seek to build a system in which DNA nanotubes could span from a start point to an end point (green and red in Figure 1a) even though these points may vary in distance and orientation. We use DNA nanotubes to study this problem; the structure and assembly mechanism of DNA nanotubes are well understood and their components can be rationally designed. In this study, we would like to demonstrate how designed nuclei could serve as starting points for nanotubes to grow, which may guide strategies for generalized linking in nanoscale systems.

Using our assembly protocol, DNA nanotubes are formed in a hierarchical self-assembly process, where individual DNA strands form DNA double-crossover molecules (DNA tiles), which then self-assemble into nanotubes via interactions of single-stranded DNA on the tile ends. Homogenous nucleation of DNA nanotubes from tiles under low supersaturation is very rare because there is an energy barrier to nucleation. Here we design nanotube nuclei and show that nanotubes grow quickly in their presence. Our nucleus design consists of a cylindrical DNA origami resembling the facet of a DNA nanotube. This design orients monomers into an assembled nanotube rather than producing a high local monomer concentration (Figure 1b). Using fluorescence microscopy, we observed a significant increase of the nanotube growth rates in the presence of these nuclei. AFM images confirm that this increase in growth rates is due to the growth of nanotubes from the nuclei (Figure 1c).

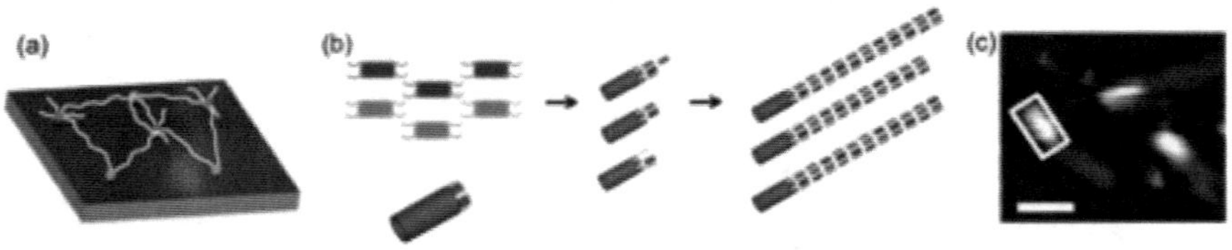

Figure 1: (a) Schematic representation of a system where DNA nanotubes connect start (green) and end (red) points; (b) Schematic showing self-assembly of DNA tiles (brown, green) on nuclei (red) to form DNA nanotubes. (c) AFM image of DNA nanotubes growing from nuclei (marked). Scale bar: 100 nm

(579b) Nano-Pipette Directed Motion of Bio-Inspired Transmembrane Channel

Meenakshi Dutt, Chemical and Petroleum Engineering, University of Pittsburgh, Pittsburgh, PA, Olga Kuksenok, Department of Chemical and Petroleum Engineering, University of Pittsburgh, Pittsburgh, PA and Anna Balazs, Chemical Engineering, Univ. of Pittsburgh, Pittsburgh, PA

Via Dissipative Particle Dynamics (DPD) approach, we study the directed motion of a transmembrane end-functionalized nanotube using a suitably functionalized nano-pipette. In our earlier work (Nanoscale 2011), we demonstrated the design and creation of biomaterials which promote controlled release by integrating end-functionalized nanotubes into lipid bilayers. Each nanotube encompasses an ABA architecture, with a hydrophobic shaft (B) and two hydrophilic ends (A). To allow controlled transport through the nanotube, we also introduce hydrophilic tethers at one end of the tube. We showed that nanotubes initially located in the outer solvent spontaneously penetrate the membrane and assume a trans-membrane position, with the hydrophilic tethers extending from the surface of the bilayer. The hydrophilic tethers can also serve as anchors for directing the motion of the inserted nanotube across the membrane. We demonstrate this process by locating a suitably functionalized nano-pipette near such an end-functionalized nanotube. The nanotube diffuses in the membrane until the tethers are close to the nano-pipette. Due to favorable interactions, the tethers anchor onto the nano-pipette. We also show that the nanotube motion can be controlled through the nano-pipette. This system can provide design guidelines for lab-on-chip applications for drug delivery.

(579c) Computational Characterization of DNA/Peptide/Nanotube Self Assembly for Bioenergy Applications

Vanessa Ortiz, Chemical Engineering, Columbia University, New York, NY

Multi-enzyme pathways have become a subject of increasing interest for their role in the engineering of biomimetic systems for applications including biosensors, bioelectronics, and bioenergy. The efficiencies found in natural metabolic pathways partially arise from biomolecular self-assembly of the component enzymes in an effort to avoid transport limitations.

The ultimate goal of this effort is to design and build biofuel cells with efficiencies similar to those of native systems by introducing biomimetic structures that immobilize multiple enzymes in specific orientations on a bioelectrode. To achieve site-specific immobilization, the specificity of DNA-binding domains is exploited with an approach that allows any redox enzyme to be modified to site-specifically bind to double stranded (ds) DNA while retaining activity. Because of its many desirable properties, the bioelectrode of choice is single-wall carbon nanotubes (SWNTs), but little is known about dsDNA/SWNT assembly and how this might affect the activity of the DNA-binding domains.

Here we evaluate the feasibility of the proposed assembly by performing atomistic molecular dynamics simulations to look at the stability and conformations adopted by dsDNA when bound to a SWNT. We also evaluate the effects of the presence of a SWNT on the stability of the complex formed by a DNA-binding domain and DNA.

(579d) Benchmarking the Scavenging of Reactive Oxygen Species Using Inorganic and Carbon-Based Nanoparticle Systems in Isolated Chloroplasts From Spinacia Oleracea

Ardemis A. Boghossian[1], Selda Sen[1,2], Brenna Gibbons[3], Fatih Sen[1,2], Juan Pablo Giraldo[1] and Michael S. Strano[1], (1)Department of Chemical Engineering, Massachusetts Institute of Technology, Cambridge, MA, (2)Department of Chemistry, Middle East Technical University, Ankara, Turkey, (3)Department of Chemistry, Dartmouth College, Hanover, NH

The chloroplast contains densely stacked arrays of light-harvesting proteins ideal for harnessing solar energy. Over time, the photoactive proteins within the chloroplast become photodamaged due to reactive oxygen species (ROS) generation. In vivo, photoautotrophs address issues of photodegradation by applying a self-repair cycle that dynamically repairs and replaces photodamaged components. Outside the cell, ROS-induced photodegradation contributes to limited chloroplast stabilities. The incorporation of chloroplasts into light-harvesting devices therefore requires regenerative ROS scavenging mechanisms to prolong chloroplast photoactivity outside the cell. For the first time, we have studied the ROS generating behavior of isolated chloroplasts directly interfaced with nanoparticles, including dextran-wrapped nanoceria (dNC) that has demonstrated remedial ROS scavenging mechanisms. We quantitatively examine the effect of dNC, along with cerium ions, fullerenol, and DNA-wrapped single-walled carbon nanotubes (SWNTs), on the ROS generation of isolated chloroplasts using the oxidative dyes, 2',7'-dichlorodihydrofluorescein diacetate (H_2DCF-DA) and 2,3-bis(2-methoxy-4-nitro-5-sulfophenyl)-2*H*-tetrazolium-5-carboxanilide sodium salt (XTT).Chloroplast photoactivity in the presence of the nanoparticle additives was also measured using chloroplast fluorescence and the artificial, photosynthetic electron accepting dye, dichloroindophenol (DCPIP). We conclude that the dNC is the most effective agent for decreasing oxidizing agent and superoxide concentrations whilst preserving chloroplast photoactivity at concentrations below 5 μM. Confocal microscopy verifies that the mechanism of remedial ROS scavenging by the dNC occurs external to the isolated chloroplast. We believe that coupled to an approach for chloroplast internalization, dNC offers a promising mechanism for maintaining regenerative chloroplast photoactviity for light harvesting applications.

(579e) Biopolyester Synthesis On Peptide Nanofibers and Liposomes

Smith Sangiambut, Biological and Soft Systems Sector, Dep of Physics, University of Cambridge, Cambridge, United Kingdom, Kevin Channon, University of Cambridge, Shun Sato, National Institute of Advanced Industrial Science and Technology, Japan, Takeharu Tsuge, Tokyo Institute of Technology, Japan and Easan Sivaniah, Cavendish Laboratory, Physics, University of Cambridge, Cambridge, United Kingdom

We present two biomimetic approaches, which control enzyme arrangement at the nanoscale, for production of biopolyester. PHA synthases are enzymes which synthesize polyhydroxyalkanoates (PHAs), a class of biodegradable, biocompatible polymers with use in applications such as degradable medical implants and drug delivery, which are naturally produced by many species of bacteria and archaea. Recombinant expression of the enzyme and addition of a monomer of PHA allow us to synthesize PHAs *in vitro*, enabling simpler purification of the polymer, but raising the issue of enzyme stability outside of its natural habitat. To address this issue, we have combined PHA synthases with two *in vitro* support systems: self-assembling peptide nanofibers and giant lipid vesicles. In the first approach, PHA synthases were engineered to associate specifically to peptide nanofibers via a tag peptide, and PHA was then synthesized to coat the surface of the nanofibers. This hierarchical assembly is one of the few examples of the functionalization of peptide nanofibers with an enzyme, and allows rapid attachment of the enzyme under mild conditions, avoiding the harsh treatments of previous methods. The second approach of synthesizing PHAs within a lipid vesicle provides the enzyme with a close match to its natural environment, providing a stabilising structure and raising the potential for establishing an efficient production system which can recycle costly substrates used in PHA synthesis.

(579f) Effect of Body Geometry On Motility of Bacteria-Powered Microrrobots (BACTERIABOTS)

Ali Sahari, School of Biomedical Engineering and Sciences, Virginia Tech, Blacksburg, VA and Bahareh Behkam, Mechanical Engineering, Virginia Tech, Blacksburg, VA

EFFECT OF BODY GEOMETRY ON MOTILITY OF BACTERIA-POWERED MICRORROBOTS (BACTERIABOTS)

Ali Sahari[1], Bahareh Behkam[1,2]

[1]School of Biomedical Engineering and Sciences, Virginia Tech, Blacksburg VA 24061

[2]Mechanical Engineering Department, Virginia Tech, Blacksburg VA 24061

Abstract— Motility of micro-scale swimming robots falls in the realm of low Reynolds number, where viscous forces exerted on the robots are dominant over inertia. In this work, the importance of body geometry on the dynamics of bacteria propelled swimming microrobots (BacteriaBots) is investigated. We have experimentally demonstrated that oblong geometries enhance directionality of the micro-particles compared with spherical geometry.

<>I. INTRODUCTION

Mobile micro-robots have unique advantages such as the ability to access small spaces and the potential to be employed in large numbers as inexpensive agents of distributed systems for swarm robotic applications. Such micro-robots are envisioned to impact a diverse range of applications, including minimally invasive diagnosis and localized treatment of diseases, environmental monitoring, and homeland security [1]. Amongst the most significant obstacles to realization of mobile robots at micro-scale are the miniaturization of on-board actuators, power sources, and communication and control modules. Bio-hybrid approaches can be employed to address these challenges by integrating prokaryotic and eukaryotic cells within the robotic system [2-4]. *BacteriaBot,* a bio-hybrid micro-robot is constructed here by interfacing a microfabricated robot body with an ensemble of live engineered bacteria with the purpose of using the bacteria for actuation, sensing, communication and control. Motile behavior of bacteria is well characterized and falls under two characteristic modes of *run* and *tumble*. This random run and tumble behavior leads to the characteristic three dimensional random walk of bacteria and consequently to the stochastic motion of micro-objects propelled by bacteria.

Motility of spherical microstructures actuated by an ensemble of attached bacteria has been thoroughly characterized in previous literature [5-8] but a systematic study of the effect of microstructure geometry on propulsive behavior is currently missing. Mobile microrobots with optimal body geometries are envisioned to impact minimally invasive diagnosis, localized treatment of diseases and environmental monitoring. Limited particle diffusion and directional coefficient of drag are some of the attributes that are enhanced through such bio-hybrid systems. In this work, we have utilized a low-cost and high throughput technique to obtain non-spherical mico-particles and investigate the effect of particle shape on the motile behavior of the BacteriaBots.

<>II. Materials and methods <>A. BacteriaBot Body Fabrication

A high throughput spherical particle casting and mechanical stretching under heat treatment, as previously described in [9], is used to produce non-spherical polystyrene (PS) particles which act as the engineered synthetic body of the robots. Briefly, 6 μm PS spheres (Polysciences) are casted in a 35 μm thick polyvinyl alcohol (PVA) film. The film is uniformly stretched in one dimension to generate voids around the micro-spheres. The casted PS micro-spheres are liquefied using a bath of hot mineral oil. For making football shaped particles, 0.5% glycerol is added to the PVA film as a plasticizer. After removal from bath and cooling, particles are released from the PVA film by soaking in 30% isopropyl alcohol (IPA): DI water solution at 80°C, and are washed by centrifugation in a 30% IPA solution.

<>B. Bacteria culture <>Escherichia coli (E. coli) strain MG1655 was cultured in Luria Broth (1% tryptone, 0.5% yeast extract, and 0.5% sodium chloride). The culture was grown to an optical density OD_{600}=0.5 at 37 °C.

<>C. BacteriaBot Construction

The mixture of Poly-$_L$-lysine and micro-particles is incubated on a vortex mixer for one hour. Bacteria are centrifuged at 3000g for 5 min and then resuspended in motility medium (0.01M potassium phosphate, 0.067M sodium chloride, 10^{-4}M EDTA, 0.21M glucose, and 0.002% Tween-20). The poly-$_L$-lysine coated micro-particles are added to the bacteria allowing the bacteria to self-assemble on the particles.

D.Â Two-dimensional Single Particle Tracking

The motion of the microbeads was captured at 20 frames per second using a Zeiss AxioObserver Z1 inverted microscope equipped with an AxioCam HSm camera. The images were analyzed using a two-dimensional (2D) particle tracking routine developed in MATLAB (MathWorks, Natick, MA). Briefly, using cell segmentation and image restoration, the artifacts existing in most of the captured images were removed. This was followed by noise removal and cell boundary recognition using a border following algorithm. Finally, the nearest-neighbor method was used to link segmented cells in successive frames and to generate the trajectories.

Table 1: Fabrication parameters for manufacturing non-spherical particles

Particle	Stretching aspect ratio	Liquefaction method	Glycerol plasticizer
Spherical	Unmodified	Unmodified	No
Barrel	1.16	130°C oil bath	No
Football	1.6	130°C oil bath	Yes

<>III. Results and Discussions

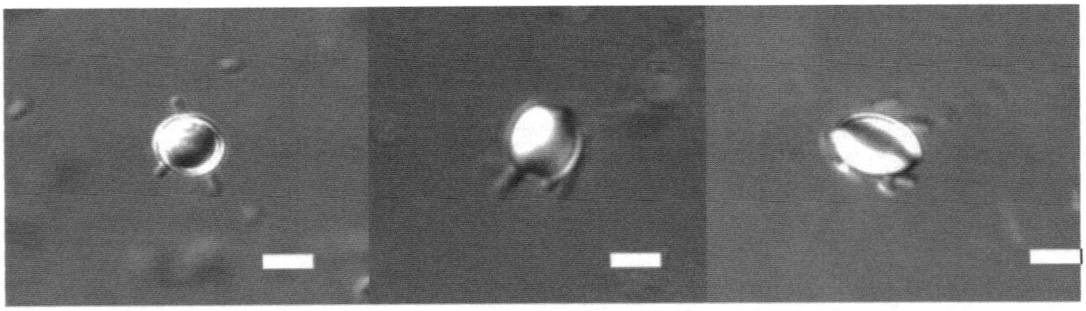

(A) (B) (C)

Fig. 2: (A) A spherical BacteriaBot with three bacteria, (B) a barrel BacteriaBot with three bacteria, and (C) a football BacteriaBot with five bacteria attached. Scale bars are 5 micron in length.

Two different particle geometries were fabricated using a casting and mechanical stretching method that was described earlier. Parameters used for the manufacturing and the scanning electron microscopy (SEM) images of the resulting geometries are given in Table 1 and Fig. 1, respectively. Representative images of the bacteriabots with different geometries are depicted in Fig. 2. The 2D image tracking routine was utilized to characterize the motion of the BacteriaBots. Of particular interest is how the body geometry will affect the directionality of the motion of BacteriaBot without the need for active steering. Directionality is defined as the ratio of the magnitude of the displacement vector to the total distance traveled. To prove that the bacteria attached to the mobile microbeads are the source of propulsion, a control experiment was performed. Minimal displacement of the control bead confirms that the bacteria attached to the mobile microbeads are the main source of propulsion. Therefore, any directed movement observed for this control bead would be neither due to diffusion nor due to the flow field generated by the free- swimming bacteria present in the background. The

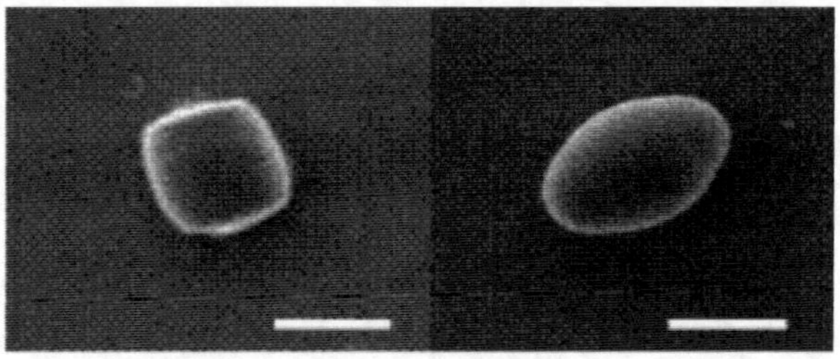

Fig. 1: Scanning electron microscopy (SEM) images of the non-spherical polystyrene geometries produced by casting and mechanical stretching of spherical particles. Scale bars are 5 micron in length.

experimental results for directionality of the bacteria- propelled bodies with spherical, barrel, and football-shaped geometries are shown in Fig. 3. It can be seen that the barrel and football shaped bodies are propelled with a higher degree of directionality compared to the spherical robots. Two representative trajectories of the BacteriaBots are shown in Fig. 4. Number of attached bacteria for all geometries varies between 1-6; however, our experimental results do not seem to be significantly dependent upon the number of bacteria attached. This is consistent with our previous observations if the attached bacteria are uniformly distributed over the body, the overall force is expected to remain largely unchanged regardless of the number of attached bacteria [7].

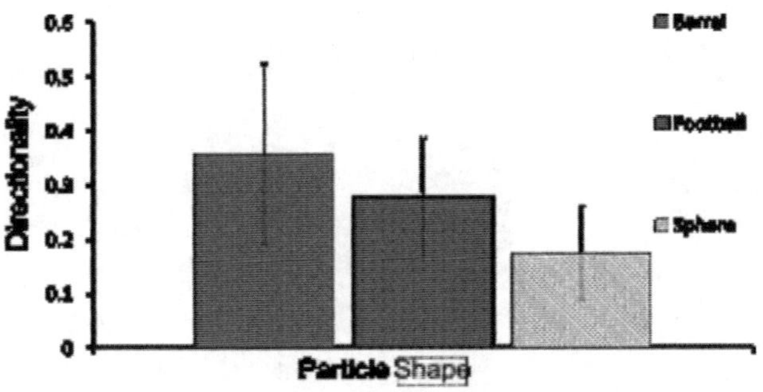

Fig. 3: Directionality as a function of BacteriaBot body shape.

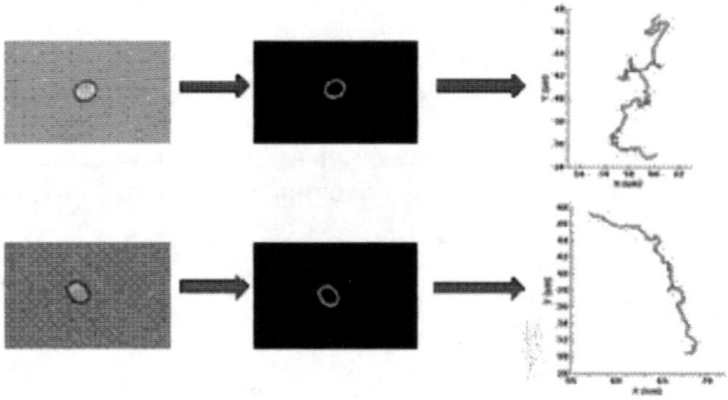

Fig. 4: Two representative trajectories of BacteriaBots provided by the

It should be noted that more than the number of bacteria attached to the microbead, the location of attachment is expected to affect the overall behavior. If the areal attachment density becomes significantly nonuniform, we expect to see a change in the average net resultant force and consequently observe a change in overall speed and directionality [3]. <>IV. Conclusion

Variation in motile behavior of BacteriaBots due to their body geometry can be very complex and can only be determined experimentally. These complexities are due to: (1) non-spherical geometries have varying coefficients of drag depending on their aspect ratio and the direction of motion, and (2) the varying radius of curvature on the surface of non-spherical geometries leads to preferential bacterial adhesion in certain locations. By utilizing a high throughput PS micro-particle manufacturing method, we characterized the motile behavior of BacteriaBots with spherical, barrel and football shaped bodies. It was shown that for BacteriaBots with uniform areal attachment density, body shape strongly affects the directionality of the motion.

We are currently examining the behavior of other geometrical shapes such as bullets. We are also investigating if the particles size will have an impact on the observed trend.

Acknowledgment

The authors would like to thank Professor Birgit Scharf in Biological Sciences Department at Virginia Tech for gifting the bacteria. This work was in part supported by the National Science Foundation (IIS-117519).

References

[1] B. Behkam, M. Sitti, "Bacteria Integrated Swimming Micro-robots", in M. Lungarella, et al., (eds.), 50 years of AI, Festschrift, Lecture Notes in Artificial Intellignece 4850, pp. 154-163, 2007.

[2] S. Martel, *et al.*, "Controlled manipulation and actuation of micro-objects with magnetotactic bacteria," *Applied Physics Letters,* vol. 89, pp. 233904-233904-3, 2006.

[3] A. A. Julius, M. S. Sakar, E. Steager, U. K. Cheang, M. Kim, V. Kumar,and G. J. Pappas, "Harnessing bacterial power for micro scale manipulation and locomotion," IEEE International Conference on Robotics and Automation (IEEE, Kobe, Japan, 2009).

[4] D. B. Weibel, *et al.*, "Microoxen: Microorganisms to move microscale loads," *Proceedings of the National Academy of Sciences of the United States of America,* vol. 102, p. 11963, 2005.

[5] B. Behkam and M. Sitti, "Bacterial flagella-based propulsion and on/off motion control of microscale objects," *Applied Physics Letters,* vol. 90, p. 023902, 2007.

[6] B. Behkam and M. Sitti, "Effect of quantity and configuration of attached bacteria on bacterial propulsion of microbeads," *Applied Physics Letters,* vol. 93, p. 223901, 2008.

[7] M. A. Traoré, *et al.*, "Computational and experimental study of chemotaxis of an ensemble of bacteria attached to a microbead," *Physical Review E,* vol. 84, p. 061908, 2011.

[8] B. Behkam and M. Sitti, "Characterization of bacterial actuation of micro-objects," Proceedings of IEEE Conference on Robotics and Automation (ICRA) 2009, pp. 1022-1027.

[9] J. A. Champion, *et al.*, "Making polymeric micro-and nanoparticles of complex shapes," *Proceedings of the National Academy of Sciences,* vol. 104, p. 11901, 2007.

Morphological evolution of polyfurfuryl alcohol derived carbon spheres assisted by emulsion polymerization of furfuryl alcohol

Carbon materials with controlled porosity derived from polymer precursors have been used in a wide range of applications such as catalysis, gas adsorption, membrane separation and electrochemical capacitors [1-8]. A key feature in these carbons which affects their performance in different applications is the surface area and pore size distribution. However, control of morphology and external properties and tailoring size and shape of porous carbonaceous materials is as important as the internal properties. In applications like catalysis, the rate of mass transfer of gases or liquids in the pores can be enhanced by controlling the size of the primary particles of the catalyst support. Emulsion polymerization is used as a common approach to synthesize polymeric particles at nano scale. In this method, a surfactant usually acts as the structure directing agent. Type of the surfactant and the monomer, composition of the mixture to start with and the synthesis condition affect the morphology and size of the final material [9-12].

In this investigation, we report a systematic study on the formation of carbon spheres using emulsion polymerization of furfuryl alcohol in the presence of Pluronic F-127 as the surfactant, followed by pyrolysis of resultant polyfurfuryl alcohol structure. Carbon spheres with the average size of 50 nm to few microns were synthesized using this approach. The pseudo-ternary phase diagram of furfuryl alcohol/surfactant/solvent mixture was developed to outline other possible morphological structures of the synthesized carbon. The effect of different synthesis parameters like polymerization initiator (HCl) concentration, surfactant and monomer concentration and solvent composition (water/ethanol) on the micelle formation, polymerization and polymer crosslinking and finally the size and morphology of the final carbonaceous structure was studied.

References

[1] Wang, Y.; He, C.; Brouzgou, A.; et al., Journal of power sources, 2012, 200, 8.
[2] Strano, M. S.; Wyre, J.; Foley, H. C.; Industrial & Engineering Chemistry Research, 2005, 44, 6414.
[3] Stevens, M. G.; Anderson, M. R.; Foley, H. C.; Chemical Communications, 1999, 5, 413.
[4] Rajagopalan, R.; Ponnaiyan, A.; Mankidy, P. J.; et al., Chemical Communications, 2004, 21, 2498.
[5] Ermete, A.; Applied Catalysis B: Environmental, 2009, 88, 1.
[6] Shah Tapan, N.; Foley, H. C.; Zydney, Andrew L.; Biotechnology Progress, 2007, 23, 1157.
[7] Merritt, A.; Rajagopalan, R.; Foley, H. C.; Carbon, 2007, 45, 1267
[8] Strano, M. S,; Zydney, A. L.; Barth, H.; et al.; Journal of Membrane Science, 2002, 198, 173
[9] Ovanda-Medina, V. M.; et. al.; Colloid & Polymer Science, 2011, 289, 759.
[10] Ya-Qing, Z.; et. al.; Powder Technology, 2010, 201, 146.
[11] Li, L.; et. al.; Journal of Applied Polymer Science, 2009, 113, 4023.
[12] Guo, J. S.; et. al.; Journal of Polymer Scienece: part A: polymer chemistry, 1992, 30, 691.

(590b) POSS-Templated Asymmetric Molecular Nanocages

Zhongliang Shen[1], Jongsik Kim[2], Christopher Downing[3], Harold H. Kung[1] and Mayfair C. Kung[1], (1)Chemical and Biological Engineering, Northwestern University, Evanston, IL, (2)Chemistry, Dong-A University, Busan, South Korea, (3)Chemistry, Northwestern University, Evanston, IL

Nanocages between 1-10 nm are interesting structures as their dimensions are similar to that of enzyme cavaties. The synthesis of the nanocage was accomplished using polyhedral oligomeric silsesquioxane (POSS) as templates. The template was constructed by derivatizing a commercially available POSS (octakis(dimethylsilyl)-POSS) to create an architecture dense in periphery functional vinyl groups. The forty eight periphery vinyl groups were linked to form a 3-D shell composed of carbosilane linkages. Dynamic light scattering measurements indicated that this shell cross linked structure had a diameter between 6.4 and 7.4 nm. NMR spectroscopy indicated that 86% of the vinyl groups were engaged in forming the shell and this number was consistent with the molecular weight measurements derived from GPC results. Since carbosilane linkages are stable in base we were able to preserve the integrity of the shell when we removed the template using tetraethyl ammonium hydroxide to generate a cavity containing 16 silanol groups. The walls of the nanocages obtained were asymmetric as the interior of the cages with the silanol functionalities were hydrophilic and the exterior surfaces with the carbosilane linkages were hydrophobic.

(590c) Block Copolymer-Mediated Anisotropic Self-Assembly of Semiconducting Metal Oxide Nanoparticles

Junzheng Wang, Ayae Sugawara-Narutaki, Atsushi Shimojima and Tatsuya Okubo, Department of Chemical System Engineering, The University of Tokyo, Tokyo, Japan

We present a facile method to prepare anisotropic nanostructure of titanium dioxide (TiO_2) and tin dioxide (SnO_2) through anisotropic self-assembly of the corresponding nanoparticles. Monodisperse TiO_2 and SnO_2 nanoparticles with size less than 10 nm assemble into a chain-like nanostructure in aqueous system with Pluronic block copolymer F127 at proper pH condition controlled by a dialysis or acid addition method. The effects of dispersion pH, F127 concentration, and heating time on the anisotropic assembly in aqueous system have been investigated systematically. The obtained nanostructure through anisotropic assembly of nanoparticles is characterized by scanning electron microscopy (SEM), transmission electron microscopy (TEM), and X-ray diffraction (XRD). The results show that highly anisotropic nanostructure is achieved. Importantly, well-defined crystallographic orientations between neighboring nanoparticles are observed in TiO_2 nanochains. Such chain-like nanostructures may reduce the grain boundaries and hence enhance electron transport, which would improve the energy storage and conversion efficiency of the solar cells that are based on TiO_2 or SnO_2.

Acknowledgement

A part of this work was supported by the New Energy and Industrial Technology Development Organization (NEDO).

(590d) Electrostatic 'catalysis' of 'nano-Molecules'

David A. Walker, Chemical and Biological Engineering, Northwestern University, Evanston, IL and Bartosz A. Grzybowski, Chemical Engineering, Northwestern University, Evanston, IL

While there has been significant progress in controlling the size, shape and properties of nanomaterials, their controlled assembly into nanostructures remains a significant challenge. As many researchers begin to look at the assortment of nanoparticles (NPs) around them as a new type of periodic table, the new standard of nanoscience is the manipulation of these NPs into structures with the same finesse that an organic chemist possess to manipulate atoms into molecules. Towards this end, we have used charged rod-like NPs, which act as a template, to effectively 'catalyze' the assembly of oppositely charged reactive NPs into 'nanomolecules'. At the crux of this system is the delicate interplay of electrostatic interactions which one can use to control the rates of reaction during the assembly process and selectively assemble different types of NPs. In this process, the rod-like particles are able to help mediate the strong electrostatic repulsions and enable the 'nano-reaction' to proceed. A mixture of experimental as well as theoretical results will be presented.

(590e) Integration of Large Nanoparticles within Vesicle Bilayers Via Adaptive Surface Chemistry

Hee Young Lee, Penn State University, University Park, PA, Sean Lewis, Chemical Engineering, Penn State, University Park, PA and Kyle J. M. Bishop, Chemical Engineering, Penn State University, University Park, PA

The incorporation of inorganic nanoparticles into lipid bilayers has important implications for medical imaging and nanoparticle actuated vesicles for controlled drug release. Depending on the material, the nanoparticle core imparts additional functionalities such as enhanced fluorescence, plasmonic excitation, or magnetic actuation. The magnitude of these effects scale strongly with particle diameter *D*. Consequently, it is often desirable to use larger particles with dimensions exceeding the thickness of lipid bilayers (typically, ~4 nm). Existing methods for incorporating NPs into bilayer structures are limited to small particles (less than ~6 nm) functionalized with hydrophobic surfaces that "fit" within the hydrophobic core of the membrane. Larger particles do not fuse with lipid bilayers but rather form lipid covered NP micelles as described by thermodynamic models. Regardless of their size, hydrophobic particles are also difficult to incorporate into *preformed* aqueous vesicles (or living cells) without the use of stabilizing detergents that must subsequently be removed. By contrast, NPs functionalized with mixed monolayers containing both hydrophobic and hydrophilic (cationic) ligands have been shown to penetrate the cell membrane and still dissolve easily in water. Such particles have the interesting ability to adapt their surface chemistry in response to environmental cues as further evidenced by the formation of amphiphilic, Janus-type NPs at liquid interfaces. Recent theoretical results suggest that environmentally-responsive particles containing binary mixtures of mobile ligands – one hydrophobic, one hydrophilic – can penetrate and fuse with lipid bilayers even when the particle dimensions (~10 nm) greatly exceed that of the membrane. Here, we confirm this prediction experimentally and demonstrate that large (up to 10 nm) gold nanoparticles functionalized with amphiphilic mixed monolayers incorporate spontaneously into the walls of surfactant vesicles.

(590f) A Modular, Biomimetic Strategy for the Synthesis of Nanoscale Inorganic Materials

Alia P. Schoen, Materials Science & Engineering, Stanford University, Stanford, CA and Sarah C. Heilshorn, Materials Science and Engineering, Stanford University, Stanford, CA

In natural systems, a single protein can be used to perform multiple functions by interacting with other proteins at well-defined epitope binding sites to form a variety of functional co-assembled structures. Inspired by the biological use of epitope recognition to enable diverse yet specific functions, we developed Template Engineering Through Epitope Recognition (TEThER), a strategy that takes advantage of noncovalent, molecular recognition between different components to achieve functional versatility from a single protein template. Engineered TEThER peptides are rationally designed to serve as molecular bridges between protein epitope binding sites and inorganic materials in a localized, specific, and versatile manner. These bifunctional peptides include protein recognition sequences that noncovalently bind to specific sites on the protein scaffold and inorganic recognition sequences that serve as sites for localized bio-enabled nucleation and growth of inorganic materials.

Here we functionalized self-assembled clathrin protein cages at specific sites through co-assembly with designed TEThER peptides to achieve nanostructures with three tailored chemical functionalities: the bio-enabled synthesis of gold, cobalt oxide, and anatase titanium dioxide nanoparticles in aqueous solvents at room temperature and ambient pressure. Compared with previous demonstrations of bio-enabled inorganic synthesis, the TEThER strategy relies solely on defined, noncovalent interactions without requiring any genetic or chemical modifications to the protein template. Therefore, this design strategy represents a mix-and-match, biomimetic approach to achieve versatile and site-specific functionalization that can be broadly applied to other protein templates to generate structures with a range of functionalities.

(590g) Stable Sequestration of Single-Walled Carbon Nanotubes in Self-Assembled Aqueous Nanopores

Meagan Mauter, Harvard Kennedy School, Cambridge, MA, Menachem Elimelech, Yale University, New Haven, CT and Chinedum Osuji, Chemical Engineering, Yale University, New Haven, CT

Many properties of anisotropic nanomaterials are contingent upon the nanomaterial's orientation, isolation, and stabilization in surrounding media. In this work we demonstrate the ability to stably sequester and align individual single-walled carbon nanotubes (SWNTs) within self-contained nanometer-scale aqueous volumes arrayed in an organic continuum. Large areal densities of 4 × 109 cm-2 are readily achieved. SWNTs are incorporated into a surfactant mesophase which forms 2.3 nm diameter water channels by lyotropic self-assembly. Near-infrared fluorescence spectroscopy demonstrates that the SWNTs exist as well dispersed tubes that are stable over several months and through multiple cycles of heating and cooling. Absence of physical distortion of the mesophase suggests that the SWNTs are stabilized by adsorbed surfactants that do not extend considerably from the surface. Our findings have important implications for templated assembly of carbon nanotubes using soft mesophases and the development of functional nanocomposites, such as the development of nanotube-based systems often envisioned for size- and chemo-selective transport.

J. Am. Chem. Soc., 2012, 134 (9), pp 3950–3953, **DOI:** 10.1021/ja209847u

(617a) Colloidal Strontium Doped Calcium Phosphate Nanocomposites for Gene Therapy

Razieh Khalifehzadeh, Chemical Engineering , University of Washington, Seattle, WA and Hong Shen, Department of Chemical Engineering, University of Washington, Seattle, WA

Colloidal strontium doped calcium phosphate nanocomposites for gene therapy

Razieh Khalifehzadeh, Hong Shen*

Department of Chemical Engineering, University of Washington, Box 351750, Seattle, WA 98195

Recent advancements in nano-biotechnology are gaining increasing attention toward the intracellular delivery of various therapeutic and diagnostic agents. Some biomolecules, such as nucleic acids (e.g., DNA or siRNA), are both charged and large molecules. It is difficult for them to cross the cell membrane and elicit their desired effects. Therefore, an appropriate delivery system is necessitated to assist their cellular uptake and trafficking. Calcium phosphate (CaP) minerals, similar to inorganic minerals of natural bone, are highly biocompatible and biodegradable. Their inherent properties such as pH-dependent solubility provide advantages in delivery and controlled release of bioactive agents at various cellular and intracellular environments. Strontium stimulates pre-osteoblast differentiation and formation of new bone. In this work, we developed a one-step method to prepare strontium-doped calcium phosphate (Sr-CaP) nanoparticles, using a mineralizing solution prepared based on the modified simulated body fluid (SBF). In order to absorb and preserve the DNA molecules, the mineralization was conducted at 37 °C and physiological pH. The size and morphological analysis by scanning electron microscopy have shown that the nanoparticles are spherical and mono-dispersed. Adjusting the strontium content enabled us to tune the physical and chemical properties of DNA-incorporated nanoparticles and mediate the gene transfer in human fetal osteoblastic cell line (hFOB1.19).

(617b) Engineering Particle Shape to Enhance Cellular Targeting

Sutapa Barua, Chemical Engineering, University of California, Santa Barbara , Santa Barbara, CA, Jinwook Yoo, College of Pharmacy , Chosun University, Gwangju, South Korea; Chemical Engineering, University of California, Santa Barbara, Santa Barbara, CA, Poornima Kolhar, Biomolecular Science and Engineering, University of California, Santa Barbara, Santa Barbara, CA and Samir Mitragotri, Chemical Engineering and Biomedical Engineering, University of California, Santa Barbara, Santa Barbara, CA

Nanoparticulate drug carriers have enormous potential to deliver macromolecular therapeutic agents such as proteins, nucleic acids and antibodies to disease sites. However, the therapeutic efficacy of particulate-based systems largely depends on their adhesion to the cell surface and intracellular uptake. Numerous attempts have been made to improve the cell membrane attachment ability and internalization of particulate-based drugs by optimizing nanoparticle size and surface decoration using ligands such as monoclonal antibodies. However, the effect of nanaoparticle shape on cell-membrane binding, cellular internalization and subsequent therapeutic activation using particulate-based drugs has not received much attention. Here we demonstrate that the specificity and cellular internalization of particulate-based drugs can be dramatically improved by engineering nanoparticle shape. We studied cell-membrane attachment and uptake of monoclonal antibody-modified spherical and rod-shaped nanoparticles in various cell lines. Antibody-coated rod-shaped nanoparticles exhibited high cell-membrane adhesion and intracellular uptake *in vitro* compared to that using spherical and disk-shaped nanoparticles. Mechanisms of shape-induced enhancement will be discussed. These results highlight the importance of the interplay between nanoparticle shape and antibody interaction with target cells for imaging, diagnostic and therapeutic drug delivery.

(617c) Activatable Virus Nanoparticles for Cancer Targeting

Justin Judd, Bioengineering, Rice University, Houston, TX, Jonathan J. Silberg, Biochemistry & Cell Biology, Rice University, Houston, TX and **Junghae Suh**, Department of Bioengineering, Rice University, Houston, TX

Adeno-associated virus (AAV) nanoparticles show great promise as gene delivery vectors due to its high safety profile, ability to transduce dividing and non-dividing cells, and its low immunogenicity. Successful implementation of cancer gene therapy will require a high degree of targeting to minimize destruction of healthy tissue. A significant body of work has demonstrated that AAV can be retargeted via cell surface receptors. However, it may be difficult to distinguish cancer cells from normal tissue by the presence of a single biomarker. We present an adeno-associated virus nanoparticle designed to integrate multiple signals in the tumor microenvironment, which influences the virus's decision to activate cell-binding and ultimately mediate transduction. The mutant displays drastic changes in heparin affinity and transduction behavior after *in vitro* treatment by MMP-7. Heparin affinity chromatography indicates the vector activates to bind its primary cell surface receptor, heparan sulfate proteoglycan (HSPG), after treatment with cancer marker MMP-7. Western blot analysis confirms the virus is specifically cleaved by this protease. Strikingly, transduction assays reveal a consistent increase in gene delivery after treatment with MMP-7 in 5 different cell lines: 293T, Cos-7, BT-549, MDA-MB-231, and HeLa. The greatest increase was in 293T cells at over a log fold increase in GFP positive cells. No difference was observed for *wt* capsids in any assay. This work is the first demonstration of an AAV nanoparticle engineered to respond to enzymatic markers of disease. Combined with the redesign of AAV2's cell binding domain, this strategy may provide the critical advancement necessary to achieve high specificity for cancer therapy.

(617d) pH Induced Drug Release From Carbon Nanohorn Based Bionanoparticles

Huang Wei[1], Jianfei Zhang[2], Harry Dorn[2] and Chenming Zhang[1], (1)Biological Systems Engineering, Virginia Tech, Blacksburg, VA, (2)Chemistry, Virginia Tech, Blacksburg, VA

Nanoparticle mediated drug delivery plays an increasingly important role in the treatment of cancers and inflammatory diseases. However, how to minimize basal drug release and thus the accompanying side effect is still a challenging problem. In this study, we have synthesized carbon nanohorn (CNH) supported liposome particles. A model drug, paclitaxel, was loaded into CNH, which displays a much higher loading capacity that that of liposomes. Lipids with different pH sensitivity were assembled into the lipid bilayer. Our initial study showed that under serum free condition, the drug release profiles of the assembled particles were highly pH dependent. At neutral pH, there was minimal release. When the pH was reduced, the release increased significantly. However, the presence of serum clearly destabilized the lipid bilayer as the bionanoparticles and liposomes showed no distinctive drug release behavior. Efforts were taken to optimize the lipid formulation to reduce the basal drug release under physiological conditions.

(617e) Grafted Polymeric Nanoscale Hydrogels for the Oral Delivery of Chemotherapeutics

Amey Puranik, Chemical Engineering, The University of Texas at Austin, Austin, TX and Nicholas Peppas, Dept. of Chemical Engineering, The University of Texas at Austin, Austin, TX

Grafted Polymeric Nanoscale Hydrogels for the Oral Delivery of Chemotherapeutics

Amey S. Puranik[1] and Nicholas A. Peppas[1,2,3]

Department of Chemical Engineering[1], Department of Biomedical Engineering[2] and Division of Pharmaceutics[3], The University of Texas at Austin, Austin, TX , 78712

Introduction: Intravenous administration of hydrophobic chemotherapeutics can result in potentially toxic systemic concentrations of therapeutics, causing several deleterious side effects and deterioration of patient quality of life. To remedy these shortcomings, we seek to investigate carriers that can enhance the therapeutic effect of hydrophobic chemotherapeutics, while minimizing side effects and be delivered orally. However, oral delivery of hydrophobic therapeutic agents involves a number of challenges that must be circumvented such as loading hydrophobic chemotherapeutics in conventional hydrophilic polymer carriers, protecting the therapeutic agent from the low pH and degradative enzymes of the stomach, and finally, absorption and transport across the lumen wall to desired cancerous location. In this work, we report the development of grafted polymeric nanoparticles that exhibit increased loading and release of the hydrophobic chemotherapeutic doxorubicin. These nanoscale hydrogels are composed of a pH-responsive hydrophilic poly(methacrylic acid – grafted – ethylene glycol) (P(MAA-g-EG)) hydrogel copolymerized with the hydrophobic monomer tert-butyl methacrylate that can associate with hydrophobic therapeutics.

Materials and Methods: Synthesis of the nanoparticles was achieved using an emulsion polymerization technique by combining methacrylic acid, tert-butyl methacrylate, poly(ethylene glycol methyl ether methacrylate) (PEGMMA; MW ~2080g/mol), tetraethylene glycol dimethacrylate (TEGDMA), 4-(2-hydroxyethoxy)phenyl-(2-hydroxy-2-propyl)ketone (Irgacure 2959) and water into a small round bottom flask, followed by probe sonication, nitrogen purging and UV-initiated free radical polymerization for 2.5 h. The resulting nanoparticles were dialyzed and lyophilized. Doxorubicin, an intravenously administered drug for a wide range of cancers, was chosen to be a model hydrophobic chemotherapeutic to evaluate the loading and release capabilities of these nanoparticles.

Results: A set of polymers with varying cross-linking density and core hydrophobicity was synthesized. Nanoparticles demonstrated reversible, pH-responsive swelling with a swollen diameter in the range of 105-125 nm at pH 7.4. In addition, SEM imaging (Figure 1A &1C) was also used to determine the dry diameters of the particles which were in the range of 50-70nm. As the crosslinking density of the nanoparticles was increased, their ability to swell decreased (Figure 1B).

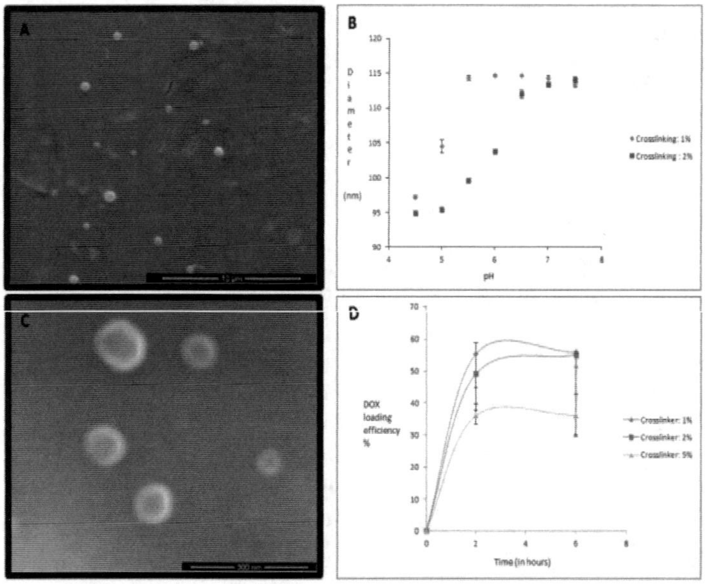

Figure 1: P(MAA-co-tBMA-g-PEG) nanoparticles: A) & C) SEM images of particles, B) Increase in diameter of nanoparticles with change in pH as determined by DLS, D) doxorubicin loading efficiency of particles with different crosslinking densities

Particles with the lowest degree of cross-linking demonstrated the highest loading of doxorubicin at an overall efficiency of 55% in 2 hours at pH 7.4 (Figure 1D). Release studies of the loaded particles were carried at pH values maintained constant (at 2 as well as 7.4), in addition to varying the pH values from 2 to 7, to better emulate the passage of the nanoparticles through the gastrointestinal tract. The physically encapsulated doxorubicin was completely released at pH 7.4 while still being retained at pH values representative of conditions in the stomach. The physically encapsulated doxorubicin was completely released at pH 7.4 while still being retained at pH values representative of conditions in the stomach. Particles with lower cross-linking displayed the fastest release kinetics with doxorubicin.

Conclusions: Our findings suggest that the P(MAA-co-tBMA-g-PEG) nanoscale hydrogels are suitable candidates as carriers that prevent release of chemotherapeutics at the low pH of the stomach during transit in the gastrointestinal tract, while still being able to preferentially load hydrophobic chemotherapeutics. Ongoing and future studies include determining the cytotoxicity of these nanoscale hydrogels against Caco-2 cell lines in addition to evaluating their mucoadhesion capacity.

Acknowledgement: This work was supported by the NIH/NCI Center for Oncophysics (CTO PSOC U54-CA-143837).

(617f) Sensitizing Cancer Cells to TRAIL Mediated Cell Death by Mitoxantrone Loaded Theranostic Micelles

Taraka Sai Pavan Grandhi[1], Thrimoorthy Potta[2], David J. Taylor[2], Yanqing Tian[3] and Kaushal Rege[4], (1)Biomedical Engineering, Arizona State University, Tempe, AZ, (2)Chemical Engineering, Arizona State University, Tempe, AZ, (3)Arizona State University, Tempe, AZ, (4)Chemical Engineering, Biomedical Engineering, Arizona State University, Tempe, AZ

Abstract

TRAIL (Tumor necrosis factor-related Apoptosis Inducing Ligand) has been shown to induce death selectively in various cancer cells. However, several cancer cells are either resistant or can develop resistance to TRAIL-induced cell death. As a result, strategies that overcome this resistance are currently under investigation. We have recently discovered several FDA-approved drugs with TRAIL sensitization activity against prostate, breast and pancreatic cancer cells. Here we demonstrate successful encapsulation of a previously unknown TRAIL sensitizing chemotherapeutic drug, mitoxantrone, in PEG-DSPE micelles in order to facilitate delivery of the drug to cancer cells. Mitoxantrone was successfully encapsulated in three different PEG-DSPE micelles with methoxy, amine, and carboxyl termini, respectively. The different PEG-DSPE micelles were characterized for drug loading, size, zeta potential and cytotoxicity. The therapeutic efficacy of encapsulated mitoxantrone was investigated as a single agent treatment, as well as in combination with TRAIL in several cancer cell lines. The micelles were also conjugated with an MRI contrast agent in order to facilitate tumor imaging. Drug-loaded micelles were found to maintain mitoxantrone's sensitization activity in both single agent and sequential drug-protein treatments in inducing synergistic apoptotic activity. Our results suggest that micelle-encapsulated mitoxantrone can be employed simultaneously for synergistic treatments and imaging applications in cancer diseases.

Keywords: TRAIL, PEG-DSPE, micelles, mitoxantrone, FDA, cancer.

(617g) Guanosine Prodrug Incorporated Polymeric Nanocarriers for Suicide Gene Therapy

Alicia Jane Sawdon, Chemical Engineering, Michigan Technological University, Houghton, MI and Ching-An Peng, Department of Chemical Engineering, Michigan Technological University, Houghton, MI

The most prominent suicide gene therapy is the use of herpes simplex virus thymidine kinase (HSV*tk*) in conjunction with a variety of guanosine-based prodrugs (ganciclovir and acyclovir). Currently, introduction of the gene for a foreign enzyme and the prodrugs are administered in a two-step process. First, the foreign gene is delivered into the targeted cancer cells, expressed and released into the cytoplasm. Then the prodrug is administered and activated to its cytotoxic form by the gene-expressed exogenous enzyme. Many gene delivery systems have been proposed for suicide gene therapy most focusing on the use of viral vectors to deliver the prodrug activation gene. In this study, a one-step approach was harnessed to deliver gene and prodrug through polymeric nanocarriers.

Guanosine-based GCV and ACV were used as the initiators in ring-opening polymerization of ε-caprolactone to form hydrophobic GCV-PCL and ACV-PCL, which were then grafted to hydrophilic chitosan to form amphiphilic copolymers for the preparation of stable micellar nanoparticles. The synthesized amphiphilic copolymers were validated by ^1H NMR and FTIR. Self-assembly behavior of micellar nanoparticles was determined by TEM, particle size and charge and critical micelle concentration (CMC). Polymeric nanocarriers complexed with HSV*tk* plasmids were cultured with colorectal HT-29 cancer cells and toxicity measured. ^1H NMR analysis shows successful synthesis of guanosine-based prodrugs to PCL. FTIR spectra reveal characteristic absorption peaks associated with chitosan and PCL are simultaneously present in amphiphilic copolymers - GCV-PCL-Chitosan and ACV-PCL-Chitosan. CMC of ACV-PCL-Chitosan copolymer was measured to be 0.0056 mg/ml, while size measured by dynamic light scattering was detected to be 150-220 nm, with a zeta potential of +20.14 mV. Toxicity results demonstrate that GCV-PCL-Chitosan/HSV*tk* and ACV-PCL-Chitosan/HSV*tk* nanocarriers are a feasible approach to kill HT-29 cells. The anti-cancer efficacy was further enhanced by encapsulated SN38 within the hydrophobic core of guanosine prodrug incorporated polymeric micelles.

(651a) Toward Design of Semiconductor Ternary Quantum Dots with Optimal Optoelectronic Function

Sumeet C. Pandey, Department of Chemical Engineering, University of Massachusetts, Amherst, MA, Xu Han, Chemical Engineering, University of Massachusetts Amherst, Amherst, MA and **Dimitrios Maroudas**, Chemical Engineering, University of Massachusetts Amherst, Amhest, MA

Heterogeneous semiconductor quantum dots (QDs), such as core/shell QDs, are semiconductor nanocrystals with non-uniform compositional distribution that have attracted significant technological interest in optoelectronic and photovoltaic device fabrication. Compositionally, it is very common for such QDs to be ternary semiconductor nanocrystals; random alloys of such ternary QDs exhibit size-dependent excitonic properties and allow for exceptional tunability of their band gaps by modulation of the nanocrystal composition. Band structure engineering based on such controllable parameters enables the development of nanomaterials with optimal function for photovoltaic and light-emitting devices. Synthesis of such semiconductor nanocrystals has followed various colloidal chemistry routes.

In this presentation, we investigate the possibility of rigorous single-step synthesis routes to minimally strained core/shell-like QDs with optimal optoelectronic function by studying systematically the distribution of atomic species in ternary QD systems, emphasizing on surface segregation as a potential means for self-assembly of core/shell-like semiconductor QDs. Toward this end, we have combined atomic-scale computations of full relaxation of semiconductor nanocrystals, first-principles electronic band structure calculations, and continuum mass transport theory and numerical modeling. The computational results determine the thermodynamic stability of core/shell QDs synthesized by two-step synthesis routes (growth of a shell on a core QD), elucidate the effects of compositional distribution on the electronic band structure of ternary QDs, and provide interpretations to X-ray photoelectron spectra (XPS) and photoluminescence (PL) emission spectra of ternary QDs.

We have computed the equilibrium concentration profiles in QDs of $ZnSe_{1-x}Te_x$, $In_xGa_{1-x}As$, and $ZnSe_{1-x}S_x$ according to a computational analysis of full nanocrystal relaxation and report results over a broad range of x and QD diameter (from ~1 to ~6 nm). The computational procedure consists of coupled compositional, structural, and strain relaxation of the semiconductor nanocrystals. The analysis is based on Monte Carlo (MC) simulations according to extended classical valence-force-field descriptions of interatomic interactions, which have been parameterized according to first-principles DFT calculations of segregation energies using slab supercells. The equilibrium profiles reflect the different tendencies of the QD constituents to segregate on the QD surface: $ZnSe_{1-x}Te_x$ and $In_xGa_{1-x}As$ form core/shell-like structures with Te- and In-deficient cores, respectively, and Te- and In-rich shells, respectively, characterized by compositionally graded core/shell interfacial regions, while the equilibrium compositional distribution in $ZnSe_{1-x}S_x$ QDs resembles a random alloy with only a weak tendency for Se surface segregation. We have determined the electronic structure of $ZnSe_{1-x}Te_x$ and $ZnSe_{1-x}S_x$QDs for various concentration distributions (core/shell, reverse core/shell, and random alloy) over the entire compositional range ($0 \leq x \leq 1$). We have predicted the dependence of the band gap as a function of x in all cases and demonstrated that, for the materials examined, the thermodynamically stable concentration distributions lead to QDs with tunable band gaps.

In the dilute limit of low x, the computed equilibrium concentration profiles are explained fully by a phenomenological species transport theory that we have developed, which accounts for Fickian diffusion of species in ternary QDs and drift due to the thermodynamic driving force for species surface segregation. The predictions of the continuum theory provide a powerful tool for the design of novel ternary QDs for applications in optoelectronics and photovoltaics. We have derived analytical and numerical solutions to the boundary-value problems for species transport in ternary QDs at equilibrium, steady state, and during transient periods. We have used the predictions for the time dependence of the near-surface-region concentration to fit XPS data for ternary $ZnSe_{1-x}S_x$ QD near-surface compositional changes during annealing at temperatures ~100 °C from ZnSe/ZnS core/shell initial configurations. The theoretical predictions provide excellent fits to the XPS data and can be used to evaluate diffusion coefficients of species interdiffusion in the QDs and the strength of the thermodynamic driving force for species surface segregation, which are the two fitting parameters used in this analysis.

(651b) Surface Mineralization and Characterization of Palladium Nanoparticles On Genetically Engineered Tobacco Mosaic Virus (TMV) Templates

Alexander Freer, Lucas Guarnaccio, Kristin Wafford, Johanna Smith, Jayne Steilberg, James Culver and Michael Harris, Chemical Engineering, Purdue University, Lafayette, IN

Certain innate properties of the tobacco mosaic virus (TMV) make it ideal as a biotemplate for nanoparticle formation. Genetic modifications of the TMV coat proteins allow for an increase in the selective deposition and controlled growth of different metals onto the surface of the virus. In the current process, TMV is coated with uniform layers of palladium metal particles in aqueous solution under very mild conditions. Palladium nanowires of 300nm in length and 30-40nm in diameter have been created with this process. Tunneling electron microscopy (TEM) and small angle x-ray scattering (SAXS) are used to charaterize the thickness and uniformity of the metal surface. Procedures to control the higher order assembly of the TMV help to limit the uncontrolled aggregation of individual nanorods and arrange the TMV into nanostructures to be used in multiple applications ranging from catalysts to sensors to battery electrodes and other devices in nanoelectronics. This presentation will focus on the aqueous coating procedure and the characterization of the resulting nanowires with TEM and SAXS.

(651c) Infrared Active Photocatalysis Through Morphological Control of Semiconductor Nanoparticles/Metal Tip Heterostructures

Doh C. Lee and Chaewon Pak, Department of Chemical and Biomolecular Engineering, Korea Advanced Institute of Science and Technology (KAIST), Daejeon, South Korea

Semiconductor nanocrystals in strong quantum confinement regime exhibit interesting size-dependent bandgap. The bandgap tunability offers a new opportunity for bandgap engineering; for example, exciton dynamics can be tailored by designing dimensions of heterostructure semiconductor nanocrystals. We synthesized PbSe/CdSe/CdS core/shell/shell nanocrystals with varying morphology, e.g., pyramid and tetrapod. These nanocrystals show varying electron-hole exchange interactions and radiative recombination lifetimes depending on their dimensions. For instance, compared to PbSe nanocrystals, PbSe/CdSe/CdS heterostructure nanocrystals show nearly 2 orders of magnitude longer radiative recombination lifetime. Using this semiconductor light absorber may help increase quantum efficiency in case an exciton has to be dissociated into an electron and a hole. By growing metal tips at the corners of the structures, we prepared semiconductor/metal nanojunction and used the nanomaterials as photocatalysts for reduction of various dyes. Interestingly, the morphology change leads to the difference in the tip growth pattern, and we demonstrate infrared active photocatalysis using the colloidal heteronanostructures.

(651d) Tunable Localized Surface Plasmon Resonances in Tungsten Oxide Nanocrystals

Karthish Manthiram[1,2] and A. Paul Alivisatos[2,3], (1)Department of Chemical and Biomolecular Engineering, University of California, Berkeley, CA, (2)Materials Sciences Division, Lawrence Berkeley National Laboratory, Berkeley, CA, (3)Department of Chemistry, University of California, Berkeley, CA

Free charge carriers, which are dielectrically-confined to a nanoparticle, participate in resonant, collective oscillations known as localized surface plasmon resonances (LSPRs). Although the vast majority of investigations of LSPRs have been conducted on noble metals, LSPRs have also been observed in heavily-doped semiconductors, in which the LSPR energy may be tuned by adjusting the stoichiometry. Transition metal oxides, such as tungsten oxide, are interesting candidates for LSPR hosts as they exhibit fascinating properties arising from the unique character of their outer-d valence electrons. We demonstrate that nanoscale oxygen-deficient tungsten oxide supports a strong LSPR mode, based on theoretical prediction of the LSPR energy from Mie-Gans theory and experimental measurement of the LSPR energy as a function of the refractive index of the medium. We also demonstrate that the LSPR energy may be tuned by heating the tungsten oxide nanoparticles in air. The identification of a tunable LSPR mode in tungsten oxide opens up the possibility of rationally designing plasmonic tungsten oxide nanoparticles for photocatalysis, bioimaging, and gas sensing.

Laser- Drawn Features on Nanoparticle Films

Sanjeev Kumar Kandpal, Aimee Co, Kody Allcroft, Michael D. Mason,

Douglas W. Bousfield, David J. Neivandt

Department of Chemical and Biological Engineering

The University of Maine, Orono, Maine -04473

Introduction

The present work explores the silver (Ag) nanoparticles film-Laser interaction based applications in the field of printed electronics, sensors, anti-counterfeiting, optical grating. Ag has the highest conductivity among metals even in its oxide form, making it the most sought after material for providing conductivity in electronics. Ag in nanoparticle size range show interesting optical, physical and electronic properties like melting point depression, large absorption and scattering cross section when compared to bulk silver, have a color that depends upon the size and the shape of the particle, high surface to volume ratio, tunable surface Plasmon resonance (SPR) peak wavelength, metal enhanced fluorescence (MEF) (Geddes *et al.*, 2003), surface enhanced Raman scattering (SERM) (Li & Peng, 2010), antibacterial application, diagnostic application, enhanced thermal and conductivity application. In the present work, we fabricated three-dimensional features on the Ag nanoparticles film using a laser and the resulting structures showed interesting fluorescent properties. We are also looking into fabricating structure on polymer-Ag nanoparticles films and characterize them for the conductivity, optical, fluorescence properties. These structures might serve as conductive paths on polymer-nanoparticle film, paving the way for an alternative method of fabricating printed circuits.

We are looking into the mechanisms behind the structure formation and our initial investigation into the process suggests that it is a combination of multiple processes like liquid melt ejection and sintering. The structure formation starts when the film is exposed to a laser beam (using the objective to focus the laser beam) and the Ag particles in the film absorb the light which is converted to heat due to the efficient photo-thermal conversion property of metal nanoparticles. The heat generated in nanoparticles starts dissipating away from the particles into the surrounding matrix. If this heat is sufficiently high, it can melt the surrounding matrix and surface of metal nanoparticles. To give a feel of melting point depression; a 20nm Ag nanoparticle has a melting point of 150°C where as bulk Ag has 960°C (Buffat & Borel, 1976). If we subject the nanoparticle film with a large flux then we can create a small volume where the temperature is much higher than the melting point of the film material, leading to its vaporization in that focal volume. These hot vapors try to expand in all directions but are unable to do so at the film surface, which leads to pressure buildup on the focal volume in the film. The buildup of pressure is released by pushing the liquid melt around the focal volume on the film surface, where it solidify. This buildup around the focal volume gives raise to structure formation if we move the film under the laser; the structure pattern formed will be the pattern in which the film was moved under the beam. If the flux used is not very high and particles are very close to each other then the heat dumped into the film will melt the surface of the particle and these melted surface run into each other and solidify, the process is known as sintering.

Materials and Methods

Synthesis

Silver nanospheres were made using single-pot redox chemical techniques according to published material (Turkevich et al., 1951; Pillai and Kamat, 2004). In this we make use of a soluble metal salt (silver nitrate), a reducing agent (sodium citrate) and a stabilizing agent (excess sodium citrate). Synthesis starts with the nucleation step, followed by nanoparticle growth and the stabilizing agent which caps the particle leaving a negatively charged surface helping to avoid aggregation (Rivas *et al.*, 2001). We start with cleaning all the glassware with concentrated nitric acid, 600mg of sodium citrate dissolved in 160 ml of 18MΩ pure water in a 500 ml round bottom flask. This was then brought to a temperature of 95°C in an oil immersion bath. Simultaneously, 40mg of silver nitrate was dissolved in 40 ml of 18MΩ ultrapure water in a glass beaker. This was then added to the round bottom flask containing sodium citrate solution and heated to the same temperature (95°C) under constant stirring. The temperature was maintained until the reaction was complete, usually less than 1 hr. The heating, stirring is stopped and solution is allowed to cool to room temperature. To impart electrostatic stability to nanoparticles, excess of stabilizing agent was added. The size of the nanoparticles are measured using a Dynamic light scattering technique (Malvern, Zetasizer Nano-ZS), a Transmission electron microscope, and UV-Vis Spectroscopy. Fig. 2a shows a transmission electron micrograph (TEM) image of the freshly prepared silver nanoparticles. From the image data the average diameter was estimated to be approximately 100 nm.

Nanoparticles Film Making

Nanoparticles solution synthesized using the method described above, is very dilute in nature. To increase the number density of particles, the solution was subjected to rotation vaporization using Rotary Evaporator (Rotovap) instrument (BUCHI Rotavapour R-205). The solution was concentrated from 0.013wt% to 0.102wt%; the solution was then characterized using DLS and UV-Vis spectroscopy. After this was done, 100µl of this solution was drop casted on 22x22mm cover slip and allowed to air dry.

Exposing film to laser and characterization them

The drop casted film was then subjected to 532nm continuous wave laser, the film was scanned under the laser in a raster pattern using an x-y scanning stage and the scan stage was controlled using LabVIEW software. Laser exposed films were characterized using an optical microscope, CLSM (Confocal Laser Scanning Microscope), SEM (Scanning Electron Microscope), AFM (Atomic Force Microscope), XRD (X-ray Diffraction), and EDAX (Energy-dispersive X-ray spectroscopy).

Result & Discussion

Characterization of particles

The nanoparticle solution was characterized using DLS, UV-Vis spectroscopy and TEM. The DLS graph shows that the freshly made Ag particles are around 50nm in radius and Rotovap concentrated are around a micron. These numbers are complemented by TEM images and they also tell us about the shape of the particles. In UV-Vis graph the flat red curve shows the aggregation effect of particles and black curve is the typical UV-Vis curve of 100nm Ag particles. These data suggests a tremendous aggregation effect in Ag particles when they are concentrated using the Rotovap.

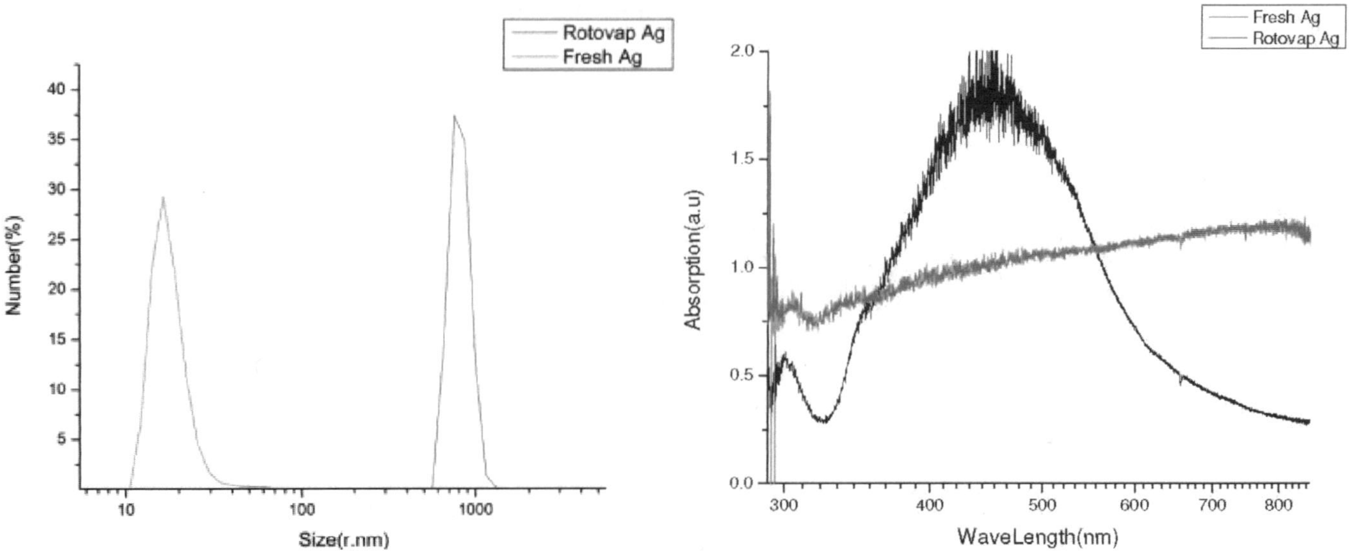

(a) **(b)**

Figure1: Comparison of DLS (a), UV-Vis spectrum (b) of freshly made and Rotovap concentrated silver.

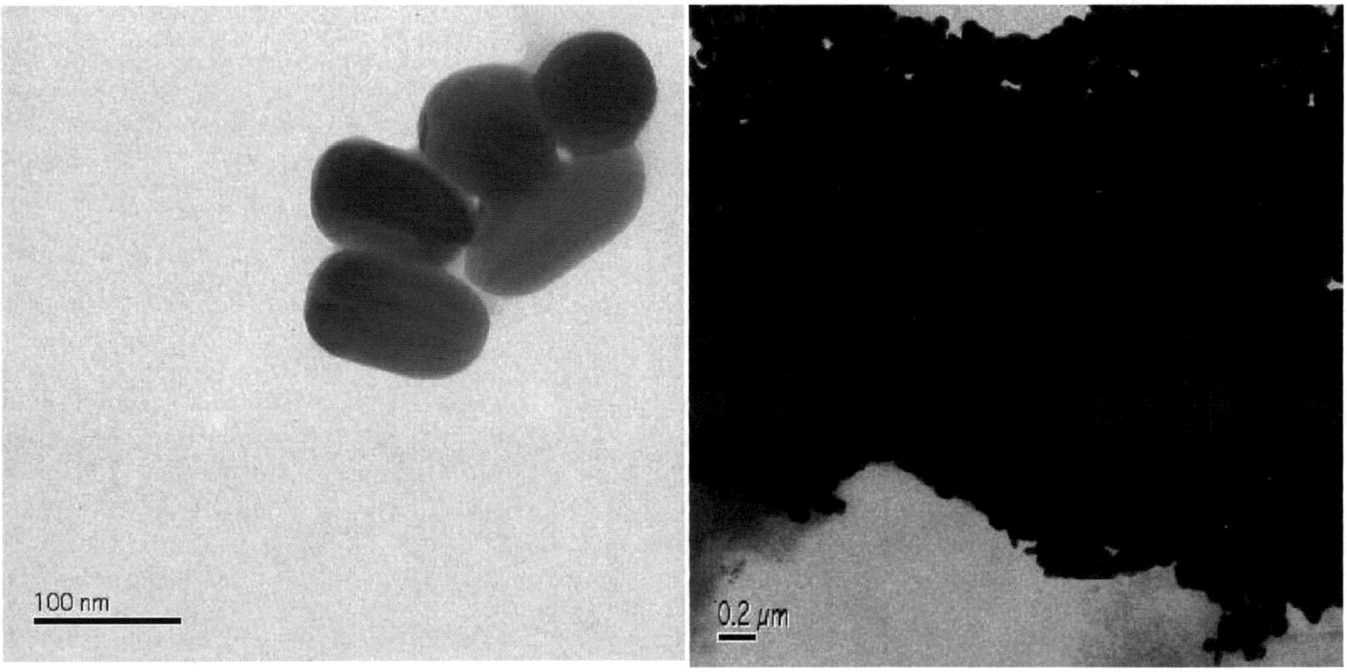

(a) **(b)**

Figure2: TEM image of freshly made silver (a), TEM image of Rotovap concentrated silver (b).

Drop Casting and Exposing

The Figure: 3a shows the digital image of 100μl drop cast of Rotovap concentrated Ag solution. The interesting feature of this drop cast film is formation of an inverse coffee ring. The film was latter exposed to 532nm CW DPSS laser using Figure: 3b optical step-up, which has provision to expose the film to laser and record the video simultaneously.

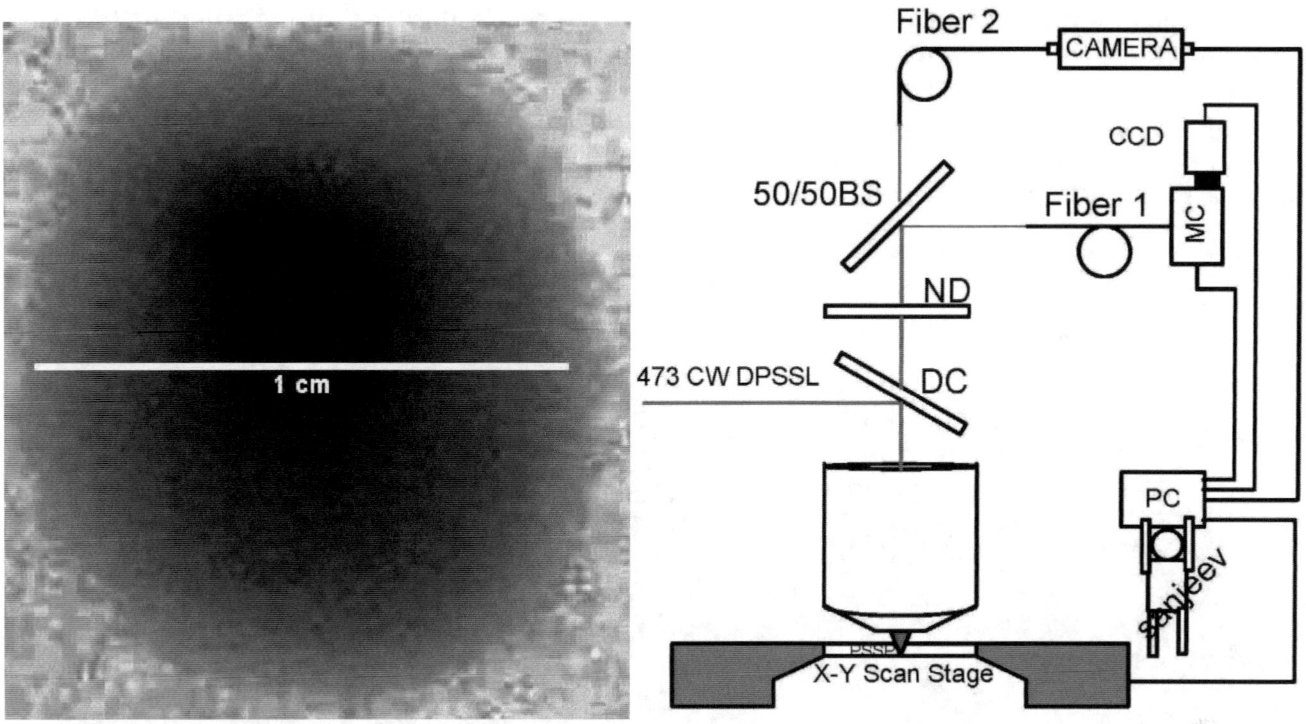

(a) (b)

Figure3: Drop cast image of silver (a) , Optical setup (b) , Transmission mode optical image of laser exposed silver drop cast image (right).

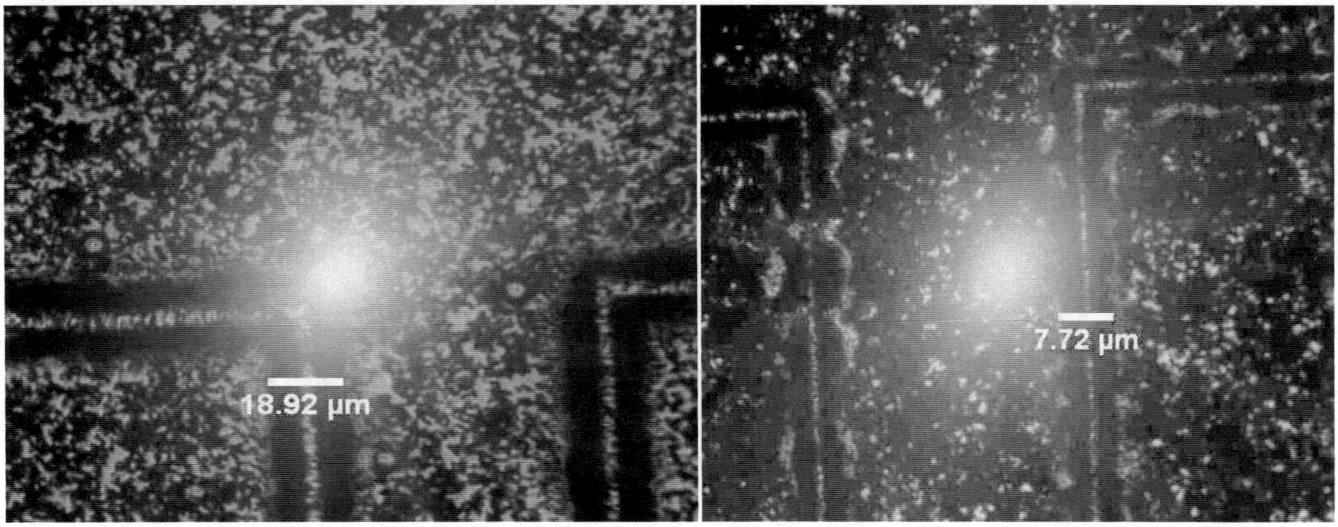

(a) (b)

Figure4: Transmission mode optical image of laser exposed silver drop cast film, using 0.4NA Objective (a), 0.65NA Objective (b).

In Figure: 4 we see a black and green speckle pattern, the black regions are the Ag particles and the green regions are the empty space between the particles. The reasons they look green are because of transmission mode imaging and camera setting. The raster patterns in the images are the regions which were exposed to the laser. It is observed that there is vacant region (bright green region) at the center of the pattern and ridges to the sides (dark regions). We observed a decrease in the width of the Laser Induced Structures (LIS) when we used a higher numerical aperture (NA) objective, which is due to the higher NA objective having the ability to focus the light to become very small volume. The resolution we were able to achieve was 10 times better than most traditional printing techniques.

Drop Cast Film on Paper Substrate

To explore the compatibility of LIS with a paper based substrate, we tried two different papers (copy paper, magazine paper). 100μl of concentrated Ag solution was drop cast onto the copy paper, allowed to air dry, and then subjected to 532nm CW laser.

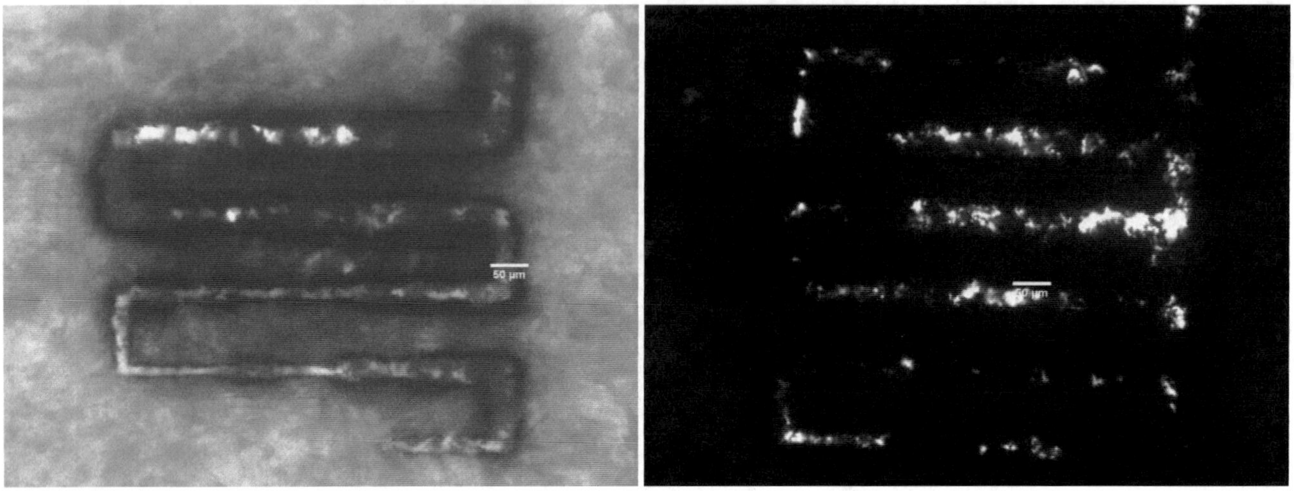

(a) (b)

Figure5: Laser exposed Copy paper, Plain (a), with Ag particles (b)

We observed the damage of the copy paper in the areas where laser was exposed (Figure5a) and the major damage when the paper was coated with Ag particles. In the Figure5, the much brighter regions are the laser damaged regions and they look bright because of light passing through that area effectively (transmission mode imaging). The reason for the enhanced damage of the Ag particle coated paper is the low optical yield of Ag particles. The light absorbed by the particles is effectively converted into heat and this heat is transferred to the surrounding material, which causes the damage to the paper substrate. We then tried coated paper (magazine paper) to overcome the damage and then we compared the coated paper without Ag particle to the coated paper with Ag particle. We observed that there was no damage to coated paper without Ag particles (Figure6a) but there was damage to the coated paper with Ag particles, however the damage was less than the uncoated copy paper. Hence, coated paper is resistant to laser damage.

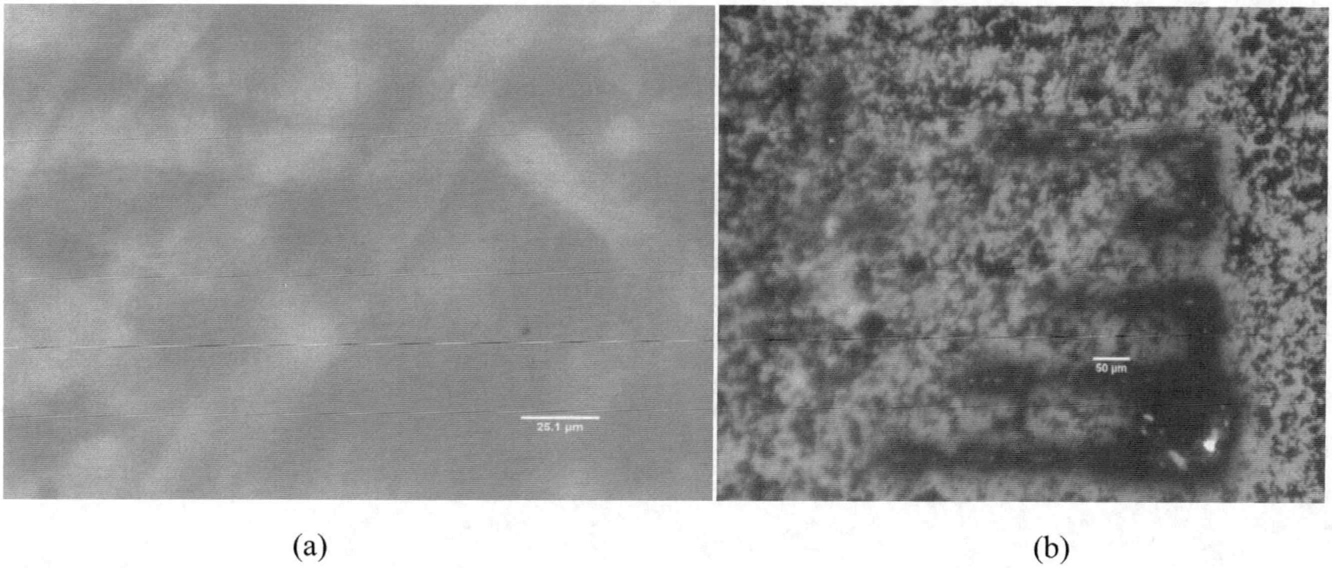

Figure6: Coated Paper without Ag particles (a), Coated Paper with Ag particles (b)

Characterization of LIS

AFM Characterization

To know the exact dimensions of these structures in three different directions, we took an AFM image of the LIS. We made 'PSSP' LIS on concentrated Ag drop cast film on glass as our sample to do characterization.

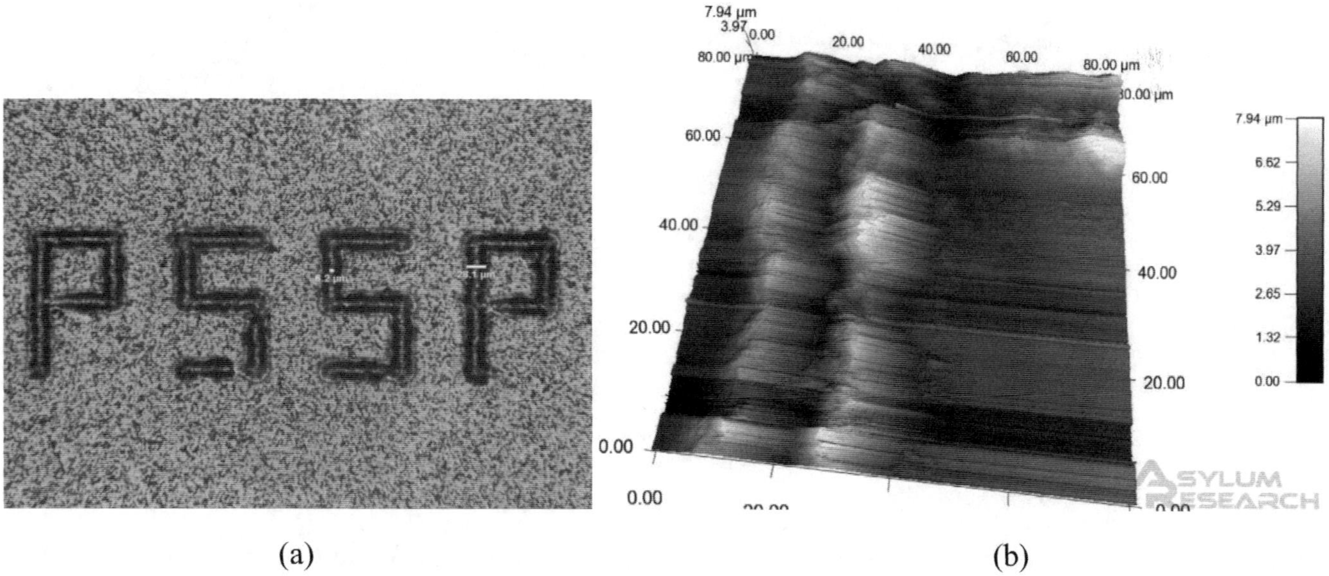

Figure7: 'PSSP' LIS (a), AFM image of tail of letter 'P'

The LIS was characterized using AFM non-contact mode and we found that the structures are 8µm high, 25µm in width and there is a cavity at the center of the structure and ridges to the sides. The AFM data confirms that there is a physical change happening under the laser.

Confocal Characterization

In this class of characterization, ‘PSSP’ LIS was excited with certain wavelengths of laser and the signal from the structure was collected using photomultiplier detector.

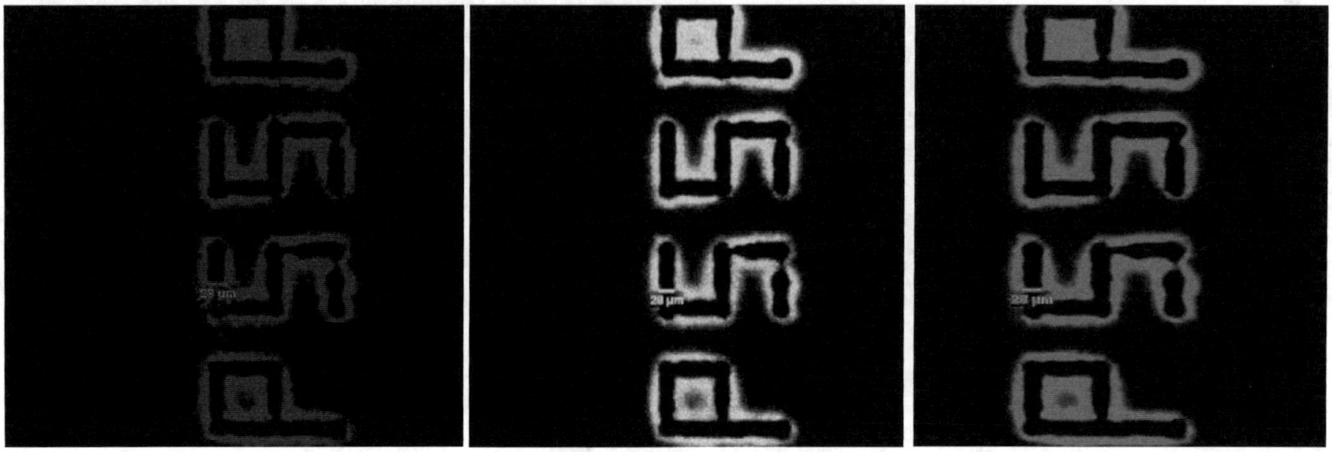

(a) Exi=488nm, Coll: 500-530nm (b) Exi=543nm, Coll: 560-620nm (c) Exi=633nm, Coll: 650-750nm

Figure8: Confocal images of LIS.

We observed a strong fluorescent signal at the edges of the LIS, and the fluorescent was over a broad range of spectrum. To quantify the fluorescent signal, we took the spectrum of the signal using a nitrogen cooled CCD camera attached to monochromator with the optical setup shown in figure3b.

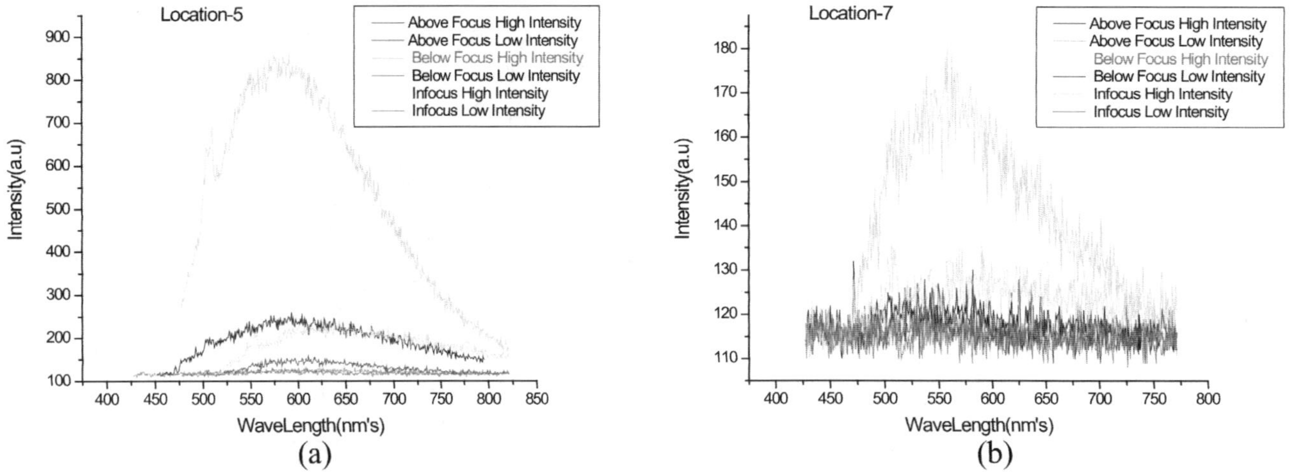

Figure9: Fluorescent spectrum at the edge of the LIS (a), on the surface of the LIS (b)

The various curves in the Figure9 are from spectrums taken at various levels of focus planes. We got maximum signal when we were on the surface of the film at edge of the LIS. We got many magnitudes less signal compared to the signal at the edge and on the Ag film surface. This spectrum also quantifies the broad range spectrum conclusion we made out of the confocal image data.

Conclusion

- Observed inverse coffee ring effect in concentrated Ag drop cast film.

- Successfully able to draw 8μm size feature on silver nanoparticle film.

- Able to go down 10x times smaller in size compared to typical printing techniques.

- Coated paper shows resistant to laser damage.

- Laser drawn structures were fluorescent at the edges.

References

- Buffat, PH, & Borel, J-P. 1976. Size effect on the melting temperature of gold particles. *Physical review A*, 13(6), 2287–2298.
- Geddes, Chris D, Haishi Cao, Ignacy Gryczynski, Zygmunt Gryczynski, Jiyu Fang, and Joseph R Lakowicz. "Applications of Indocyanine Green to in Vivo Imaging". *J. Phys. Chem. A 107* (2003): 3443-3449.
- Li, Wenbing, Yanyan Guo, and Peng Zhang. "SERS-Active Silver Nanoparticles Prepared by a Simple and Green Method." *The Journal of Physical Chemistry C* 114, no. 14 (2010): 6413-6417.
- Gold, Colloidal, John Turkevich, Cooper Stevenson, and James Hillier. "A study of the nucleation and growth processes". *Discussions of the Faraday Society* 11 (1951): 55-75.
- Rivas, L, S Sanchez-Cortes, and JV Garcia-Ramos. "Growth of silver colloidal particles obtained by citrate reduction to increase the Raman enhancement factor." *Langmuir* 17, no. 3 (2001): 574-577.
- Pillai, ZS. "What factors control the size and shape of silver nanoparticles in the citrate ion reduction method?" *The Journal of Physical Chemistry B* 3, no. 10 (2004): 945-951.

(651f) Controllable Synthesis of Cu_{2-X} E (E=S, Se) Nanocrystals with Localized Surface Plasmon Resonance and Investigation of Conductivity of NC Thin Films

Xin Liu, Xianliang Wang and Mark T. Swihart, Chemical and Biological Engineering, University at Buffalo (SUNY), Buffalo, NY

Inorganic nanocrystals (NCs) are of great interest due to their attractive optical, electronic and optoelectronic properties. Metal nanoparticles (NPs) exhibit localized surface plasmon resonance (LSPR) when photons induce oscillation of the free electrons in conduction band at resonant frequencies. However, in semiconductors, most electrons are confined to the filled valence band and LSPR is not observed. Non-stoichiometric $Cu_{2-x}S$ and $Cu_{2-x}Se$ NCs have a high concentration of free holes in the valance band. Recently, LSPR was observed in these heavily self-doping semiconductor NCs. Here, we demonstrate a facile method for controllable synthesis of $Cu_{2-x}S$ and $Cu_{2-x}Se$ NCs with superior LSPR. Moreover, we studied the electrical conductivity of solution-processed $Cu_{2-x}S$ and $Cu_{2-x}Se$ NCs thin film in ITO-NC film-Al devices. After chemical post-treatment to displace the long-chain surface ligands attached to the particle surfaces, the nanocrystal thin films show surprisingly high electrical conductivity. As a result, they show promise for use in solution-processed field-effect transistors.

(653a) Growth of Nanowires On Bulk Metal Foils: Opportunities and Challenges towards High-Throughput, Catalyst-Free Nanowire Production

Tobias Hanrath, Chemical and Biomolecular Engineering, Cornell University, Ithaca, NY

Silicon and germanium nanowires are promising building blocks for a variety of proposed nanomaterial-enabled technologies. Recently, nanowires have drawn significant interest as a high performance anode materials for next-generation lithium ion batteries. Challenges between high-throughput production, ease of applicability, and cycling performance present restriction to bring this promise towards technological fruition. We have studied high-throughput nanowire growth directly on bulk metal films. This approach obviates the need for noble metal seed catalyst particles and facilitates direct integration of nanowires on macroscopic contacts. This configuration has significant advantages for the potential application of nanowire-based anodes since the direct contact to the current collecting metal foil avoids the need for a conducting binder and thereby reduced overall weight of the battery cell. This synthesis method has been observed to produce high-yields of good quality nanowires that are epitaxially attached to a conductive surfaces, and could be adapted to a roll to roll process--addressing the key restrictions to implementation.

In contrast to nanowires grown from metal seed particles via the vapor-liquid-solid mechanism, the growth mechanism underlying the formation of nanowires on bulk metal foils is poorly understood. A better mechanistic understanding of the nanowires grown on bulk foils is required to optimize the nanowire yield and to explore opportunities for controlled synthesis of nanowires with predefined growth direction, diameter and length. We report our investigation of the fundamental growth mechanism based on nanowire growth on metal films with controlled composition and thickness. We investigated the composition of the metal seed foil using X-ray photoelectron spectroscopy depth profiling. The physiochemical properties of the nanowire surface are a critical determinant of their electrical properties; this is particularly critical in the formation of stable solid-electrolyte interfaces in the application of nanowire-based anodes. We investigated chemical and physical surface modification techniques and established the relationship between surface chemistry and electrochemical performance.

(653b) Hydrothermal Preparation of Ceramic Nanowires

Xinjie Zhang[1], Zhilong Wang[1] and Anthony Allegrezza[2], (1)Novarials Corporation, Woburn, MA, (2)Congress Street Consulting LLC, MILFORD, MA

One-dimensional nanostructured materials are gaining significant attentions in academic and industrial research communities. The most important and extensively studied one, carbon nanotubes, has found dozens of applications. Inorganic nanowires, own to their wide choice of elements and versatile properties, are predicted to have much wider applications than carbon nanotubes. At least dozens of kinds of inorganic nanowires have fabricated so far. Because most inorganic nanowires are currently prepared at high cost via often complicated preparation procedures, many researchers are still focusing on their high-value end applications such as in electronics, sensors, displays etc.

Hydrothermal growth has demonstrated to be a powerful procedure for preparing high quality and precisely engineered inorganic nanowires. In this talk, we present our recent progresses on the hydrothermal growth of several important inorganic nanowires including titania nanowire, alumina nanowire, magnesia nanowire, molybdenum oxide nanowire, and vanadium oxide nanowires etc. The growth parameters such as precursor choice, precursor concentration, temperature, duration, and additives are demonstrated.

The commercial availability of these premium inorganic nanowires opens up a wide opportunity for fabricating novel products and devices using these premium basic building blocks.

(653c) Growing Titania Nanowires Using Sol-Gel Chemistry in Supercritical CO2

Paul A. Charpentier[1], Nasrin Farhangi[2], Serge Ayissi[2] and Qasem Alsharari[2], (1)Department of Chemical and Biochemical Engineering, Western University, London, ON, Canada, (2)Chemical & Biochemical Engineering, Western University, London, ON, Canada

Growing Metal Oxide Nanowires using Sol-Gel Chemistry Using

Supercritical CO_2

P. A. Charpentier*, N. Farhangi, Qasem Alsharari, S. Ayissi

Department of Chemical and Biochemical Engineering, Faculty of Engineering,

University of Western Ontario, London, Ontario Canada, N6A5B9.

Supercritical carbon dioxide ($scCO_2$) provides a new approach for the synthesis of nanomaterials and nanodevices due to its low viscosity, high diffusivity, and "zero" surface tension [1]. This paper will review our methodology and new synthetic and density functional theory (DFT) modelling data for understanding the directed self-assembly of titania and other metal alkoxides in $scCO_2$ using sol-gel chemistry.[2] This methodology has allowed the formation of well defined nanostructures of TiO_2, and modified titania with ZrO_2, Fe, N and other dopants by this template-free scalable method. Well defined nanospheres, wires, tubes and other unique morphologies have been formed, with diameters ranging from 5 to 50 nm. Other unique properties including high surface areas, tunable morphologies, and 3-dimensional monolithic structures have been enabled. Nanowires have advantages over other nanostructures in terms of direct electron transport (to100 X) to electrodes compared to other morphologies and reduced recombination. [3] Tri and multi- metallic nanowires were successfully prepared using this methodology with high metal oxide dispersion, which is very challenging to be achieved using other methods, such CVD. Metal dispersion is a critical parameter for providing not only high surface area but also more active catalytic sites for reactions to take place.

The formation of these structures has been monitored by *in situ* FTIR along with offline mass-spectrometry, Raman, NMR, XRD, XPS and other techniques in which the titania alkoxides were found to form hexamer crystal structures that react and self-assemble by sol-gel and supramolecular chemistry through the acetate ligands.[2] Also, uniform TiO_2 nanowires and TiO_2/ZrO_2 nanotubes were grown from the surface of graphene sheets, which acted as a template for sol-gel growth.[4] Doped TiO_2 nanowires showed smaller crystal size, higher visible absorption, surface area, thermal stability, photocatalytic activity and higher efficiency solar cells compared to similar materials without graphene.[5] The directed self-assembly and interactions between graphene and TiO_2 were comprehensively studied using the Vienna ab-initio Simulation Package (VASP)[6] based on Density Functional Theory (DFT). Molecular, Rutile and Anatase TiO_2 were decorated on the graphene sheets in 3 different positions corresponding to physical adsorption phenomena (physisorption): the top, the bridge and the hollow sites. An additional position corresponding to a chemical adsorption phenomenon (chemisorption) was created by the organic functionalization process of graphene: the carboxylate site. In all cases Rutile (Ti_2O_4) and Anatase (Ti_4O_8) showed higher stability than a molecular TiO_2 while adsorbed on graphene. In the specific case of functionalized graphene, the carboxylate site showed a much higher binding energy value confirming the significant difference between physisorption (~2 eV) and chemisorptions (~5 eV).

1) R. Sui, Paul A. Charpentier*, Synthesis of Metal and Silicon Oxides Nanostructures by Direct Sol-Gel Reactions in Supercritical Fluids, *Chem. Rev.*, (2012) In Press.

2) Ruohong Sui, John M. H. Lo, Nasrin Farhangi, and Paul A. Charpentier, Self–Assembly of 1D Nanostructures by Linear Condensation of Metal–Acetate Complexes in Supercritical Carbon Dioxide, Accounts of Chemical Research (2012), Submitted.

3) Oudel, P., & Qiao, Q. (2012). One dimensional nanostructure/nanoparticle composites as photoanodes for dye-sensitized solar cells. Nanoscale, 4(9), 2826-2838.

4) Nasrin Farhangi, Yaocihuatl Medina-Gonzalez , Rajib Roy Chowdhury, and Paul A. Charpentier, Using Supercritical CO_2 for growing TiO_2/Graphene Nanoassembles: Synthesis and Photocatalytic Effect, *Nanotechnology* (2012), In Press.

5) Nasrin Farhangi, Rajib Roy Chowdhury, Yaocihuatl Medina-Gonzalez, Madhumita B. Ray and Paul A. Charpentier Visible light active Fe doped TiO2 nanowires grown on Graphene using Supercritical CO_2, Applied Catalysis B: Environmental 110 (2011) 25– 32.

6) G. Kresse, D. Joubert, Phys. Rev. B 59 (1999) 1758

(653d) A Generalized Kinetic Model for the Growth of Single-Walled Metal Oxide Nanotubes

Sankar Nair, Chemical & Biomolecular Engineering, Georgia Institute of Technology, Atlanta, GA, Ipek Yucelen, School of Materials Science and Engineering, Georgia Institute of Technology, Atlanta, GA, Dun-Yen Kang, School of Chemical & Biomolecular Engineering, Georgia Institute of Technology, Atlanta, GA and Haskell W. Beckham, Materials Science and Engineering, Georgia Institute of Technology, Atlanta, GA

The synthesis of 1-D nanoscopic objects (such as nanotubes and nanowires) on a larger scale (e.g., kilogram or ton scales) is necessary for their technological application. A quantitative multiscale understanding of the growth of such objects *via* detailed growth models, is critical to be able to predict and control key parameters such as the length distribution and yield. Such a model can then be used to design liquid-phase reactors for scale-up of nanotube or nanowire synthesis. In this talk, we present a generalized kinetic model to describe the reactions leading to formation and growth of single-walled metal oxide nanotubes, together with detailed experimental data on the evolution of single-walled nanotube populations in a batch reactor. This model is capable of explaining the evolution of nanotube populations as a function of kinetic parameters. It also allows considerable insight into meso/microscale nanotube growth processes. For example, it shows that two different mechanisms operate during nanotube growth: (1) growth by precursor addition, and (2) by oriented attachment of nanotubes to each other. Furthermore, the predictive capabilities of the model are illustrated by comparison of experimental and simulated nanotube length distributions during the synthesis. We conclude with a discussion on the use of such models in the design of reactors for production of nanotubes (and by corollary, nanowires) with complex structures.

(679a) Dynamics of DNA Release From Polyplexes Induced by Inter-Polyelectrolyte Exchange

Yi Zou[1], David Oupicky[2] and Guangzhao Mao[1], (1)Chemical Engineering and Materials Science, Wayne State University, Detroit, MI, (2)Pharmaceutical Sciences, Wayne State University, Detroit, MI

Polymeric gene delivery systems are attractive alternatives to virus-based systems due to safety concerns. A major problem of polymeric gene carriers is their short circulation half-life *in vivo*. We carried out a mechanistic study of DNA release induced by inter-polyelectrolyte exchange, which is believed to be one of the main causes of instability for polyplexes in extracellular environment. Bioreducible hyperbranched poly(amido amine) (RHB) has shown improved gene delivery performance in previous studies, and its depolymerization-induced DNA release process has been visualized at the molecular level. In this study RHB as well as poly(ethyleneimine) (PEI) were used as polycations to condense plasmid DNA. Heparin and poly(sodium 4-styrenesulfonate) (PSS) were used as competing polyanions to DNA to induce the inter-polyelectrolyte exchange reaction. The DNA release process was studied at the single polyplex level using *in situ* real-time AFM imaging. The AFM results were correlated with fluorescence data as well as cell transfection data. A critical polyanion concentration was determined and found to be a function of polycation charge density, chain architecture, and amino to phosphate (N/P) ratio. AFM was used to capture the morphological release pathway with distinct polyplex structures. These characteristic morphologies include core-shell, toroid, nanoparticle decorated toroid, and loose chains held by a compact core structure. This study contributes to basic understanding of DNA release dynamics from single polyplex particles in simulated physiologic conditions and provides synthetic strategies to improve gene delivery efficiency of polymeric vectors.

(679b) Effects On Macrophage Activation Profiles Induced by Adsorption of Serum Proteins Onto Polyanhydride Nanoparticles

Julia Vela Ramirez[1], Amanda Ramer-Tait[2], Michael J. Wannemuehler[2] and Balaji Narasimhan[1], (1)Department of Chemical and Biological Engineering, Iowa State University, Ames, IA, (2)Department of Veterinary Microbiology and Preventive Medicine, Iowa State University, Ames, IA

Biodegradable polymers have shown promising characteristics as adjuvants and/or delivery vehicles by enhancing antigen presentation compared with soluble protein. The chemical and physical properties of these materials can be tailored to enhance their interactions with antigen presenting cells (APCs). In particular, polyanhydride nanoparticles exhibit desirable characteristics as adjuvants such as immunomodulation, sustained antigen release, and stabilization of protein antigens. Upon in vivo administration, serum proteins coat the surface of nanoparticles, potentially affecting their interactions with immune cells. For example, opsonization of the particles may be affected and this enhances cell phagocytosis by the stimulation of different receptors (e.g., Fcγ, mannose) and by the activation of the complement cascade. Analysis of these interactions is necessary to rationally select and shape the desired immune response. It is known that complement receptor 3 (CR3) is required to induce phagocytosis through its interaction with iC3b and modulation of IL-12 production by APCs.

In this work, the effect of serum protein adsorption onto polyanhydride nanoparticles following stimulation of CR3-deficient ($CR3^{-/-}$) mice macrophages on the internalization and activation of these cells was investigated. Polyanhydride nanoparticles based on sebacic acid (SA), 1,6-bis-(*p*-carboxyphenoxy)hexane (CPH), and 1,8-bis-(*p*-carboxyphenoxy)-3,6-dioxaoctane (CPTEG) were used in this study. The profile of adsorbed serum proteins on the surface of the nanoparticles was a function of polymer chemistry. This resulted in different activation phenotypes of the macrophages. The results also showed that the CR3-mediated pathway was important for the uptake of polyanhydride nanoparticles by macrophages. These studies have important implications for the in vivo performance of nanoparticle-based systems for vaccine delivery.

(679c) Click Chemistry Induced Nanomedicine Cancer Targeting

Vahid Mirshafiee, Chemical and Biomolecular Engineering, Uni of Illinois at Urbana-Champaign, Urbana, IL, Jianjun Cheng, Department of Materials Science and Engineering, University of Illinois at Urbana-Champaign, Urbana, IL and Mary L. Kraft, Chemical & Biomolecular Engineering, University of Illinois at Urbana-Champaign, Urbana, IL

Click Chemistry Induced Nanomedicine Cancer Targeting

Vahid Mirshafiee, Jianjun Cheng, Mary L. Kraft

Nanoparticulate delivery vehicles for cancer targeting and therapy are routinely prepared by incorporating a targeting ligand, such as an antibody or aptamer, to the surface of nanoparticles. The interaction between the targeting ligand and specific receptor on the cell membrane is anticipated to drive the nanoparticles to preferentially accumulate in tumor tissues, which enhances antitumor efficacy. However, this design has one major drawback; nanoparticles containing targeting ligands usually have substantially enhanced, undesirable retention in the spleen and liver as compared to unmodified nanoparticles. This undesirable biodistribution of the modified nanoparticles prohibits their *in vivo*targeting, resulting in increased immune response and reduced anticancer efficacy. To address this problem, we are investigating the use of click chemistry for targeted drug delivery.

We have developed nanoparticles that are functionalized with highly-strained and highly-reactive cycloalkynes. These functionalized nanoparticles undergo spontaneous, reagent-free covalent reaction with metabolically incorporated azido-sugars on the cell surface, which is anticipated to promote nanoparticles internalization through endocytosis. This approach is expected to allow high cellular uptake of the nanoparticles while triggering a lower immune response than nanoparticles functionalized with protein- or aptamer-based targeting ligands. Here we report our progress towards achievingthese goals.

(679d) Liposome-Nanoparticle Assembly Controlled Release

Matthew Preiss[1], Arijit Bose[2] and Geoffrey D. Bothun[2], (1)Chemical Engineering, University of Rhode Island, Kingston, RI, (2)Department of Chemical Engineering, University of Rhode Island, Kingston, RI

Development of new drug formulations is plagued by high attrition rates. Only about 11% of new promising therapeutic compounds in clinical development are eventually approved. Nearly 70% of drug failures are attributed to poor pharmacokinetics, efficacy, toxicology, clinical safety, and formulation.[1] Liposome-nanoparticle assemblies (LNAs) can help to overcome these deficiencies by providing a delivery vehicle capable of triggered and controlled release. Decorated LNAs (D-LNAs) are formed by embedding hydrophobic nanoparticles in the lipid bilayer. Controlled and triggered release can be achieved by heating the nanoparticles with alternating current electromagnetic fields (EMF) at radiofrequency. Heating causes the lipid bilayer to melt increasing permeability and releasing encapsulated drugs.

Encapsulation of drugs and release from D-LNAs exposed to EMF is well established. However, the release mechanism of encapsulates remains misunderstood. Our group has demonstrated controlled release from D-LNAs exposed to EMF of smaller drug models (< 600 MW).[2] Previous studies have also established that release is likely due to more than just the change in permeability of the bilayer due to heating.[3] In this research, we investigated the encapsulation and release of large drug models (>3000 MW). Release of larger drugs will provide insight into the release mechanisms of EMF heated D-LNAs. The encapsulation and stability of D-LNAs will also be investigated to determine the feasibility of D-LNAs as a delivery vehicle for large MW drugs.

---

[1] Kola, I., and Landis, J. (2004). Opinion: Can the pharmaceutical industry reduce attrition rates? Nature Reviews Drug Discovery 3, 711-716.

Leeson, P.D., and Davis, A.M. (2004). Time-Related Differences in the Physical Property Profiles of Oral Drugs. Journal of Medicinal Chemistry 47, 6338-6348.

[2] Chen, Y.J., Bose, A., and Bothun, G.D. (2010). Controlled Release from Bilayer-Decorated Magnetoliposomes via Electromagnetic Heating. ACS Nano 4, 3215-3221.

[3] Bothun, G.D., and Priess, M.R. (2011). Bilayer heating in magnetite nanoparticle–liposome dispersions via fluorescence anisotropy. J Coll Inter Sci 357, 70-74.

(679e) siRNA-Dendrimer Nanocarriers for the Lung Epithelium: In Vitro Transfection and Aerosol Formulation

Denise S. Conti, Daniel Brewer, Jordan Grashik and Sandro R. P. da Rocha, Chemical Engineering and Materials Science, Wayne State University, Detroit, MI

RNA interference (RNAi) is a gene therapy approach that works by silencing the expression of target genes in a post-transcriptional manner. Small interfering RNA (siRNA) therapeutics have several advantages over small molecules drugs and proteins, as for example the rapid identification and optimization of lead compounds, and the potential to inhibit all targets, including those considered 'non-druggable'. There are many potential opportunities in formulating siRNA therapeutics in oral inhalation (OI) devices, as it may provide for a direct and non-invasive means of targeting the lung tissue. However, there are many challenges in the development of OI formulations for the efficient delivery of siRNA to the lungs, and in the design of siRNA-based nanocarriers capable of overcoming the extra and intracellular barriers present before siRNA reach the cytosol in relevant cell populations in the lungs.

In this work we report (i) the preparation of siRNA-dendrimer complexes and their characterization by dynamic light scattering, SEM, TEM and AFM, (ii) the formulation of those nanocarriers in portable inhalers, and (iii) their gene silencing efficiency in lung epithelial cell models. Airway (Calu-3) and alveolar (A549) cell lines that stably express eGFP were developed and the gene knockdown ability of the nanocarriers as a function of relevant variables such as N/P ratio evaluated on those cell lines. Our studies show that with the appropriate particle engineering technology, siRNA-dendrimer nanocarriers can be formulated in portable OI devices that possess aerosol characteristics conducive to deep lung deposition.

The relevance of this work stems from the fact that siRNA-based gene therapy is a promising approach to treat a variety of medically relevant pulmonary disorders, including lung cancer, pulmonary tuberculosis, cystic fibrosis, and asthma, and also due to the fact that portable inhalers are the most widely used devices for the non-invasive, regional targeting of such lung diseases.

Precise gene delivery for cell reprogramming with nanochannel based electroporation (NEP)

Junyu Ma[1], Daniel Gallego-Perez[1], Xinmei Wang[1], Yun Wu[1], Keliang Gao[1], Pouyan E. Boukany[1], L. James Lee[1,2,3]

[1]Nanoscale Science and Engineering Center for Affordable Nanoengineering of Polymeric Biomedical Devices, The Ohio State University, 174 West 18th Avenue, Room 1012, Columbus, Ohio 43210, United States
[2]Division of Pharmaceutics, College of Pharmacy, The Ohio State University, 500 West 12th Avenue, Columbus, Ohio, USA)
[3]William G. Lowrie Department of Chemical and Biomolecular Engineering, The Ohio State University, 140 West 19th Avenue,Room 125A, Columbus, Ohio 43210, United States

Cellular reprogramming has presented great promise for regenerative medicine and tissue engineering. Pluripotent stem cells can be generated from mouse and human somatic cells by enforced expression of defined factors. However, clinical uses of current virus methods were limited by integration of transduction factors to the host genes, which may result in tumorigenicity. Meanwhile, even current nonviral methods successfully avoid exogenous DNA integration, low transfection efficiencies and poor reproducibility are still main problems to be solved.

Our lab developed a nanochannel based electroporation device to deliver trancription factors, which may achieve dosage control and minimize side effect of electroporation and result in high reprogramming efficiency and quality.

In this study, mouse embryonic fibroblast (MEF) cells were loaded into microchannels via rotation. The pCAG2LMKOSimO vector containing complementary cDNAs of Oct4, Klf4, Sox2, c-Myc and m-Orange was introduced into mouse embryonic fibroblasts (MEFs) by NEP method under various conditions. Expression level of transduction factors was followed and characterized by image analysis and RT-PCR method. High percentages of cells expressed red fluorescence after 24h of cultures, which indicated successful transfection of pCAG2LMKOSimO plasmid. Compared to bulk electroporation method (Neon, invitrogen), MEF cells transfected by NEP method also showed longer red fluorescence expression time and more homogeneous red fluorescence distribution which indicated longer gene expression time and gene delivery dosage control.

Our newly developed nanochannel based electroporation (NEP) technique was used to transfect MEFs with the pCAG2LMKOSimO vector and achieve a successful precise dosage control. The results greatly suggested NEP a potential alternative delivery method which may increase cell reprogramming efficiency and avoid tumorigenicity.

(679g) Poly(amino ether)-Gold Nanorod Assemblies for shRNA Delivery

James Ramos, Biomedical Engineering, Arizona State University, Tempe, AZ and Kaushal Rege, Chemical Engineering, Biomedical Engineering, Arizona State University, Tempe, AZ

Poly(amino ether)-Gold Nanorod Assemblies for shRNA Delivery

James Ramos, Kaushal Rege

As the field of nanotechnology continues to grow, many novel nanomaterials are showing strong potential for use in the biomedical field. Gold nanorods (GNRs) have emerged as one such nanomaterial with the capacity for use in biosensing, imaging, photothermal therapy and gene delivery as a cancer therapy. Using polymers from a novel poly(amino ether) library synthesized in our laboratory, we have synthesized poly(amino ether) coated GNRs (PAE-GNRs) by depositing polyelectrolyte multilayers on cetyltrimethyl ammonium bromide (CTAB)-based gold nanorods. These PAE-GNRs exhibit an increase in the nanoparticles stability as well as the capacity to electrostatically bind to nucleic acids. We have previously shown the PAE-GNRs ability to deliver plasmid DNA to two different prostate cancer cell lines for enhanced transgene delivery when compared to GNRs deposited with polyethylenimine, a current standard for polymer mediated gene delivery. Here we show the capability for the PAE-GNRs to delivery shRNA for gene knockdown. By loading the PAE-GNRs with shRNA targeted against the luciferase reporter gene and using them to treat cancer cell lines that constitutively express the luciferase, we demonstrate the ability of the PAE-GNRs to deliver the shRNA for gene silencing. This is confirmed by the observed reduction in luciferase expression levels compared to no treatment and delivery of non-silencing nucleic acids. By optimizing the shRNA loading amount, treatment times, and PAE-GNR concentrations we have determined the ideal treatment conditions for maximum gene silencing. These conditions can then be translated for delivery of shRNA targeted against a therapeutically relevant target as a gene therapy to combat disease. Additionally, due to the unique optical properties of the GNRs, we also demonstrate the ability to image the nanoassemblies after delivery *in vitro* using two-photon induced luminescence of the GNRs. This can be translated as a labeling/detection application for PAE-GNR assemblies modified with moieties targeted to a specific cell type. Our results indicate that the engineering of the PE-GNRs leads to a gold nanoassembly that combines high stability with low cytotoxicity and the capacity to mediate shRNA delivery for gene silencing, as well as be used for TPL imaging. These PE-GNRs also have the potential to be used for photothermal ablation, photothermal enhanced delivery, and carrying of other small anionic molecules for delivery making them a powerfully potential single platform theranostic nanomaterial.

(711a) Lead-Free Nanosolders and Their Application for Nanowire Assembly and Joining

Fan Gao and Zhiyong Gu, Chemical Engineering, University of Massachusetts Lowell, Lowell, MA

Due to the requirements for high-performance interconnects in advanced packaging, the development of nanotechnology has to meet the dramatic scaling down of the electronics packaging size from micron-scale to nanoscale. Solder-based connections have great potential and provide ample opportunities to form robust joints between individual nanocomponents and integrate them into a nanosystem with desired functionality and geometry for device applications. Prior study has shown that nanosolders can be used to assemble and join various metallic and metal oxide nanowires. In this study, the approach of nanosoldering is shown to be useful for the construction of functional nanostructures from nanomaterials by utilizing conventional reflow process with proper flux selection and utilization. The interaction between lead-free tin-based nanosolders onto metallic nanowires and reactive substrates such as copper are studied. The melting and wetting behavior of nanosolders on a reactive surface provide primary model of nanoscale joining formed on 2-dimensional planar. Such 1-D cylindrical nanowire structure provides the critical information to process the joining techniques in micro/nano electronics packaging to devices in a much smaller scale.

(711b) Electromechanically Driven Nonlinear Dynamics of Voids in Metallic Thin Films Under Anisotropic Mechanical Stress

Dwaipayan Dasgupta, Georgios I. Sfyris and Dimitrios Maroudas, Chemical Engineering, University of Massachusetts Amherst, Amhest, MA

Void migration and morphological evolution, driven by the combined action of mechanical stresses and electric fields, are the leading mechanisms of failure in metallic thin films, which are used widely in microelectronics. This electromechanically driven void dynamics is extremely rich and complex and it depends strongly on the void size, the strength of the applied electric field, mechanical loading conditions, as well as diffusional anisotropy parameters for diffusion on the void surface. In this presentation, we report results for the effects on the current-driven nonlinear void dynamics of biaxial mechanical loading of the film, in the form of a general applied stress tensor.

We have studied the electromechanically driven surface morphological evolution of voids in single crystalline thin films of face-centered cubic metals for <100>-oriented film planes, characterized by four-fold symmetry of surface diffusional anisotropy. The thin film is subjected to an electric field and a mechanical load expressed by a transversely isotropic biaxial stress tensor. We have investigated all the possible loading modes, i.e., purely tensile loading, purely compressive loading, as well as a more general loading where the applied stress components can be either tensile or compressive, resulting in a mixed (tensile + compressive) loading mode in addition to the loading anisotropy. We have conducted self-consistent dynamical simulations of the driven void surface morphological response according to a well validated, two-dimensional, and fully nonlinear model. The model describes driven mass transport on the void surface and accounts for curvature-driven and stress-driven surface diffusion, electromigration driven by the electric field, and surface diffusional anisotropy. In the simulations, a Galerkin boundary-integral method for the solution of the electrostatic and elastostatic boundary-value problems is combined with a front tracking method for monitoring the shape evolution of the void surface.

We have analyzed the effects of the various mechanical loading modes on the surface morphological evolution of the void for electric-field strengths that are above and below a critical level, which corresponds to a Hopf-bifurcation transition from steady states to time-periodic states in the unstressed film; these asymptotic states of voids are solitons that migrate along the electric-field direction at constant speed. We have spanned the entire range of mechanical loading anisotropy, from uniaxial loading to isotropic biaxial loading. We have mapped the void stability domain boundaries and found that under isotropic or transversely isotropic biaxial loading that is either purely tensile or purely compressive, both steady and time-periodic states can be stabilized in the driven void morphological response; transitions between these states occur at supercritical Hopf bifurcation points. Upon increasing the stress level beyond a certain limit, the void tip extends across the film width causing film failure. Under the same loading anisotropy and stress level, purely compressive and purely tensile stresses result in very similar void morphological responses in terms of oscillation amplitude at the time-periodic states, ranges of stress that result in stable steady and time-periodic states, and level of stress that the film can sustain prior to failure. However, under mixed tensile and compressive biaxial loading, we found that the stable steady states in the void morphological response become completely suppressed and only time-periodic states can be stabilized for stress levels below the critical stress for film failure. Moreover, under such mixed loading, the levels of stress that the film can sustain prior to its failure are much lower than those it can sustain under purely tensile or purely compressive loading. We conclude that such mixed mechanical loading modes make metallic films with voids more vulnerable to electromechanical failure.

(711c) Atomic Layer Deposition Enabled Synthesis of Nanostructured Composite $BiFeO_3/CoFe_2O_4$ Thin Films for Multiferroic Applications

Calvin D. Pham, Department of Chemical and Biomolecular Engineering, University of California Los Angeles, Los Angeles, CA and Jane P. Chang, Department of Chemical and Biomolecular Engineering, University of California, Los Angeles, Los Angeles, CA

Multiferroic materials, that can either exist as single-phase materials or multi-phase composites, exhibit two or more forms of ferroic order such as (anti)ferroelectricity, (anti)ferromagnetism, or ferroelasticity and have been proposed for use in future non-volatile memory technology. Atomic layer deposition (ALD) is proposed as a scalable approach to synthesize multiferroic thin films and to enable the synthesis of multiferroic composites which utilize conformal deposition onto 3-D nanostructures. Challenges that must be overcome in the ALD of multiferroic materials is the amorphous nature of as-deposited films and the difficulty in attaining the desired crystallinity and structure that would enable multiferroic properties to emerge from these materials.

In this work, multiferroic $BiFeO_3$ was deposited by ALD as a single-phase multiferroic thin film as well as the ferroelectric component in a composite multiferroic using a ferrimagnetic $CoFe_2O_4$ mesoporous template that was synthesized using an evaporation induced di-block copolymer self-assembly technique. The ALD process used the metallorganic precursors $Bi(tmhd)_3$ (tmhd = 2,2,6,6-tetramethylheptane-3,5 dione) and $Fe(tmhd)_3$ alongside oxygen atoms produced from a coaxial waveguide microwave powered atomic beam source. A variety of ALD process conditions were studied, such as the effects of process temperature, precursor pulsing time, and precursor pulsing ratio on film composition, growth rate, and crystallization. The ALD films were able to be grown with a composition ratio Bi:Fe close to unity and with a controlled nanostructure and growth rate of ~0.7 Å/cycle. In order to achieve the desired crystalline material after rapid thermal processing, the composition and nanostructure of the as-deposited films must first be controlled via the ALD process to fit within narrow windows.

To compare the performance of the multiferroic ALD films to more well established synthesis methods, measurements of magnetic and ferro/piezoelectric properties were accomplished using SQUID magnetometry and piezoresponse force microscopy, respectively. Magnetic measurements showed that the out-of-plane remnant magnetization of a composite film at room temperature was approximately 66.4 emu/cm^3 while the coercive field was approximately 1950 Oe which was comparable to epitaxial films grown by other methods such as PLD. The magnetoelectric coupling effects in the composite films were studied to assess the effectiveness of the nanostructured material approach.

(711d) Atomic Layer Deposition of AlN Thin Films As Gate Dielectrics for Wide Bandgap Semiconductors

Ya-Chuan Perng and Jane P. Chang, Department of Chemical and Biomolecular Engineering, University of California, Los Angeles, Los Angeles, CA

Aluminum nitride (AlN) is a potential dielectric layer for wide bandgap semiconductor based power electronic devices, such as those demanded in radio frequency, high-speed and high-temperature communication, because of its wide bandgap and high dielectric constant. In particular, for 4H-SiC, AlN is also a promising interfacial layer due to their similar atomic arrangement, small lattice mismatch (1.3%) and comparable thermal expansion coefficients. Although various deposition techniques have been investigated to synthesize AlN thin films with atomic controllability over a large substrate remains a challenge. Atomic layer deposition (ALD) was thus used in this work to grow AlN thin films.

AlN deposition was performed in an ultra-high vacuum chamber with base pressure of 10^{-7} Torr using trimethylaluminum (TMA) and ammonia (NH_3) as precursors. It was discovered that ALD of AlN is possible only when the minute amount of moisture in NH_3, which competed with and inhibited the nitride growth, was completely eliminated. The ALD window was found to be 500-570 °C with a growth rate of 1.5 Å/cycle. The deposited film composition was evaluated via *in-situ* x-ray photoelectron spectroscopy (XPS) with Al/N determined to be 1.2. *In-situ* reflective high-energy electron diffraction (RHEED) measurements showed as-deposited AlN films were crystalline, which was confirmed by x-ray diffraction (XRD). AlN/4H-SiC MIS capacitors were fabricated to examine the electrical properties with the dielectric constant of AlN determined to be 8.3 and a leakage current density of 10^{-3} A/cm^2 at 4.3 MV/cm. The ALD AlN passivated AlGaN/GaN hetero-structure demonstrated an increase in the carrier density and a 3% decrease in mobility compared to those of non-passivated hetero-structure. With amorphous Al_2O_3 surface passivation, the mobility decreases by 22% with increased carrier density, showing that the crystalline AlN providing a superior property on passivating the hetero-structures.

Title : Observations, Formation Mechanisms and Reduction Methods of 3-Dimensional Cross-Slip Dislocations in Silicon Single Crystal

Paper number : ID#270201:

Author : Do Won Song, Kyo Kim†

LG Siltron Inc, Gyeonggi-do, Korea
† University of Seoul, Seoul, Korea

As part of Materials Engineering and Science Division
Session: Nanoelectronic Materials and Devices II
scheduled for Thursday, November 1, 2012: 1:50 PM - 2:07 PM, Westmoreland East (Westin)

1. Introduction

1) Requirements of Si Wafer

Conceptual Model of Wafers

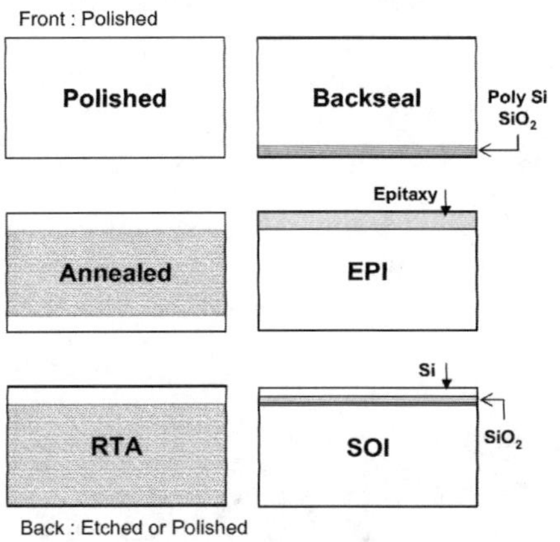

Requirements for Wafer

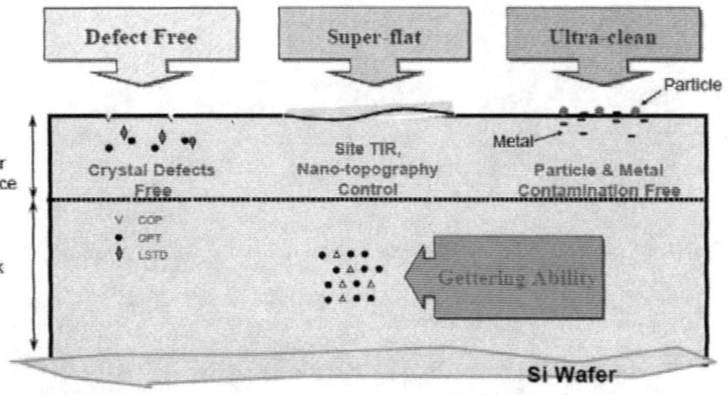

EPI : Epitaxial, RTA : Rapid Thermal Anneal, SOI : Silicon On Insulation, COP : Crystal Originated Particle, OPT : Oxygen Precipitates, LSTD : Laser Scattering Tomography Defect

SILTRON

1. Introduction

2) Defects & Analyzing Method in Silicon Crystal

Various Crystal Defects

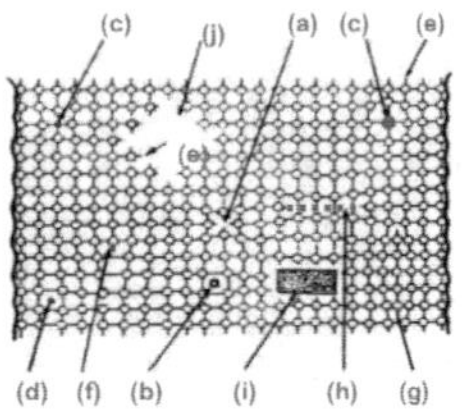

Point defect
- intrinsic
  - vacancy (a), self interstitial atom (b)
- extrinsic
  - substitutional impurity atom (c)
  - interstitial impurity atom (d)
  - dangling bond (e)

Line defect
- dislocation : edge- (f), screw-

Plane defect
- stacking fault
  - interstitial- (g), vacancy-type (h)
- twin, grain boundary

Volume defect : precipitate (i), void (j)

Defects of Si Wafer & Analyzing Method

Species	Vacancy Related Defects			Interstitial Related Defects		
Dimension	Volume Defects	Plane Defects	Point Defects	Point Defects	Plane Defects	Volume Defects
Etching + Visual Light	**COP** FPD					**LDP** EPD
Heat Treat + Etching + Visual Light		**O-band** (OiSF)			**T-zone**	
Heat Treat + Electrical Measurement	V-Cluster	O-Band	**VDP**	**IDP**	T-zone	I-Cluster
Eye	Bubble (Air Pocket)					

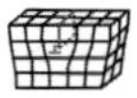

edge-dislocation

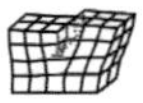

screw-dislocation

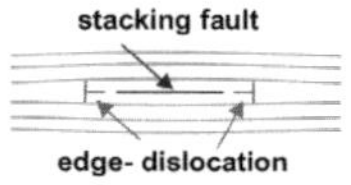

OISF : wet oxidation at 1100℃ for 60min
(Wright 5min, 10,000X, by SEM)

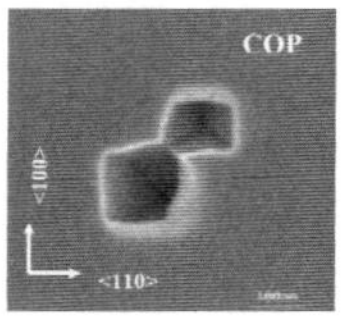

FPD : Flow Pattern Defect, LDP : Large Dislocation Pit, EPD : Etch Pit Defects, OiSF : Oxidation induced Stacking Fault, VDP : Vacancy Dominated Point Defect, IDP : Interstitial Dominated Point Defect

SILTRON

2. Observations of Line Defects

1) Defects in Wafer

Particle Counter

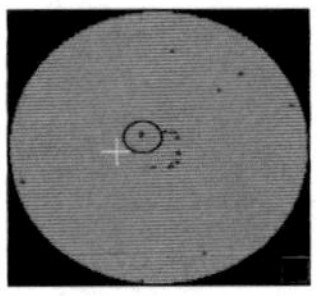

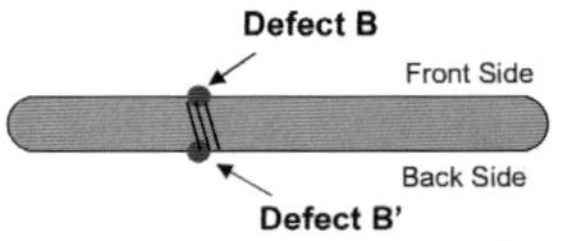

CSD Position (r,z)

Wafer 700 pcs

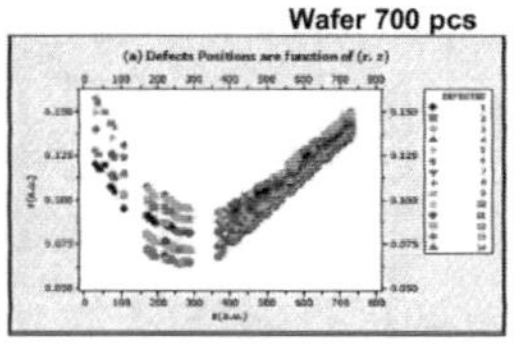

CSD Position (r,Θ)

Wafer 700 pcs

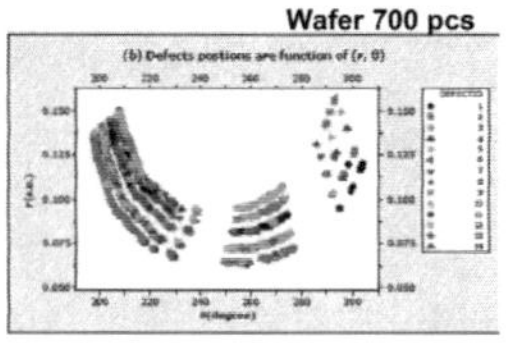

EPD Test Wright etching 5 min

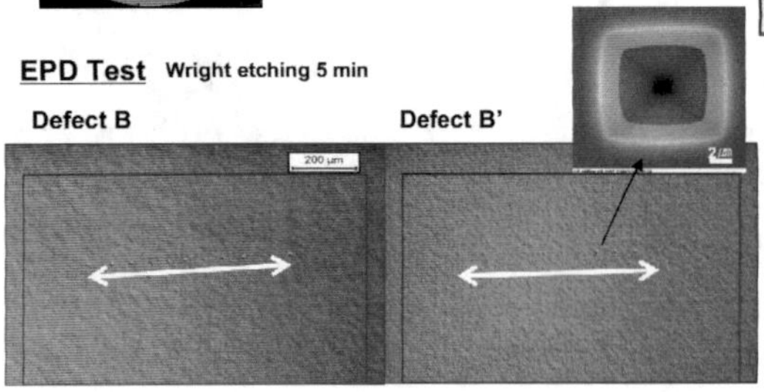

Line Defects in Ingot

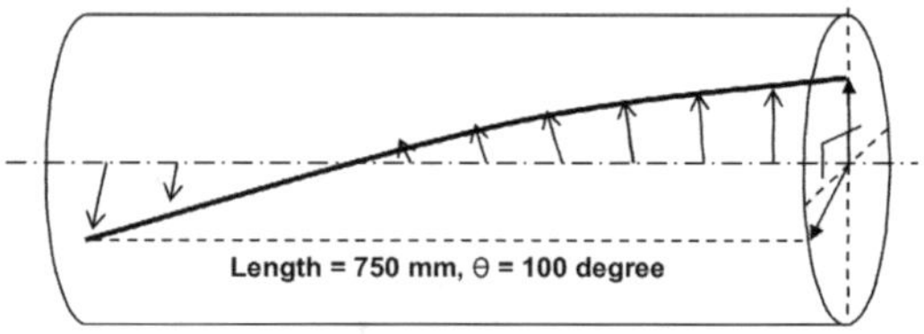

EPD : Etch Pit Defect, CSD : Cross-Slip Dislocation

SILTRON

2. Observations of Cross-Slip Dislocation

2) HR-TEM Images

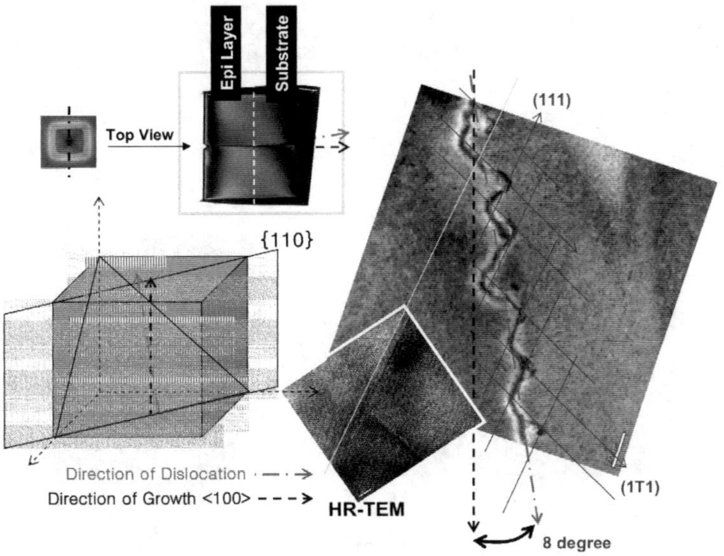

TEM Sample : PT DEPO about 800~900 Å (metal sputter), Etched by Usr3 Beam

3. Effects on Gate-Oxide Insulation

BV Fail Rate(%)

Gate Oxide Thickness : 80A

Electric field	<4	<8	<10	<12
Fail Mode	A	ΣAB	ΣA~C	ΣA~C+
Reference	0.0	0.0	0.0	0.6
CSD 1	0.0	0.6	1.2	1.5
CSD 2	0.0	0.0	0.6	0.9

(MV/cm)

BV Fail Full Map

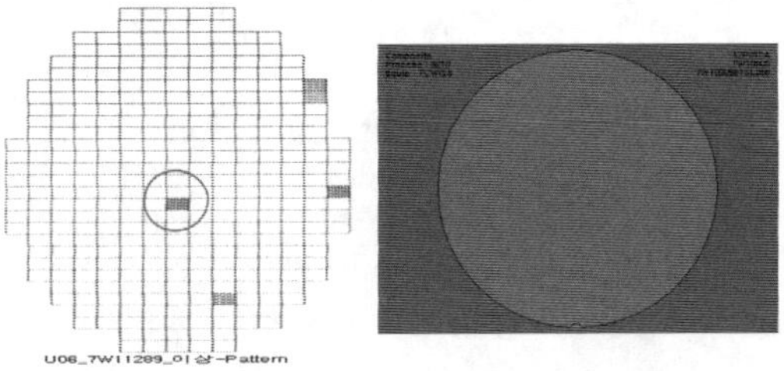

Full MAP

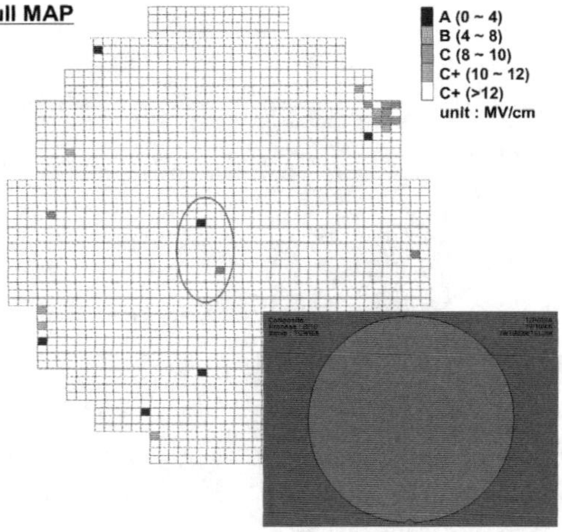

4. Formation Mechanisms

1) Screw Dislocation Growth Model

☑ **Gibbs-Thomson Equation on the Interface**
(Added Constitutional, Capillary and Kinetic Undercooling)

HMCZ Melt Flow → **Screw Dislocation**

$$T^{I} = T_m\left(1 + 2H\frac{\gamma}{L_{\upsilon}}\right) + mC - \mu^{-1}V_n$$

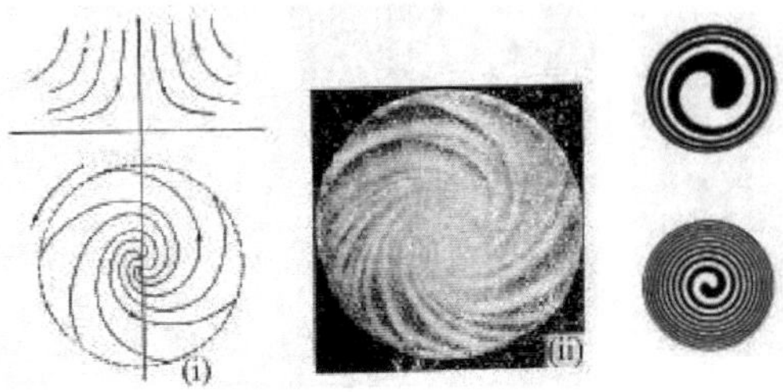

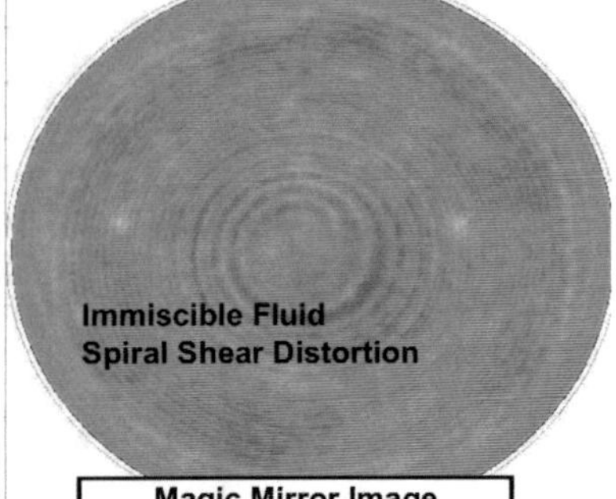

Magic Mirror Image

※ W.A. Tiller "The Science of Crystallization" (1991)

4. Formation Mechanisms

2) Climb & Movement of a Cross-Slip

☑ **Climb & Movement of a Cross-Slip; Sessile Frank Partial**
(Jogged Screw-Dislocation)

Interface Height → **Thermal Stress** → **Jogged Screw Dislocation** → **Cross-Slip**

Scanning Infra-Red Depolarization Image

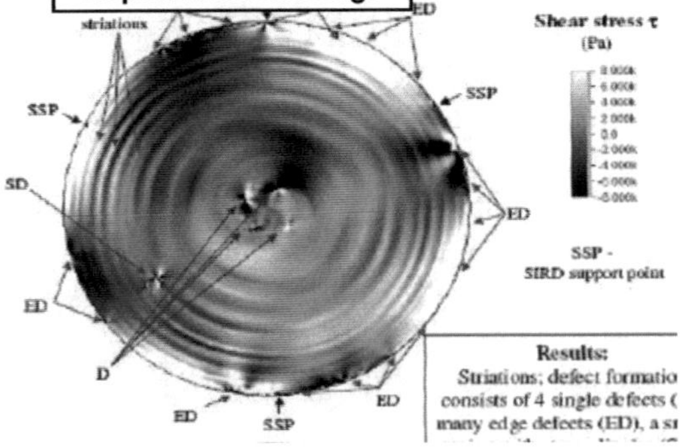

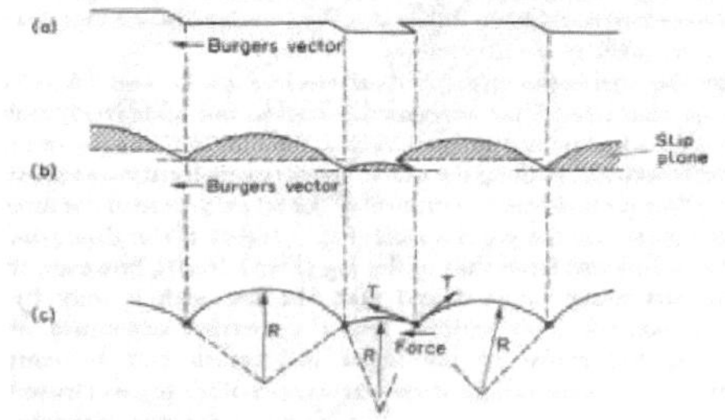

※ D. Hull & D.J. Bacon "Introduction to Dislocation" (1984)

4. Formation Mechanisms

3) Conclusions

Asymmetric Melt Flow → **Spiral Shear Distortion** → **Mechanical & Thermal Stress** → **Screw Dislocation** → **Jogged Screw Dislocation** → **Cross-Slip**

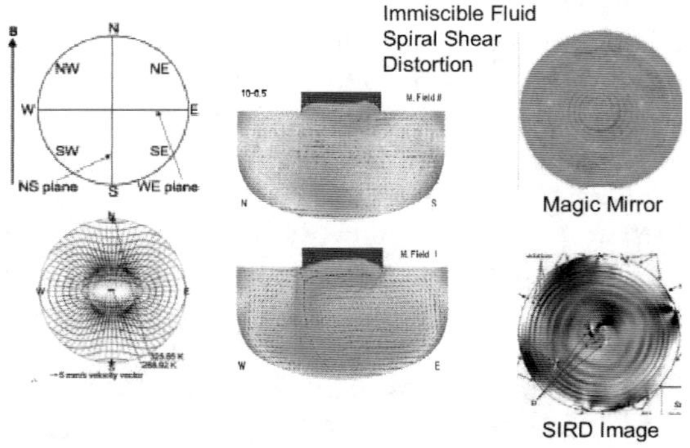

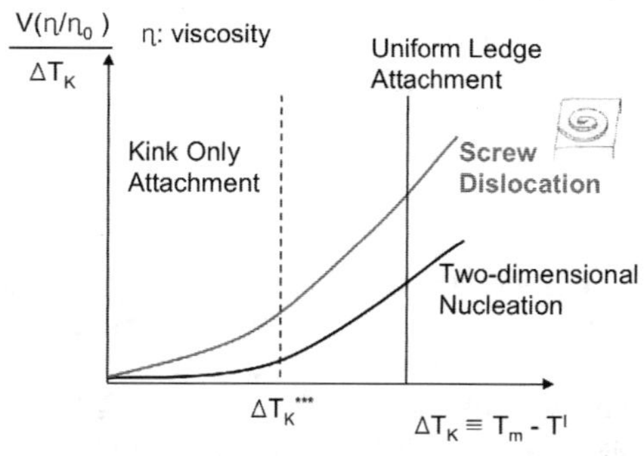

※ International Journal of Heat and Mass Transfer 48 (2005)

※ W.A. Tiller "The Science of Crystallization" (1991)

SIRD : Scanning Infra-Red Depolarization

SILTRON

5. Reduction Method

Decrease Curvature → **Increase Gradient** → **Decrease Spiral Shear Distortion**

$$T^I = T_m\left(1 + 2H\frac{\gamma}{L_\upsilon}\right) + mC - \mu^{-1}V_n$$

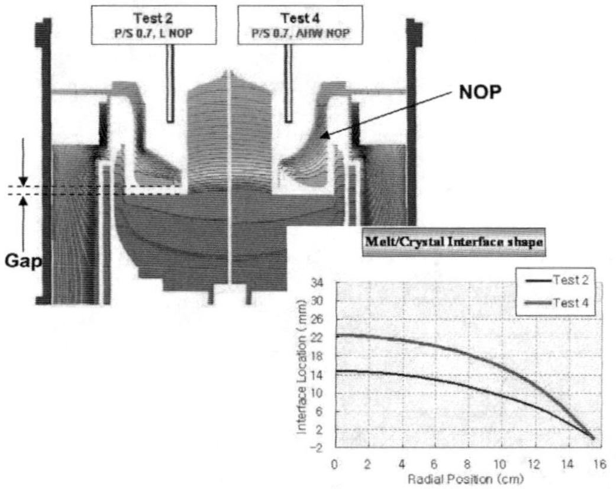

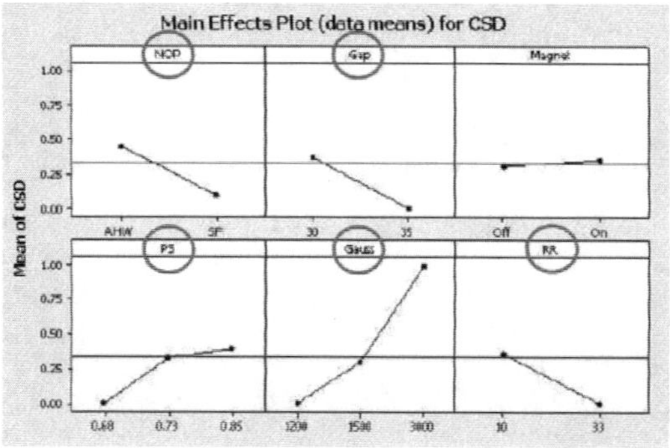

PS : Pulling speed of crystal, RR : Rotation ration between crystal and crucible, NOP : Apparatus of heat shield, Gap : distance between NOP and melt

SILTRON

(711f) Numerical Methods for Solving Mode-Related Problems in Photonic Devices

Meng-Mu Shih, Electrical and Computer Engineering, University of Florida, Gainesville, FL

This work demonstrates a numerical process of solving the mode-related problems in photonic devices with periodic structures. The periodic structures can make the photonic devices behave quite differently than without. In quantum photonics, this periodicity can cause the perturbation effect and show a stop band in the frequency and wave-vector diagram. This effect also can make the propagation constant of the photonic device become a complex number. A visualized method to approach the solution effectively and efficiently is therefore of the central interest.

This work will use the light-emitting devices with hybrid materials of semiconductor and metal as the computational example. This photonic device can emit the light at telecommunication wavelength. This process will cover several sub-processes such as material selection and optical properties, material and optical matching, mathematical issues such as boundary value, linear algebra, and numerical methods. The numerical method has to consider the trade-offs between the precision of the solution and the time required to get the solution. The above results obtained by the photonic method will be compared with the results by the optical method. The results by both methods are close. This photonic method can be applied to photonic devices with different materials and structures for wide applications.

(711g) Analyzing the Dynamics of the Horizontal Ribbon Growth Process for Solar Silicon

Parthiv Daggolu, Andrew Yeckel and Jeffrey J. Derby, Chemical Engineering and Materials Science, University of Minnesota, Minneapolis, MN

Horizontal ribbon growth (HRG) promises the growth of crystalline silicon at rates that are orders of magnitude greater than vertical ribbon growth technologies. If successful, this process would enable the production of higher-quality, near-single-crystalline silicon wafers at fraction of the cost of current production techniques. This fascinating process was first conceived by Shockley in late 1950's for silicon growth and was practiced by Bleil on the late 1960's for germanium growth. Large-scale development efforts were subsequently carried out by Kudo in Japan in the late 1970's and by the Energy Materials Corporation in the US in the early 1980's. However, after encouraging early results, experimental advances and process development efforts stalled, and this technique was abandoned in favor of growth methods that were easier to develop.

Unlike vertical meniscus-defined crystal growth processes, such as edge-defined film-fed growth (EFG), which are inherently stable, there are many failure modes that must be avoided in the HRG process. We argue that its successful operation will rely on a thorough understanding of system design and control–issues that are perhaps only feasibly addressed via computational modeling of the system. Towards these ends, we present a comprehensive thermal-capillary model based on finite-element methods to study the coupled phenomena of heat transfer and interfacial phenomena (solidification and capillarity) in the HRG process. Bifurcation analysis coupled with transient computations using this model reveals process limitations consistent with known failure modes and suggests operating windows that may allow for stable process operation.

(711h) Melt Growth of CZT with Convex Interfaces VIA A Bell-Curve Gradient Freeze Profile

Nan Zhang, Andrew Yeckel and Jeffrey J. Derby, Chemical Engineering and Materials Science, University of Minnesota, Minneapolis, MN

The availability of large, single crystals of cadmium zinc telluride (CZT) with uniform properties is key to improving the performance of gamma radiation detectors. The quality and cost of CZT substrate is primarily determined by crystal growth outcomes. However, the growth of these crystals is still not well understood, and most modifications to crystal growth have been driven by intuition obtained from growth experience with traditional semiconductor systems that are very differently from CZT. We aim to advance the practice of crystal growth by applying physics-based, computational models to understand the most important factors affecting these processes.

In this presentation, a novel, bell-curve furnace temperature profile is presented and predicted to achieve macroscopically convex solid-liquid interface shapes during melt growth of CZT in an electrodynamic gradient freeze (EDG) furnace. A strategy is also presented to dynamically adapt this furnace profile so that uniform, convex interface shapes are maintained through an entire CZT growth run. This approach represents a significant advance over traditional gradient-freeze profiles, which always result in concave interface shapes, and furnace heat transfer modifications, such as redesigned ampule support pedestals, that can achieve convex interfaces, but only over only a small portion of the growth run.

Realizing a convex solidification interface via this adaptive bell-curve furnace profile is postulated to result in better crystallinity and higher yields than conventional CZT growth techniques. Importantly, this strategy may be applied to any Bridgman configuration that utilizes multiple, controllable heating zones.

(721a) Response of Breast Tumor Cells to Hybrid Polymer-Peptide Self-Assembled Nanoparticles

Esmaiel Jabbari, University of South Carolina, Columbia, SC

Introduction: A novel strategy to circumvent the short half-life, limited solubility, and improve uptake of anti-cancer drugs is encapsulation in colloidal self-assembled nanoparticles (NPs). Hybrid polymer-peptide NPs can selectively target the tumor over normal cells while enhancing intracellular uptake of the drug. The objective of this work was to investigate the effect of charged head-group of the peptide Cys-Val_6-Y_2 (CY_6Y_2), where Y is arginine or lysine, conjugated to acrylated poly(L-lactide) (PLAA-CV6Y2) on uptake and migration of tumor cells when loaded with Doxorubicin (Dox) chemotherapeutic agent.

Methods: PLA was conjugated to CY_6Y_2 peptide, where Y was lysine (CV_6K_2) or arginine (CV_6R_2), and the polymer-peptide conjugate was self-assembled into NPs as described [1]. Control PLAA NPs (without charged peptide) stabilized with poly(ethylene oxide) (PLAA-EO) were synthesized as described [1]. Size and distribution of the NPs were measured with light scattering. For cell uptake experiment 4T1 (mouse breast tumor) cells were incubated with NPs encapsulating FITC (2 mg/mL, 2% loading) for 24 hours. At every time point, the supernatant and cells were collected and analyzed by a fluorescent plate reader. For invasion and migrations studies, 4T1 cells at a density of 1.5×10^4 cells/well in 24-well Transwell plates were exposed to empty and Dox-loaded NPs (5 μM Dox for loaded particles, 5% loading). Free Dox was used as a positive control. The cells were allowed to migrate through the transwell membrane in the absence (migration) or presence (invasion, 5% wt gel in media) of Matrigel in the upper chamber. After 24 h, the membranes were fixed, stained with eosin-Y and azure dyes, imaged, and the number of migrating cells were counted with an inverted microscope.

Results and Conclusions: NPs had a narrow size distribution between 50-150 nm. Release kinetics showed sustained Dox delivery for 25-35 days. After 24 h incubation of 4T1 tumor cells with self-assembled PLAA-CV_6R_2 and PLAA-CV_6K_2 NPs, 60 and 50% of the NPs were taken up by the cells, respectively, compared with only 30% for PLAA-EO NPs. Migration experiments indicated that Dox-loaded PLAA-CV_6R_2 NPs were more effective in retarding the migration of 4T1 cells than PLAA-CV_6K_2, PLAA-EO, or the free Dox. Our results demonstrate that the arginine head group was more effective with respect to NPs uptake and retarding the migration of tumor cells than the lysine head group

(721b) Designing Nanostructured Biomaterials Via the Self-Assembly of Functionalized Nanotubes and Lipids

Meenakshi Dutt, Chemical and Biochemical Engineering, Rutgers University, Piscataway, NJ

Our objective is to design nanostructured biomaterials using the self-assembly of functionalized nanotubes and lipid molecules. In this presentation, we summarize the multiple control parameters which direct the equilibrium morphology of a specific class of nanostructured biomaterials. Individual lipid molecules are composed of a hydrophilic head group and two hydrophobic tails. A bare nanotube encompasses an ABA architecture, with a hydrophobic shaft (B) and two hydrophilic ends (A). We introduce hydrophilic hairs at one end of the tube to enable selective transport through the channel. The dimensions of the nanotube are set to minimize its hydrophobic mismatch with the lipid bilayer. We use a Molecular Dynamics-based mesoscopic simulation technique called Dissipative Particle Dynamics which simultaneously resolves the structure and dynamics of the nanoscopic building blocks and the hybrid aggregate. The amphiphilic lipids and functionalized nanotubes self-assemble into a stable hybrid vesicle or a bicelle in the presence of a hydrophilic solvent. We demonstrate that the morphology of the hybrid structures is directed by factors such as the temperature, the rigidity of the lipid molecules, and the concentration of the nanotubes. We characterize the equilibrium morphology of the hybrid aggregate in terms of the properties of its components.

(721c) Designed DNA Nanotube Architectures

Rebecca Schulman and Abdul M. Mohammed, chemical and biomolecular engineering, Johns Hopkins University, Baltimore, MD

Designed DNA Nanotube Architectures

Rebecca Schulman and Abdul M. Mohammed

Chemical and Biomolecular Engineering

Johns Hopkins University

The cytoskeleton and extracellular matrix consist of semiflexible biopolymers that are dynamically assembled into precise topologies and continuously reassembled based on inputs from cells and environmental signals. This assembly process leads to specific interactions of the resulting materials with cells that enable tissue formation and can lead to properties such as the capacity for the polymer matrix to self-heal or to reassemble continuously, producing motion of the structure. Recapitulating the same assembly mechanisms synthetically is a current challenge that would lead to new materials with novel biological interactions and programmable, active reformation.

We report initial steps toward designing mechanisms to synthetically control the topology of DNA nanotube architectures. DNA nanotubes consist of units called tiles, approximately 14x4 nm structures that self-assemble via DNA hybridization. Tubes have a persistence length that depends on the radius; for the structures we assemble it is estimated to be 10-20 microns. We show that the nucleation and termination of tubes can be efficiently directed at specific locations, with the result being that we can create nanotube asters or point-to-point nanotube links. Both nucleation and termination are guided by designed DNA complexes, folded structures that emulate the facet of a growing DNA nanotube. We characterize controlled nanotube growth using both fluorescence light microscopy to track growth and atomic force microscopy to characterize the structure of individual DNA nanotubes.

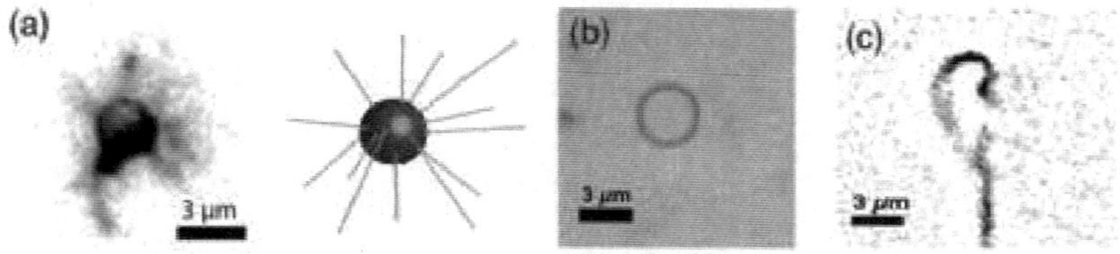

Figure 1: Nanotube networks. (a) Fluorescence micrograph (left) and cartoon (right) of a DNA nanotube aster growing from a bead coated with nucleating complexes. (b) Without nucleating complexes, nanotubes do not grow; bead location is shown by an orange circle. (c) A nanotube can link a bead coated with nucleating complexes and a bead coated with end capture complexes.

(721d) Sequential Self-Assembly of Engineered Protein Building Blocks in 3D Matrices

Won Min Park and Julie Champion, Chemical and Biomolecular Engineering, Georgia Institute of Technology, Atlanta, GA

Self-assembly is an attractive tool to manipulate physical properties of bio-macromolecules by organizing them into supramolecular structures at physiologically relevant conditions. Naturally derived protein motifs present useful self-assembling properties, and combination of multiple motifs in a self-assembly system can provide new opportunities to control the self-assembly process. We use two protein building blocks with different characteristic self-assembling motifs. The first assembly component is a diblock copolypeptide that consists of an elastin-like polypeptide (ELP) block, which undergoes inverse phase transition and aggregation, and a leucine zipper that forms coiled-coils. Its self-assembly partner is fluorescent protein mCherry, acting as a globular model protein, fused with a leucine zipper motif. The lower critical solution temperature (LCST) of the ELP leucine zipper fusion protein ranged from 25°C to 40°C at micromolar concentrations. Consistently, upon a temperature increase from 25°C to 37°C, the diblock copolypeptide rapidly aggregated into nano- and micro-particulates. The leucine zipper domain fused with ELP allows for further assembly with mCherry leucine zipper fusion protein by formation of high-affinity, heterodimeric coiled-coils. The second step of assembly significantly influenced the phase transition, colloidal stability, and dynamic behavior of aggregated particles in solution, primarily controlled by the ratio of leucine zipper building blocks. These sequential multi-step assembly processes spontaneously occur in physiological environments, where no further stimulus is necessary, and can be applied as a new system for protein delivery. In a new strategy, non-equilibrium *in situ* self-assembly of the building blocks resulted in self-assembled nano- and micro-structures in 3D matrices, which promote substantial control of molecular transport of the model protein. Dynamic behavior of the self-assembled structures in 3D matrices was characterized by fluorescence microscopy to quantify the parameters that govern protein retention in tissue. This simple, all protein, self-assembling system can be applied to control delivery and extend retention of therapeutic proteins in natural tissue or artificial matrices.

(721e) Future of the Layer-by-Layer Assembled Nanomaterials

Nicholas Kotov, Chemical Engineering, University of Michigan, Ann Arbor, MI

Layer-by-layer assembly (LBL) is a very established field of materials engineering with multiple demonstrations of its utility and creative resolution of many interesting technological problems. It is important to understand better what could be the main directions of its future development. The key advantage of the technique is that it offers the possibility of ***multiscale*** engineering of nanocomposite materials based on sequential adsorption of nanometer scale layers of polymers and inorganic colloids. Using this property LBL can resolve hard challenges of materials science related to mechanical, electrical, optical, and biological properties. One of them is the design of materials with hard-to-reach combinations of electrical and mechanical properties necessary for energy conversion and storage characteristics. In this presentation, the principles of property engineering of these materials will be described. The focus technological areas will be (1) Flexible solar cells and solid state light emission devices; (2) High capacity lithium batteries; and (3) Biomedical implants with energy storage requirements. Manufacturing of such composites in large scale will also be addressed.

LBL films exhibit exceptional mechanical performance. This property becomes very essential in preparation of transparent conductors for flexible solar cells. One of the key hurdles on their mass utilization is replacement of expensive indium tin oxide coating, which also reveals deficient performance in bending, which is essential for flexible electronic devices. It will be demonstrated that the LBL films made from carbon nanotubes and appropriate polymers reveal cumulative figure of merit taking into account both electrical and mechanical performance higher than standard ITO glass. Surface conductivity of the coatings was as low as 80 Ohm/square.

Mechanical performance is essential for lithium ion batteries which is the key for attaining safety parameters. They are considered to be the main challenge for battery technologies in automotive and solar energy industries. It will be demonstrated that LBL films can achieve the combination of several characteristics enabling the new type of flexible Li^+ batteries.

Biomedical implants must have many parameters similar to those needed in energy conversion with additional requirements for biocompatibility and biodegradability. LBL composites can satisfy all these stringent set of conditions. Preparation and functionality of a new type of implantable electrodes using LBL multilayers from carbon nanotubes and other materials will be demonstrated.

As a part of the future of LBL-made materials, the need to develop scaled-up versions of LBL deposition techniques is immediate and is expected to dominate the practical aspects of its realization. Some of the potential approaches to the resolution of this challenge and considerations for future devices in this area will be included in the presentation.

Relevant References.

1. Zhu, Jian et al, *J. Am. Chem. Soc.* **2011**, 133 (19), 7450–7460

2. Andres, Christine. M.; Kotov, Nicholas A. *J. Am. Chem. Soc.,* **2010**, 132(41), 14496-14502

3. Shim, Bong Sup, et al., *ACS Nano*, **2009**, *3* (7), 1711–1722.

4. Podsiadlo Paul., et al , *Science,* **2007**,318, 80-83.

5. Mamedov, Arif. et al. *Nature Materials*, **2002**, 1 190-194;

6. Tang, Zhiyong.; et al. *Nature Materials*, **2003**, 2, 6, 413-418.

(721f) Microrheological Measurements of the Effects of Cholesterol On the Mechanical Properties of Lung Surfactant Monolayers

Kyuhan Kim, Chemical Engineering, Univerisity of California, Santa Barbara, Santa Barbara, CA, Todd M. Squires, Chemical Engineering, University of California, Santa Barbara, Santa Barbara, CA and Joseph A. Zasadzinski, Department of Chemical Engineering and Materials Science, University of Minnesota, Minneapolis, MN

Cholesterol plays a significant role in biological interfaces, especially in the organization of membrane rafts in cell membranes and in clinical lung surfactants used to treat neonatal respiratory distress. We are particularly interested in lung surfactant (LS) monolayers, where the role of cholesterol remains unclear. Different clinical surfactants used for treatment of premature infants have widely different cholesterol contents; one of the main difficulties is that even though a small mole fraction of cholesterol (5-10%) has been reported in extracts of mammalian LS, there is still a question whether cholesterol originates from the LS or from cell debris. We show, using a novel, micron-scale surface rheometer, that small fractions of cholesterol dramatically change the elastic and viscous properties of LS monolayers. As little as 2 mol% cholesterol decreases the viscosity of a LS monolayer by two orders of magnitude, while likely enhances the spreading of LS as the alveolar area changes during respiration. 5 mol% cholesterol induces an elastic response in the monolayer, which may help retain LS in the deep lung by resisting the Marangoni flow caused by the surface tension gradient between the deep lung and bronchi. Using a combination of atomic force microscopy and optical fluorescence microscopy, we show the change in monolayer morphology induced by cholesterol provides important clues to the dynamic processes. We conclude that a cholesterol content of about 4 mol% is a critical concentration around which rheological properties can vary dramatically. Subtle changes in composition around this concentration can enable both a low viscosity to improve spreading during respiration, as well as a high elasticity to minimize Marangoni flow out of the lungs.

(721g) Immobilization Scheme Utilizing a Photopolymerizable Crosslinker and Biotinylated Fusion Proteins for Neural Tissue Engineering Applications

Aleesha M. McCormick[1], Asanka Wijekoon[2] and Nic D. Leipzig[1], (1)Department of Chemical and Biomolecular Engineering, The University of Akron, Akron, OH, (2)ARL BioPharma Inc., Oklahoma City, OK

Recombinant proteins have propelled tissue engineering into new directions through the inclusion of specific moieties that precisely direct cell and tissue behavior. Nerve growth factor (NGF) and semaphorin3A (Sema3A) are two proteins that aid in nervous system development and repair. Implementation of these bioengineered proteins into neural tissue engineering strategies is a novel approach for axon guidance, pruning and regeneration. For these studies, *E. coli* was transformed with pET-21a+ plasmids incorporating the fusion proteins NGF or Sema3A, and an AviTag sequence, which can be biotinylated using birA enzyme. NGF and Sema3A expression and isolation was successfully preformed using Ni-NTA affinity chromatography and further purified using fast protein liquid chromatography. N-(2-mercaptoethyl)-3-(3-methyl-3H-diazirine-3-yl) propanamide (N-MCEP-diazirine), a heterobifunctional crosslinker, was synthesized for the specific linkage of biotinylated NGF and Sema3A to chitosan films. Biotinylated proteins were also tagged with 5-(and-6)-carboxyfluorescien succinimidyl ester and attached to chitosan films utilizing N-MCEP-diazirine. DRGs were isolated from chicken embryos and placed inside immobilized regions containing either NGF or Sema3A in serum-free growth medium. Controls contained either soluble protein or adsorbed protein directly onto chitosan films. DRGs were imaged under brightfield, and neurite number and extension was quantified with ImageJ software. The amount of protein immobilized to the films compared to adsorbed groups was fluorescently quantified using a spectrophotometer. Our immobilization approach showed that we could specifically control tethering of both proteins at concentrations needed to control cell responses. Engineering and specifically immobilizing these proteins through inserted moieties via a photopolymerizable crosslinker, affords better axon outgrowth utilizing a minimal amount of protein. Our approach for the specific attachment of fusion proteins can be used in a wide variety of tissue regenerative applications. In addition, multiple recombinant proteins can be patterned onto the same substrate through the employment of various protein affinity pairings or bioconjugation techniques.

(721h) Structure, Properties, and Stromal Cell Response of Self-Assembled Micellar Star Poly(ethylene oxide-co-lactide) Hydrogels

Esmaiel Jabbari, University of South Carolina, Columbia, SC

Introduction: PEG hydrogels, due to their inert and non-fouling nature, provide enormous flexibility in designing and controlling the cell microenvironment. However, the use of PEG gels in tissue regeneration is limited by their persistence in the site of regeneration. Copolymerization of PEG with lactide has been used to impart degradability to PEG macromer. However, the covalently crosslinked copolymer networks, due to phase separation and entrapment of functional groups, had significantly lower stiffness than the PEG gels. The objective of this work was to investigate self-assembly, micelle structure formation of star acrylated poly(ethylene oxide-co-lactide) (SPELA) macromonomers in aqueous solution with lactide segment length, and the response of marrow stromal cells (MSCs) to the self-assembled hydrogels.

Methods: A two-step procedure was used to synthesize SPELA macromonomer. In the first step, star poly(ethylene glycol-co-lactide) macromers were synthesized by melt ring-opening polymerization with star PEG. In the second step, the terminal hydroxyl groups of SPEL macromers were reacted with acryloyl chloride to produce SPELA. Gelation kinetics of the macromonomer in aqueous solution and modulus of the hydrogel was measured by rheometry. Water content and degradation was measured in PBS as described. The effect of lactide segment length on micelle formation and micelle size was simulated by molecular dynamic simulation using the Mesocite module of the Materials Studio software (Accelrys). For evaluation of cell response, MSCs were mixed in the polymerizing precursor solution and UV crosslinked. The cell-laden gels were cultured in osteogenic media for 21 days. At each time point, homogenized samples were analyzed by DNA content, extent of mineralization, and mRNA analysis.

Results and Conclusions: As the number of lactide groups on each arm of the SPELA macromonomer was increased from 2 to 10, the diameter of the self-assembled micelles increased from 2 to 5 nm, respectively, but the micelle size was independent of SPELA concentration in aqueous solution. Interestingly, due to aggregation and micelle formation by the lactide segments, the inter-acrylate distance decreased, leading to increase in modulus with lactide segment length. For a given time, degradation of SPELA gel increased with lactide segment length up to segment length three, followed by a decrease in mass loss for higher values. The addition of only one lactide per macromonomer arm increased mass loss to 50% after 6 weeks. Lactide segment length did not have a significant effect on the viability of embedded MSCs. MSCs encapsulated in SPELA gels differentiated to osteogenic lineage when cultured in osteogenic media and formed a mineralized matrix. Results demonstrate that the inert non-fouling self-assembled SPELA gels are potentially useful as a degradable cell delivery matrix in tissue regeneration.

Acknowledgement: This work was supported by research grants to E. Jabbari from the National Science Foundation under grant Nos. CBET0756394, CBET0931998, and DMR1049381.

Reference: [1] Mercado, A.E. et al. J. Contr. Rel. 2009, 140:148.

(732a) Click-Functionalized Antibody-Nanoparticle Conjugate Drug Delivery System for Cancer Treatment

Emily Smith, Chemical Engineering, University of Cincinnati, Cincinnati, OH and Joo-Youp Lee, School of Energy, Environmental, Biological, and Medical Engineering, University of Cincinnati, Cincinnati, OH

Breast cancer is one of the leading causes of cancer patient death in this country. Current chemotherapy treatments use combinations of passive and targeted drug delivery systems, but no such system exists which combines the two. Chemo drugs can be encapsulated into nanosized carriers which passively target tumor cells via the EPR effect. To take this type of passive system to being an actively targeted system we propose using copper-catalyzed azide alkyne cycloaddition to attach an anti-HER2 antibody to the surface of the particles since this chemistry is efficient, regioselective, and can be performed in the benign solvents that proteins can tolerate. Our particles are made using fluorescent conjugated polymer blends with PLGA-PEG via nanoprecipitation. A flow cytometer has been used to measure the fluorescence present on MCF7 breast cancer cells which have been incubated with the modified particles to confirm conjugation of the antibody to the particle and to show that the antibodies retain their biological activity post modification. Results show that the chemistry sequence is successful in creating actively targeting particles.

(732b) Eliposomes: Novel Drug Delivery Vehicles

Marjan Javadi, Department of Chemical Engineering, Brigham Young University, Provo, UT, William G. Pitt, Department of Chemical Enginering, Brigham Young University, Provo, UT and David Belnap, Department of chemistry and Biochemistry, Brigham Young University, Provo, UT

An eLiposome is a liposome encapsulating an emulsion nanodroplet, and can be used for drug delivery. For example, therapeutic agents are encapsulated inside the eLiposomes, and application of ultrasound can cause the emulsion droplet to change from liquid to gas, thus increasing the volume inside the vesicle, causing rupture and release of the drug.

In this novel method, emulsion droplets were made of perfluorohexane (PFC6) or perfluoropentane (PFC5) and liposomes were made separately and then they were mixed using ultrasound; thus we call this the "ultra" method. Calcein was encapsulated inside the eLiposomes as a model drug. Dynamic light scattering and transmission electron microscopy were used to measure the size of the emulsions, liposomes and eLiposomes. The size of eLiposomes is appropriate for extravasation into tumors with malformed capillary beds. We hypothesize that ultrasound can be used to break open these eLiposomes. eLiposomes were constructed with folate on their surfaces to induce endocytosis. eLiposomes were used to successfully deliver calcein as a model drug to HeLa cells.

Figure 1 demonstrates that folate was successfully incorporated into PFC5 ultra eLiposomes, and that the folate was necessary to obtain calcein delivery to the cytosol of cells. Parallel experiments were conducted using eLiposomes prepared with and without DSPE-PEG2000-folate incorporated into the eLiposome bilayer. 200 uL of each eLiposome sample was added to HeLa cells in 12-well plates. The plates were incubated at 37°C for about 2 hours and then 20-kHz ultrasound was applied at 1 W/cm^2 for 2 seconds. Finally the cells were washed to remove external calcein and eLiposomes. Figure 1A shows the resulting calcein fluorescence from eLiposomes without folate, while figure 1B shows the parallel experiment having folate incorporated in the eLiposomes. Because the calcein was concentrated (30 mM) in the eLiposomes, it did not fluoresce until released. There is very little calcein signal observable in Figure 1A, suggesting that very few calcein-containing eLiposomes were endocytosed when folate was absent. In Figure 1B, the calcein fluorescence is originating from the cell cytosol, not the cell surface, suggesting that the folated eLiposomes had been endocytosed and that the calcein had been released to the cytosol. Some of these cells exhibit a granular pattern of calcein fluorescence, perhaps indicating that some calcein remains in endosomes, and while released from the eLiposome and diluted sufficiently to fluoresce, some calcein may not have been released completely from the endosome.

A B

Fig. 1. Confocal image of HeLa cells exposed for 2 hours to ultra eLiposomes containing 30 mM calcein, followed by application of 20-kHz ultrasound at 1 W/cm^2 for 2 seconds. (A) eLiposomes were not folated. (B) eLiposomes contained folate in their phospholipid membrane.

While Figure 1 shows that folate is necessary, Figure 2 shows that emulsions are also necessary for calcein delivery. Folated ultra eLiposomes with PFC5 emulsions and conventional liposomes (without emulsions) were prepared, each containing calcein at 30 mmol. 200 μL of each sample was added to HeLa cells in 12-well plates. Plates were incubated at 37°C for about 2 hours and then 20 kHz ultrasound was applied (1 W/cm^2 for 2 seconds). Figure 10A shows very little release of calcein to the cells exposed to folated conventional liposomes (without emulsion) while Figure 10B shows considerable release of calcein in the parallel experiment employing folated eLiposomes. Both these images were collected under identical optical conditions with the same gain and other settings on the confocal microscope. Most cells show calcein distributed through the cell, but again sometimes calcein appears more intensely in punctate spots, which might be unruptured endosomes.

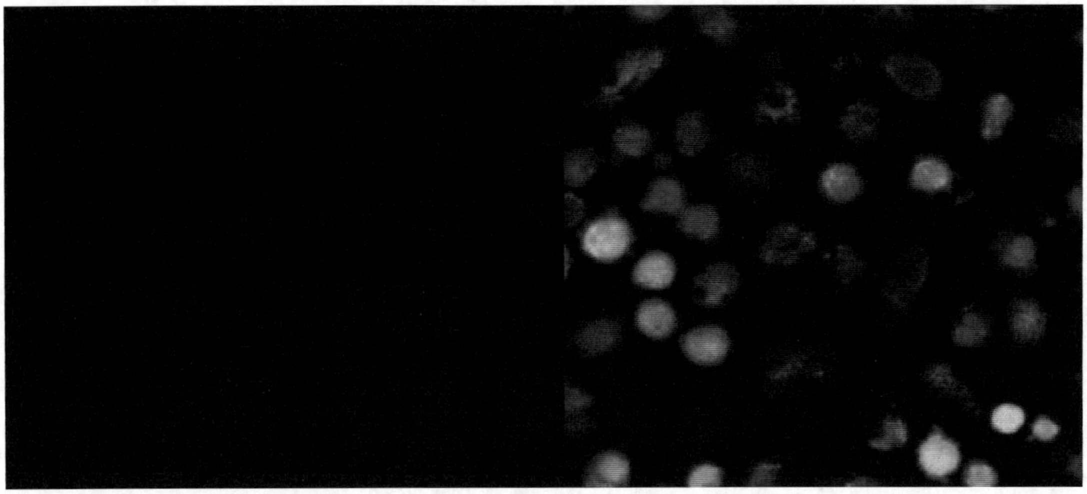

A B

Fig. 2. (A) Confocal image of HeLa cells exposed to calcein-containing folated liposomes (no emulsion within) for 2hours after which ultrasound was applied at 1W/cm^2 for 2 seconds. (B) Confocal image of the same experiment except that folated eLiposomes (containing emulsions) were used.

We propose that our eLiposomes can be used in ultrasonically activated drug delivery as follows. The eLiposomes could be loaded with drugs, such as Doxorubicin, using conventional or pH gradient techniques [1, 2]. Upon exposure to ultrasound, the local pressure oscillates above and below ambient pressure. At sufficiently high acoustic amplitudes, the local pressure decreases below the vapor pressure of the PFC emulsion inside the eLiposome. During this short time, the PFC may start to boil to a gas phase.

Lipid bilayers can only sustain about 3% area expansion before they rupture [3, 4]. For the PFC6 and PFC5 of this research, the liquid to gas expansion ratios are about 259-fold and 137-fold, respectively. For a liposome of 250-nm-diameter eLiposome containing a single 100-nm-diameter PFC6 emulsion droplet, less than 0.2% of the PFC liquid is required to vaporize to achieve the 3% area expansion for eLiposome lysis. Nucleation of the vapor phase may or may not occur on the first negative pressure cycle of the ultrasonic wave.

Although any liquid of high vapor pressure could be employed in eLiposomes, PFCs are useful for making emulsion droplets within eLiposomes because the phospholipids and PFCs are biocompatible and non-toxic [5].

[1] Y. barenholz, gilad haran, "Method of amphiphatic drug loading in liposomes by pH gradient," 1993.

[2] C. Celia, N. Malara, R. Terracciano, D. Cosco, D. Paolino, M. Fresta, and R. Savino, "Liposomal delivery improves the growth-inhibitory and apoptotic activity of low doses of gemcitabine in multiple myeloma cancer cells,"

Nanomedicine: Nanotechnology, Biology and Medicine, vol. 4, pp. 155-166, 2008.

[3] E. A. Evans, R. Waugh, and L. Melnik, "Elastic Area Compressibility Modulus of Red-Cell Membrane," *Biophysical Journal,* vol. 16, pp. 585-595, 1976.

[4] R. R. Netz and M. Schick, "Pore formation and rupture in fluid bilayers," *Physical Review E,* vol. 53, pp. 3875-3885, Apr 1996.

[5] M. Nieuwoudt, G. H. C. Engelbrecht, L. Sentle, R. Auer, D. Kahn, and S. W. van der Merwe, "Non-toxicity of IV Injected Perfluorocarbon Oxygen Carrier in an Animal Model of Liver Regeneration Following Surgical Injury," *Artificial Cells, Blood Substitutes and Biotechnology,* vol. 37, pp. 117-124, 2009.

(732c) Effect of Uptake Pathway On Non-Viral Gene Delivery in Vitro and in Vivo

Victor Wing Tat Shum, Chemical and Biomolecular Engineering, University of Illinois at Urbana-Champaign, Urbana, IL and Daniel W. Pack, Chemical & Biomolecular Engineering, University of Illinois Urbana-Champaign, Urbana, IL

Efficient non-viral gene delivery often involves the conjugation of a cell-specific ligand to the vector, which directs the vector to its intended target. However, little is known in terms of how the process of endocytosis affects the performance of the delivery vehicle. Previously, we have demonstrated that caveolin-mediated endocytosis is important to *in vitro* polyethylenimine (PEI)-mediated gene delivery in HeLa cells through the use of small molecule drugs and small-interfering RNA. The goal of this project is to elucidate the effects of cellular uptake mechanism on non-viral gene delivery *in vitro* and *in vivo* utilizing small hairpin RNA (shRNA) Tet-on system. Specifically, we are investigating the efficacy of PEI-mediated gene delivery upon internalization via clathrin-dependent endocytosis or caveolin-dependent endocytosis in HeLa and MDA-MB-231 cell cultures and xenograft tumors. We attached transferrin and folate ligands to PEI to direct internalization by clathrin- and caveolin-mediated mechanisms, respectively, and transfected HeLa and MDA-MB-231 cells with targeted and untargeted PEI in the presence of small molecule drugs that inhibit clathrin- or caveolin-mediated endocytosis. We have demonstrated that caveolin-dependent endocytosis is important to successful PEI-mediated gene delivery in both HeLa and MDA-MB-231 cells. In order to further understand the impact of endocytic pathways on gene delivery efficiency *in vitro*, we have modified HeLa and MDA-MB-231 cells to express small hairpin RNA (shRNA) that target clathrin, caveolin, or lamin A/C (negative control) protein expression (HeLa-CLTC, HeLa-CAV, HeLa-LAM, 231-CLTC, 231-CAV, and 231-LAM) in the presence of tetracycline. Using the shRNA Tet-on system, we are able to demonstrate ~90% clathrin knockdown and ~60% caveolin knockdown in both modified HeLa and MDA-MB-231 cells. Using this shRNA knockdown mechanism, we have confirmed that caveolin-dependent endocytosis is more important than clathrin-dependent endocytosis in PEI-mediated gene delivery, which corroborates the results observed in unmodified cell lines. We will also report the effect of clathrin-dependent and caveolin-dependent endocytosis on *in vivo* PEI-mediated gene delivery in a xenograft murine tumor model. Such studies of intracellular trafficking and its effect on PEI gene delivery efficacy in cancer cell lines and tumors enable researchers to design more efficient non-viral gene delivery vectors for cancer gene therapy, by synergistically combining the design of cell targeting ligands and the understanding of cellular uptake mechanism for polymeric carriers.

(732d) Cationic Polylactides for siRNA Delivery

Chih-Kuang Chen[1], Wing-Cheung Law[2], Ravikumar Aalinkeel[3], Bindukumar Nair[3], Atcha Kopwitthaya[3], Supriya D. Mahajan[3], Jessica L. Reynolds[3], Jiong Zou[1], Stanley A. Schwartz[3], Paras N. Prasad[2] and **Chong Cheng**[1],
(1)Department of Chemical and Biological Engineering, SUNY-Buffalo, Buffalo, NY, (2)Institute for Lasers, Photonics and Biophotonics, SUNY-Buffalo, Buffalo, NY, (3)Department of Medicine, SUNY-Buffalo, Buffalo, NY

Cationic polymers have been broadly studied as scaffolds for gene delivery. However, because typical cationic polymers cannot degrade readily, their long-term cytotoxicity is a serious concern. Therefore, we are interested in integrating significant degradability with cationic polymers to create novel and safe polymeric materials for applications in gene delivery. In this presentation, synthesis and characterization of cationic polylactides (CPLAs), as well as their application in siRNA delivery, will be described. A series of CPLAs with different amounts of pendent tertiary amine functionalities were synthesized by the preparation of allyl-functionalized PLA via ring-opening polymerization, followed by thiol-ene click functionalization. The CPLAs had well-controlled molecular weights and amine mol%, as well as low polydispersities. With significant hydrolytic degradability, the CPLAs exhibited very low cytotoxicity via MTS cell viability assay, indicating that they are relatively safe materials for biomedical applications. According to DLS and TEM analysis, CPLAs can readily complex with negatively charged siRNA to form CPLA-siRNA nanoplexes by electrostatic interaction. In vitro studies showed that, with ready cell uptake of the CPLA-siRNA nanoplexes, efficient siRNA transfection via the nanoplexes can be realized.

(732e) Effect of Ionizable Head Group Architecture On the Delivery Efficiency of Lipid-Based siRNA Nanoparticles

Christopher Alabi[1], Kevin Love[2], Gaurav Sahay[1], Robert Langer[3] and Daniel G. Anderson[1], (1)Koch Institute for Integrative Cancer Research, Massachusetts Institute of Technology, Cambridge, MA, (2)Chemical Engineering, Massachusetts Institute of Technology, Cambridge, MA, (3)Massachusetts Institute of Technology, Cambridge, MA

Synthetic small interfering RNAs (siRNAs) have emerged as powerful tools for the regulation and silencing of endogenous genes. To achieve efficient gene silencing, siRNAs must be encapsulated or conjugated to delivery vehicles that can ensure cytosolic delivery of intact siRNAs across the cellular membrane. So far, liposomal delivery vehicles have shown promising results for *in vivo* systemic delivery to hepatic and immune cellular targets. Using a combinatorial synthetic chemistry approach in conjunction with a lipid nanoparticle (LNP) formulation, our group recently identified a new class lipid-like structures (lipidoids) that facilitate efficient siRNA delivery. In an effort to comprehend the relationship between lipidoid structure and delivery efficiency, both *in vitro* and *in vivo*, this body of work focuses on evaluating the effect of the ionizable amine core structure of the lipidoid on the biophysical properties of the LNPs such as particle size, siRNA entrapment, LNP pKa, LNP lytic ability and extracellular LNP stability. These biophysical properties, along with the structure of the amine core are then correlated with uptake and gene silencing efficiency of the LNPs both *in vitro* and *in vivo*. The biophysical parameters obtained from this study, along with their correlation to lipidoid structure and gene-silencing efficiency, gives us some insight into possible future design strategies for the synthesis and evaluation of potent lipid structures for siRNA delivery.

(732f) A Novel Anionic Nanoparticle Delivery System for Microrna-29b Targets FLT3 and KIT Receptor Tyrosine Kinase Expression in Acute Myeloid Leukemia

Xiaomeng Huang[1], Sebastian Schwind[2], Bo Yu[1], Shujun Liu[2], Jiuxia Pang[2], Ramasamy Santhanam[2], Yue-Zhong Wu[2], Kenneth K. Chan[3], William Blum[2], Clara D. Bloomfield[2], Danillo Perroti[2], Ramiro Garzon[2], John C. Byrd[2], Natarajan Muthusamy[2], Robert J. Lee[3], Guido I. Marcucci[2] and Ly James Lee[4], (1)Nanoscale Science and Engineering Center, The Ohio State University, Columbus, OH, (2)Comprehensive Cancer Center, The Ohio State University, Columbus, OH, (3)College of Pharmacy, The Ohio State University, Columbus, OH, (4)William G. Lowrie Department of Chemical and Biomolecular Engineering, The Ohio State University, Columbus, OH

MicroRNAs, small non-coding RNA molecules, contribute to the pathogenesis of various human cancers, including acute myeloid leukemia (AML). One of these microRNAs - *miR-29b* - is frequently downregulated in AML, has antileukemic activity and was shown to target genes involved in leukemogenesis.[1] Furthermore, higher pretreatment levels of *miR-29b* were associated with improved clinical response to treatment with the anti-leukemic agent decitabine.[2] Recently we demonstrated that *miR-29b* also downregulates the expression of the receptor tyrosine kinases (RTKs) KIT and FLT3 in AML.[3,4] The aberrant activation of the RTKs KIT and FLT3, by mutations or overexpressed is frequently found in AML and much effort has been undertaken to target these RTKs. In AML, due to development of early resistance, most inhibitors of RTKs have not led to substantial clinical benefit. However, instead of targeting the activity of RTKs, decreasing their expression may be a valuable therapeutic alternative. Thus the artificial increase of *miR-29b* levels in AML blasts to downregulate the RTKs KIT and FLT3 may be of high clinical benefit. Although some miR delivery methods have been developed for solid tumors, these methods have limitations that prevent their use in leukemia therapy. Cationic lipid/polymer formulations frequently used in solid tumors have the tendency to accumulate in lungs, kidney and liver and may cause a non-specific immune response, for which the consequences in AML have not been fully explored yet. We developed a transferrin-conjugated anionic lipid-based nanoparticle (NP) system to deliver synthesized *miR-29b* mimic molecules to leukemia cells. Low-molecular-weight polyethylenimine (PEI) was used to provide a positive charge, which can easily capture the negatively charged miR molecules achieving high entrapment efficiency of more than 90%. The lipid-based carrier was made of 1,2-Dioleoyl-sn-Glycero-3-Phosphoethanolamine (DOPE), Linoleic acid and 1,2-Dimyristoyl-sn-glycerol, methoxypolyethylene Glycol (DMG-PEG). We chose linoleic acid because of its low binding affinity to small RNAs, which implies easy dissociation of miRs from the lipopolyplex core. Moreover, the NPs were protected from reticuloendothelial system clearance by DMG-PEG, which allowed long *in vivo* circulation times. Since AML blasts overexpress the transferrin receptor, we conjugated the NPs with transferrin (Tf) using a post-insertion method to increase the uptake by targeted delivery. The average size and zeta potential of empty NP and miR-loaded NP were 129.6 ± 1.02nm (± Standard Deviation [SD]) and -9.82 ± 1.53mV (± SD) and 137.6 ± 0.96nm and 22.48 ± 1.44mV, respectively. After the Tf-conjugation of the miR-loaded NP the size increased to 147.3 ± 4.74nm and the zeta potential was 5.82 ±1.87mV. To evaluate the efficiency of NP uptake we encapsulated FAM-fluorescent labeled miR and treated two AML cell lines (i.e. Kasumi-1 and MV4-11). Tf-conjugated NP (Tf-NP) treated cells showed the highest uptake by confocal microscopy and flow-cytometry (mean fluorescence intensity (MFI): 21.0 and 22.9 for Kasumi-1 and MV4-11 respectively) compared to non-Tf-conjugated NPs (MFI: 8.3 and 13.0). Next, we evaluated the efficiency of *miR-29b* mimics delivery to AML cell lines and primary patient blasts with the Tf-NP system. After treatment with *miR-29b* loaded Tf-NPs (Tf-NP-*miR-29b*) we observed an increase of intracellular levels of *miR-29b,* 420 folds increase in Kasumi-1 cells, 220 folds increase in MV4-11 cells and an average 1000 folds increase in patient blasts compared to negative control-loaded Tf-NP (Tf-NP-nc) treatment after 24 hours. The expression levels of an unrelated miR, i.e. *miR-140,* remained unchanged. To assess whether the increased *miR-29b* levels would lead to a decrease of KIT and FLT3 expression we performed western blot analysis 48 hours after treatment. Tf-NP-*miR-29b* treatment led to 3.4 and 2 folds decrease of FLT3 and 1.5 and 1.4 folds decrease of KIT in Kasumi-1 and MV4-11 cells respectively and on average 1.5 folds decrease of FLT3 and 2.5 folds decrease of KIT in cells derived from three AML patients compared to Tf-NP-nc treated cells. The Tf-NP-*miR-29b*-mediated downregulation of these targets had antileukemic effects on AML cells. Tf-NP-*miR-29b* treatment reduced growth (per day) from 32.2% to 25.3% (P=0.01) in Kasumi-1 cells and from 53.0% to 43.9% (P=0.007) in MV4-11 cells, compared to Tf-NP-nc treatment. We also observed a reduction of numbers of colony formed by 55-66% after Tf-NP-*miR-29b* treatment after 2 weeks. The average numbers of colonies formed were 143±9(±SD) and 65±6 (P<0.001) for Kasumi-1 and 213±7 and 80±5 (P<0.001) for MV4-11 of Tf-NP-nc and Tf-NP-*miR-29b* treatment, respectively. Since we recently demonstrated that higher pretreatment levels of *miR-29b* were associated with improved clinical response to decitabine treatment,[4] we also tested whether priming AML cells with Tf-NP-*miR-29b* prior to decitabine treatment would have additional therapeutic effect. Tf-NP-*miR-29b* treatment (for 48h) followed by 0.5 μM decitabine (for 48h) decreased the cell viability compared to Tf-NP-nc treatment or unprimed cells by 40% (P=0.001) and 30% (P=0.01), respectively in Kasumi-1 and by 16% (P=0.003) and 22% (P=0.001), respectively in MV4-11 cells. We confirmed these findings in three AML patient blasts, Tf-NP-*miR-29b* priming, followed by 5 μM decitabine treatment decreased cell viability for patient 1 by 20% (P=0.01), for patient 2 by 40% (P=0.003) and for patient 3 by 22% (P=0.002), respectively compared to Tf-NP-nc priming. In conclusion, we report here a novel anionic NP-based delivery system for miRs in AML. Our delivery system for *miR-29b* showed targeting activity against the important RTKs KIT and FLT3 that are associated with leukemogenesis, antileukemic effects and improved response to decitabine treatment in AML cells. The modulation of *miR-29b* levels with the here proposed delivery system as a single agent or in combination with other anti-cancer drugs - i.e. decitabine - may represent a promising new targeted treatment option for AML patients in the near future.

1. Garzon R, Liu S, Fabbri M, et al. MicroRNA-29b induces global DNA hypomethylation and tumor suppressor gene reexpression in acute myeloid leukemia by targeting directly DNMT3A and 3B and indirectly DNMT1. *Blood*. 2009;113(25):6411–6418.
2. Blum W, Garzon R, Klisovic RB, et al. Clinical response and miR-29b predictive significance in older AML patients treated with a 10-day schedule of decitabine. *PNAS*. 2010;107(16):7473–7478.
3. Liu S, Wu L, Pang J, et al. Sp1/NFκB/HDAC/miR-29b Regulatory Network in KIT-Driven Myeloid Leukemia. *Cancer Cell.* 2010; 17(4): 333-347
4. Blum W, Schwind S, Tarighat SS, et al. Clinical and Pharmacodynamic Activity of the Combination Bortezomib and Decitabine: a Phase I Trial in Patients with Acute Myeloid Leukemia. *Blood.* 2012; published ahead of print as doi:10.1182/blood-2012-03-413898.

(756a) Understanding and Controlling the Substrate Effect On Graphene Electron Transfer Chemistry Via Reactivity Imprint Lithography

Qing Hua Wang[1], Zhong Jin[2], Ki Kang Kim[3], Andrew J. Hilmer[1], Geraldine L. C. Paulus[1], Chih-Jen Shih[1], Moon-Ho Ham[4], Javier D. Sanchez-Yamagishi[5], Kenji Watanabe[6], Takashi Taniguchi[6], Jing Kong[3], Pablo Jarillo-Herrero[5] and Michael S. Strano[1], (1)Department of Chemical Engineering, Massachusetts Institute of Technology, Cambridge, MA, (2)Chemical Engineering, Massachusetts Institute of Technology, Cambridge, MA, (3)Department of Electrical Engineering and Computer Science, Massachusetts Institute of Technology, (4)School of Materials Science and Engineering, Gwangju Institute of Science and Technology, (5)Department of Physics, Massachusetts Institute of Technology, (6)Advanced Materials Laboratory, National Institute for Materials Science

Graphene has exceptional electronic, optical, mechanical, and thermal properties that make it promising for electronic, optoelectronic, and sensing applications. The chemical functionalization of graphene has been pursued to control its electronic properties and interactions with other materials. Covalent modification of graphene by organic diazonium salts has been used to achieve these goals, but because graphene is only a single atomic layer, it is strongly influenced by the underlying substrate. In this paper, we show a stark difference in the rate of electron transfer reactions with organic diazonium salts for monolayer graphene supported on a variety of substrates. Reactions proceed rapidly for graphene supported on SiO_2 and Al_2O_3 (sapphire), but negligibly on alkyl-terminated and hexagonal boron nitride (hBN) surfaces, as shown by Raman spectroscopy. We develop a model of reactivity based on substrate-induced electron-hole puddles in graphene, as the local Fermi level fluctuations directly influence the electron transfer rate between the graphene and diazonium molecule. We also develop a new patterning technique called reactivity imprint lithography (RIL) for spatially patterning chemical groups on graphene by patterning the underlying substrate, and apply it to the covalent tethering of proteins on graphene.

(756b) Analysis of Defect-Induced Amorphization of Single-Layer Graphene

Corinne Carpenter, Chemical Engineering, University of Massachusetts Amherst, Amherst, MA, Ashwin Ramasubramaniam, Mechanical and Industrial Engineering, University of Massachusetts Amherst, Amherst, MA and Dimitrios Maroudas, Chemical Engineering, University of Massachusetts Amherst, Amhest, MA

Defect engineering of graphene has been studied extensively, both experimentally and computationally, aimed at approaches for tuning graphene's electronic, optical, and mechanical properties. Ion-beam and electron-beam irradiation have been employed to introduce in the honeycomb lattice of graphene defects, such as vacancies, reconstructions with pentagon and heptagon rings, as well as larger-scale defects such as two-dimensional (2D) holes or voids. Recent experimental studies also have reported defect-induced amorphization of graphene after high electron-beam exposure. High-resolution imaging of the resulting amorphized graphene demonstrated that such irradiation-induced amorphized graphene remains a coherent single-layer 2D structure that consists of a random patch of polygons and also contains 2D holes.

In this presentation, we address some fundamental questions on the defect-induced amorphization of graphene, including how the defect-induced structure and morphology depend on the inserted defect concentration, the determination of the onset of defect-induced amorphization, and the effects of the defect density and amorphization transition on the graphene's electronic structure and properties. Our study was based on isothermal-isobaric molecular-dynamics (MD) simulations according to a reliable bond-order potential for the description of interatomic interactions in graphene and using large supercells that contain thousands of C atoms. Random distributions of single vacancies (or other types of point defects) were introduced into the graphene honeycomb lattice at a given concentration and the response of the defective graphene sheets was explored systematically over a broad range of defect concentration based on a variant of simulated annealing. The analysis of the simulation results for the relaxation of the defective graphene samples included detailed structural and morphological characterization of the relaxed configurations based on calculation of certain metrics, such as the radial distribution function, structural order parameters, and RMS surface roughness; the relaxed configurations consitute computer models of irradiated graphene. Finally, we used density-functional tight-binding calculations to compute the electronic density of states of the structurally relaxed configurations.

We predict that the onset of the amorphization transition occurs at an inserted vacancy concentration between 5 and 10%, with the transition becoming less abrupt with vacancy concentration as temperature increases, and determine the surface roughness as a function of defect concentration. Our predictions for the defect-induced amorphization transition are similar for all the types of point defects examined in our study. We find that the computed amorphized configurations are in good agreement with those observed in recent experiments of high-dose exposure of graphene to 100-keV electron irradiation. We also find that the electronic density of states of vacancy-amorphized graphene is characterized by introduction of localized states near the Fermi level of perfect single-layer graphene, consistent with those arising from random vacancy insertion in graphene at low concentrations. Our findings set the stage for exploratory work on the functionalization of defect-amorphized graphene toward useful band-gap engineering and tuning of electronic transport and optical properties.

(756c) Quantum Tunneling in Graphene-Based and Carbon Nanotube-Based Nano Electronic Devices

Meng-Mu Shih, Electrical and Computer Engineering, University of Florida, Gainesville, FL

Graphene has unique electric, optical and mechanical properties. The high electronic mobility and compact size can improve the performance of electronics. The high thermal conductivity can conduct heat easier so the heat accumulation problem can be solved in smaller electronics. Flexible and light graphene-based electronics possibly can make the products more wearable, portable, durable and affordable. With the progress in fabrication methods, graphene-based electronics can be fabricated and integrated more easily.

A modified preliminary model shows how graphene-related parameters can affect quantum tunneling in graphene-based electronic devices. Conventional quantum tunneling can be affected by parameters such as material work function, applied electric field, effective mass of material. In the modified model, the graphene-related parameters such as the number of graphene layer can affect the quantum tunneling in nano electronic devices. Quantum tunneling in carbon nanotube-based electronic devices will be also discussed and compared. Future work and possible applications will be proposed.

(756d) Facile, Controllable Graphene-Based P-N Junctions Using Self-Assembled Monolayers

Jose Baltazar[1], Hossein Sojoudi[2], Juan Vargas[3], Janusz Kowalik[3], Laren M. Tolbert[4], Samuel Graham[5] and Clifford L. Henderson[6], (1)School of Chemical and Biomolecular Engineering, Georgia Tech, Atlanta, GA, (2)Woodruff School of Mechanical Engineering, Georgia Tech, (3)Chemistry & Biochemistry, Georgia Tech, (4)School of Chemistry and Biochemistry, Georgia Institute of Technology, Atlanta, GA, (5)Mechanical Engineering, Georgia Institute of Technology, Atlanta, GA, (6)Chemical and Biomolecular Engineering, Georgia Institute of Technology, Atlanta, GA

Graphene, a two-dimensional sp^2 hybridized carbon lattice that is also the fundamental building block of graphite, has attracted significant interest recently due to its distinctive electrical and mechanical properties. An important challenge is to develop methods for controllably doping graphene, since such methods potentially give access to both p-type and n-type junctions and can allow for adjustment of the work function of graphene. In this study we investigate the use of a self-assembled monolayer (SAM) to create a p-n junction in graphene films. Previous techniques rely on charge transfer from adsorbants or electrostatic gate/potentials. Here we demonstrate that, by successfully modifying the SiO_2 surface with an aminopropyltriethoxysilane (APTES) layer, and using intrinsically p-doped transferred CVD graphene films, a well-defined junction can be achieved. Field-effect transistors and p-n junction regions are fabricated prior to graphene film transfer, in order to preserve the pristine properties of the graphene. The I-V characteristic curve indicates the presence of two thermally-controllable neutrality points. This method allows a facile, controllable and low temperature fabrication of graphene p-n junctions.

(756e) Enhanced Electrochemical Detection Performance of Multiwall Carbon Nanotubes Functionalized by Aspartame

Miao Liang, **Renliang Huang**, Rongxin Su, Wei Qi and Zhimin He, School of Chemical Engineering and Technology, Tianjin University, Tianjin, China

Carbon nanotubes (CNTs) are outstanding materials due to their exceptional properties including excellent mechanical strength, high electrical and thermal conductivities, high surface area, and high aspect ratios [Adsul et al., 2011]. However, their water insolubility is a major bottleneck in the exploitation of these excellent properties. Generally, noncovalent functionalizations of the CNTs sidewalls can preserve their desired properties, while improving their solubilities quite remarkably [Zhao and Stoddart, 2009]. Therefore, inexpensive, nontoxic, and biocompatible materials that can disperse CNTs in aqueous solutions through a noncovalent approach are highly desirable.

In this paper, we report the use of an amphiphilic dipeptide derivative, aspartame, as an effective dispersing agent in preparing highly stable multiwall carbon nanotubes (MWCNTs) suspensions under ultrasonication. In addition, we investigated the stability of the resulting MWCNT/aspartame composites at a wide range of pH values by zeta potential measurement. Furthermore, an investigation into whether the MWCNT/aspartame composites could improve the electrochemical performance of glassy carbon electrode (GCE) was conducted. The main ideas and conclusions are summarized as follows.

(1) The MWCNTs can be well dispersed in water in the presence of aspartame via brief ultrasonication. The resulting MWCNT/aspartame composites demonstrated a high degree of dispersion, which derived from the amphiphilic properties of aspartame. Specifically, L-phenylalanine of aspartame molecule gives aspartame the strong affinity property [Piao et al., 2008], which allows it to be adsorbed onto the MWCNT surface through the p-p stacking of aromatic groups. In addition, The MWCNT/aspartame composites exhibited zeta potentials as low as -36.5 mV at pH values between 6 and 12. The displayed negative zeta potential indicates that the MWCNT surface was negatively charged over a broad range of pH values, which gives the composites great promise in biological applications.

(2) The dispersion qualities and structural details of the MWCNT/aspartame composites was characterized by SEM, HRTEM and FTIR spectroscopy. SEM images showed the majority of the nanotubes exist individually, with some small aggregates of two or three tubes. HRTEM images also demonstrated smooth sidewalls was "polluted" by extra particles of a different species, which are believed to be aspartame debris, after irradiations. Therefore, the morphological observations further suggest that the amphiphilic aspartame was indeed adsorbed onto the nanotube sidewalls. Moreover, FTIR spectra also confirm that aspartame was successfully adsorbed onto the surface of the nanotubes through the p-p stacking interaction between the MWCNTs and the aromatic ring of aspartame.

(3) The MWCNT/aspartame composites could improve the electrochemical performance of GCEs. The enhanced electrochemical reversibility of the MWCNT/aspartame modified electrodes with the strongest peak currents was observed compared with that of the bare electrode. The calculated effective surface area of the MWCNT/aspartame - coated GCE was 11.68 mm^2, which is larger compared with 9.14 mm^2 for the MWCNT-coated GCE and 7.21 mm^2 for the bare GCE. Furthermore, the MWCNT/aspartame-coated electrode possesses better electrochemical detection performance for hydrogen peroxide, with its greatly enhanced electrocatalytic response and reduced potential. These findings indicate that the utilization of aspartame as a dispersing agent for CNTs provides a favorable avenue for the fabrication of CNT-modified electrodes and offers great promise for amperometric biosensors.

This work was supported by the Natural Science Foundation of China (No. 20806057 and 31071509), the Ministry of Science and Technology of China (Nos. 2012BAD29B05 and 2012AA06A303), and the Ministry of Education (No. NCET-11-0372).

References

1. Adsul M.G., Rey D. A. and Gokhale D.V. Combined strategy for the dispersion/dissolution of single walled carbon nanotubes and cellulose in water, *Journal of Materials Chemistry*, 21, 2054-2056, 2011
2. Zhao Y. L. and Stoddart J. F. Noncovalent Functionalization of Single-Walled Carbon Nanotubes, *Accounts of Chemical Research*, 42, 1161-1171, 2009
3. Piao L. Y., Liu Q. R., Li Y.D. and Wang C. Adsorption of L-Phenylalanine on Single-Walled Carbon Nanotubes, *The Journal of Physical Chemistry C,* 112, 2857-2863, 2008

(756f) Asymmetric Deposition of Manganese Oxide in Single Walled Carbon Nanotube Films As Electrodes for Flexible High Frequency Response Electrochemical Capacitors

Yuan Chen, School of Chemical and Biomedical Engineering, Nanyang Technological University, Singapore, Singapore and Jianmin Shen, Institute of Chemical and Engineering Sciences, Singapore , Singapore, Singapore

Manganese oxide (MnO_2) is a promising pseudocapacitive electrode material because of its high capacitance, abundant resource, low-cost, and environmental friendliness. However, its poor electrical and ionic conductivities and low stability hinder applications. Forming MnO_2 nanocomposites with high surface area porous metal, carbon materials, or conducting polymers is a possible solution. In this work, we have developed a facile and scalable asymmetric *in situ* deposition method to incorporate MnO_2 nanoparticles in conductive single walled carbon nanotube (SWCNT) films. The high porosity of vacuum filtrated SWCNT films accommodates pseudocapacitive MnO_2 nanoparticles without sacrificing the mechanical flexibility and electrochemical stability of SWCNT films. We exposed one side of SWCNT films to acidic potassium permanganate ($KMnO_4$) solution. The infiltrated $KMnO_4$ solution partially etches SWCNTs to create abundant mesopores, which ensure electrolyte ions efficiently access deposited MnO_2. Meanwhile, the remaining SWCNT network serves as excellent current collectors. The electrochemical performance of the SWCNT–MnO_2 composite electrodes depends on the porosity of SWCNT films, pH, and concentration of $KMnO_4$ solution, deposition temperature and time. Our optimized two–electrode electrochemical capacitor, with 1 M Na_2SO_4 in water as electrolyte, showed a superior performance with specific capacitance of 529.8 F g^{-1}, energy density of 73.6 Wh kg^{-1}, power density of 14.6 kW kg^{-1}, excellent capacitance retention (99.9%) after 2000 charge and discharge cycles, and one of the highest reported frequency responses (knee frequency at 1318 Hz). The high performance flexible electrochemical capacitors have broad applications in portable electronics and electrical vehicles, especially when high frequency response is desired.

(756g) Flexible Resistive Based Temperature Sensor to Detect Heat and Sweat Inside the Sockets of Prosthetics

Nathaniel J. Blasdel, Department of Chemical and Biomolecular Engineering, The University of Akron, Akron, OH

The most current estimate states there are approximately 1 in 190 persons living in the United States with major limb-loss while the rate of amputations completed increases each year. This necessitates an importance for understanding both quality of life rated (QOLR) issues and options to remediate prevalent problems. There are a number of important quality of life factors affecting people living with the loss of a limb and the number one issue is ambulation or replacing the functionality of the lost limb. Throughout time we have developed many different styles of prosthetics to fill this deficiency, while the socket style prosthetic has evolved to be one of the most common solutions. These socket style prosthetics facilitate a need to replace a missing limb, but with this comes problems associated with the use of the device. Proper fitting is important, along with proper care of the device and residual limb. These are issues that can be controlled by the amputee. One major, currently uncontrollable issue reported by amputees using socket style prosthesis is the combination of heat and sweat inside the socket. This is troublesome for a residual limb, because the socket can become hot and humid during even just regular use and can cause a variety of dermatological conditions if proper care is not taken. The major contributors to heat and sweat inside the socket are personal activity, and socket and liner materials of construction. Socket and liner materials of construction negatively affects the socket environment by inhibiting heat transfer away from the residual limb and just ten minutes of walking can increase the average residual limb temperature by 1.7¼C. A reduction in heat transfer causes sweat inside the socket, which can create a moist, abrasive environment between the skin and the sock or liner, coupled with pressure from the prosthetic socket during use facilitates an uncomfortable environment with the potential for sores, blisters, and rashes. The purpose of this presentation is to discuss the feasibility of monitoring heat and sweat by using a sensing composite made of nylon 6, carbon nanotubes, and polypyrrole. The sensing unit incorporates a resistive device composed of the substrate, nylon nanofibers, the bulk electron transport medium, carbon nanotubes, and the sensing layer, conductive polypyrrole. This presentation describes the design and fabrication methods of a working flexible temperature sensor. It illustrates how the fiber diameter, nanotube loading, and polymerization scheme affects the sensor current response and it will discuss the techniques used to analyze and verify the material.

(766a) Gelatin Stabilized Noble Metalic Nanoparticles for SERS Detection

Changwon Lee, Chemistry, University of Cincinnati, CINCINNATI, OH and Peng Zhang, Chemistry, University of Cincinnati, Cincinnati, OH

Gelatin reduced and stabilized noble metallic nanoparticles have been developed with one-pot, green synthesis for Surface Enhanced Raman Spectroscopy (SERS) applications. Both gold and silvernanoparticles were developed and the size of the particle was determined to be 17~ 18 nm for gold nanoparticles and 8 ~ 13 nm for silver nanoparticles by TEM. The size of the nanoparticles was dependent on the initial gelatin concentration during the synthesis.

Gelatin was chosen for both reducing and stabilizing agent for many advantages over conventional methods including easy one-pot preparation, water solubility, improved stability, providing proper gap between metallic surface and SERS substrate molecules. SERS sensitivity of gelatin gold and silver nanoparticles were compared with Rhodamine 6G (R6G) and Ru(BPy) as reporter molecules. And for silver nanoparticles, the SERS sensitivity for melamine and folic acid was shown. The concentration of the SERS substrates for the detection was as low as 0.1 μM for R6G and Ru(BPy) and 10 μM for melamine and folic acid in case of silver nanoparticles.

(766b) Coaxial Nanofibers with Aligned Gold Nanorods near the Fiber Surface for Surface Enhanced Raman Spectroscopy

Jay Hoon Park, Chemical and Biomolecular Engineering, Cornell University, Ithaca, NY and Yong L. Joo, School of Chemical & Biomolecular Engineering, Cornell University, Ithaca, NY

Molecular sensors were made using Au or Ag and polymer composite fibers fabricated by electrospinning. Firstly, Surface-enhanced Raman Scattering (SERS) template was synthesized using electrospun Au nanorod/Poly vinly alcohol (PVA) nanofibers. The electromagnetic resonance from mutual orientation of Au nanorods in PVA fiber exhibited great enhancement of 3,3′-diethylthiatricarbocyanine iodide (DTTCI) signals under Raman scope. We then directed the placement of nanorods along the surface by coaxially electrospinning Au nanorod/PVA and pure PVA as shell and core, respectively, which showed even greater DTTCI signals than the monoaxially electrospun Au nanorod/PVA fiber with the same Au nanorod concentration. This finding suggests that the loading of costly Au nanorods can be reduced up to three fold with the same sensitivity, resulting in more cost-effective free-standing SERS template. In addition, we have used coarse-grained molecular dynamics simulations to demonstrate that the extensive elongation from electrospinning attributes to the dispersion and alignment of nanorods in the cylindrical polymer matrix, and the selective placement of nanorods along the shell in coaxial electrospinning. Secondly, we will demonstrate that metal nanofibers from water-based electrospinning of highly loaded metal precursors can be served as a highly efficient SERS template. Fabrication of such cost-effective and highly productive ultrasensitive SERS template shows great potentials in advanced biomedical imaging or single-molecule detection device.

Coaxial Nanofibers with Aligned Gold Nanorods near the Fiber Surface for Surface Enhanced Raman Spectroscopy

(766c) Highly Luminescent Metal Nanoclusters (< 1 nm) for Chemical and Biological Sensor Development

Jianping Xie, Chemical and Biomolecular Engineering, National University of Singapore, singapore, Singapore and Xun Yuan, National University of Singapore, Singapore, Singapore

Ultrasmall (< 1 nm) noble metal nanoclusters (NCs) present an emerging super-molecule consisting of several to a hundred metal atoms, and with very interesting chemical and physical properties which are not observed in single metal atoms and large metal nanocrystals (> 2 nm). Due to the strong quantum size confinement in this size regime (< 1 nm), NCs possess discrete and size-tunable electronic transitions, and display unique molecule-like properties, such as quantized charging and luminescence. Strong luminescence is one of the most attractive features for these NCs due to their ultrafine size, good photostability, and low toxicity, some of which could not be realized by other luminophores such as organic dyes (*e.g.*, poor photostability) and semiconductor quantum dots (*e.g.*, toxicity concerns). This presentation will summarize our recent work in the development of novel synthetic routes for highly luminescent metal NCs, and for sensor applications for various analytes, such as toxic metal ions and biomolecules.

(766d) Oscillatory Surface Reactions of Riboflavin, Trolox and Reactive Oxygen On Single-Walled Carbon Nanotubes Probed by near Infrared Fluorescence

Fatih Sen[1,2], **Ardemis A. Boghossian**[1], Selda Sen[1,2], Zachary Ulissi[1] and Michael S. Strano[1], (1)Department of Chemical Engineering, Massachusetts Institute of Technology, Cambridge, MA, (2)Department of Chemistry, Middle East Technical University, Ankara, Turkey

Semiconducting single-walled carbon nanotubes (SWNTs) fluoresce in the near-infrared and are promising as optical sensors when suitably functionalized to enable selective analyte recognition. Their detection limit has been extended down to the single molecule, as the SWNT emission can stochastically fluctuate in a quantized manner in response to single-molecule adsorption and desorption events. In this study, we identify for the first time, the signature of the complex decomposition and reaction pathways of Riboflavin using isolated SWNT sensors dispersed onto an aminopropyltriethoxysilane (APTES) coated glass slide. The SWNT emission is quenched by Riboflavin and singlet oxygen, but increased by Trolox, which functions as a reductive brightening agent. Because Riboflavin oxidizes the Trolox and is also a photosensitizer for singlet oxygen and superoxide generation, the disparity in the reaction rates for each of these reactions creates an oscillatory fluorescence response from each isolated SWNT or bundle on the surface. This oscillatory behavior allows for the simultaneous detection of multiple species for a reactive system using SWNT emission. A series-parallel kinetic model is shown to describe the critical points of these oscillations, with predicted partition coefficients on the order of 10^{-5}.

(766e) Detecting Toxicity of Carbon Nanotubes Using Nanoindentation

Chenbo Dong[1], Michael L. Kashon[2], David H. Lowry[2], Linda M. Sargent[2] and Cerasela Zoica Dinu[1], (1)Chemical Engineering, West Virginia University, Morgantown, WV, (2)National Institute for Occupation Safety and Health, Morgantown, WV

Carbon nanotubes are being explored for broadest applications including electrical circuits, hydrogen storage, fiber optics, and conductive plastics. In recent years, their biocompatibility with small molecules, such as proteins and nucleic acids, opened up exciting biomedical applications. Thus, concerns about their potential cytotoxicity on both human and environmental have risen. It is necessary to develop significant and sensitive analysis technologies for carbon nanotubes cytotoxicity investigation. Our research aims at elucidating the mechanisms of interactions of carbon nanotubes with biological systems and evaluating the overall effects on the cell fate. To unravel the molecular mechanisms of such interaction, we need to use the latest technological advances in which both structure and function of the cells upon exposure are being analyzed. Here we employed Atomic Force Microscopy (AFM) to carry on system analysis on cellular topography and mechanical properties of immobilized cells treated with carbon nanotubes with high spatial resolution and increased sensitivity. Our results indicated that cellular exposure to carbon nanotubes lead to significant changes in cell biophysical and biomechanical properties. Moreover, such analysis provided the possibility to develop biomechanical based methods to probe the structure-function relationship in live cells and cells incubated with different nanomaterials.

(767a) Morphological and Surface Dependant Nano-Particle Behavior for Medical Imaging: Evaluation of Magnetic Nanoparticles for MRI and Gold Nanoparticles for CT-SCAN

Angel A. Galvis, Juan R. Reyes, Luis C. Serrano, Laura M. Hernandez and Watson L. Vargas, Department of Chemical Engineering, Universidad de los Andes, Bogotá, Colombia

The field of medical imaging has recently enhanced its medical potential applications based on nanotechnology; this is due in part to an extensive interest from different disciplines. Several researchers are currently focused on the development of molecular imagenology using nanoparticles as contrast media. These techniques have been outlined as the "non-invasive, quantitative and replicable imaging of targeted tissues or biological processes *in vivo* and *ex-vivo*". The future applications of molecular imaging are expected to include: more accurate prognoses, personalized medicine, the ability to follow the effectiveness of treatments, and the early detection of many diseases such as cancer [1].

Current development on gold nanostructures is related to plasmon phenomena (*e.g.* Photothermal ablation) and its appeal arises from their potential use as both therapeutic agents and contrast media for several medical imaging techniques (e.g. photoacustic imaging and CT-SCAN). On the other hand, magnetic nanoparticles have been widely studied for magnetic resonance imaging MRI due to the alterations they induce in the spin–spin relaxation time when introduced into biological tissues.

The aim of this study is twofold. First, to provide experimental evidence as to how the shape of gold nanoparticles can affect the X-rays attenuation behavior. Here, this is achieved using gold nanoshells, nanorods, hollow gold nanoparticles and colloidal gold on controlled experiments from which the evolution of X-rays attenuation is determined. Second, to evaluate the effects of different surface modifications on superparamagnetic iron oxide nanoparticles (SPION´s) with the purpose of gathering information on how these modifications can alter the spin-spin relaxation time for MRI applications. The nanoparticles studied in this case include: magnetite capped with polyvinylpyrrolidone (PVP), magnetite capped with citrate, magnetite capped with an amino-silane, and core-shell magnetite-SiO2. The introduction of different chemical groups on the surface of magnetite is expected to modify the translational diffusion of water molecules to the magnetic core, therefore affecting the spin-spin relaxation time.

Nanoparticles are characterized by Transmission Electron Microscopy (TEM), Scanning Electron Microscopy (SEM), Dynamic Light Scattering, Zeta potential, Fourier transform infrared spectroscopy (FT-IR), X-ray diffraction (XRD) and UV-VIS-NIR spectroscopy. Their properties such as X-ray attenuation and spin-spin relaxation time are measured in commercial medical devices for CT-SCAN tomography and MRI.

Keywords: Shape effects, surface effects, nanoparticle based biomedical imagenology, magnetic nanoparticles, gold nanoparticles, SPION´S.

References:

[1] Thorek D. J., Chen A. K., Czupryna J., Tsourkas A.; Superparamagnetic Iron Oxide Nanoparticle Probes for Molecular Imaging, *Annals of Biomedical Engineering* 34 1 ,p. 23-38 (2006).

(767b) Developing Novel Virus-Like Particles As Tunable, Targetable Reporter Platforms

Mark T. Smith, Chad T. Varner, Matthew A. Burnham, Jay M. Rainsdon and Bradley C. Bundy, Chemical Engineering, Brigham Young University, Provo, UT

Virus coat and envelope proteins are compelling candidates for nanoparticle technology development because they self-assemble into small, highly symmetric ensembles called virus-like particles (VLP). These monodisperse VLP particles have a conserved morphology and can be modified both genetically and chemically in a highly controlled manner, allowing for tuned adaptation for each specific application. The VLP based on bacteriophage Qβ coat proteins (CP) and its maturation protein, A2, are especially appealing for nanoparticle development. The Qβ VLP is highly stabilized by disulfide bonds, which cross-link the CP. Each VLP contains 180 copies of the CP while only incorporating a single A2.

In this work, we explore the development of customized bacteriophage Qβ VLP for targeted photoreporter platforms. Via site-directed genetic mutations and global-methionine replacement, unnatural amino acids (uAA) are inserted into the CP during cell-free protein synthesis. Use of the *in vitro* cell-free synthesis approach provides ease of access to the synthesis environment, allows for synthesis of highly cytotoxic proteins such as A2, and simplified transport of otherwise less soluble uAA. These uAA provide bio-orthogonal chemical moieties for conjugation of a variety of photoreporters after VLP assembly. The result of photoreporter attachment is a small, stable and densely populated nanoparticle, with upwards of 100 photoreporters per particle. We genetically modified the A2 protein to contain a single, site-specific insertion for uAA incorporation. This uAA also contains a bio-orthogonal moiety that acts as the targeting-ligand attachment site. Combining the modified CP and A2 proteins, we produce a platform for high density attachment of various photoreporters that can be covalently attached to targeting-ligands of choice.

(767c) Supramolecular Nano-Beacons As Protease Sensors and Imaging Agents

Lye Lin Lock, Andrew G. Cheetham, Pengcheng Zhang and Honggang Cui, Chemical and Biomolecular Engineering, Johns Hopkins University, Baltimore, MD

Molecular beacons which convert specific chemical reactions or binding events into measurable signals are essential tools for understanding cellular and sub-cellular activities at molecular level. In this context, we successfully designed and synthesized a peptide-based drug conjugated nano-beacon that could serve as a cancer detection tool as well as a carrier for targeted drug delivery in cancer treatment. An important characteristic of nano-beacons is that they can self-assemble into nanostructure under physiological condition, which enhances the cancer-cell permeability and retention (EPR effect). Here we designed and synthesized a novel nano-beacons structure conjugated with anti-cancer drug and an enzymatically degradable linker. This linker is susceptible to the cleavage of cathepsin-B, a lysosomal enzyme that is over-expressed in many cancerous cell lines, hence allowing the release of drug at cancer site. Our results showed that these cleavable drug-conjugated nano-beacons are uptake by cancerous MCF-7 cells at the lysosome before diffusing into the cell nuclei. The delivery of anti-cancer drug to MCF-7 results in cell apoptosis as well as the decrease in viability after incubation with the drug-conjugated nano-beacon.

AUTHOR INDEX

AUTHOR INDEX

AUTHOR INDEX

AUTHOR INDEX

AUTHOR INDEX

AUTHOR INDEX

AUTHOR INDEX

AUTHOR INDEX